Permian and Triassic rocks in the eastern Tethyan region form continuous marine sequences that record the waning phases of the Paleozoic and the early stages of the Mesozoic eras. This book describes and interprets these rocks, summarizing the distribution of major fossil groups in a way that will allow detailed comparison with strata of comparable age in the western Tethys and other parts of the world. The sixteen contributions by forty authors are the culmination of the five-year long International Geological Correlation Programme Project 203. The detailed information presented here is gathered from many areas in the eastern Tethyan region – from France to Australia – and will be of use in the evaluation of the major changes in the global marine biosphere known to have taken place at the end of the Paleozoic Era. The stratigraphic record for this fascinating segment of Earth history is not widespread elsewhere in the world and is most continuous in the region covered by this book.

Permo-Triassic Events in the Eastern Tethys

Permo-Triassic Events in the Eastern Tethys

Stratigraphy, Classification, and Relations with the Western Tethys

EDITED BY

W. C. SWEET, YANG ZUNYI,
J. M. DICKINS, AND YIN HONGFU

International Geological Correlation
Programme Project 203: Permo-Triassic events
of East Tethys region and their
intercontinental correlation

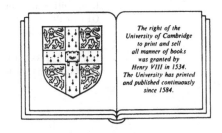

The right of the
University of Cambridge
to print and sell
all manner of books
was granted by
Henry VIII in 1534.
The University has printed
and published continuously
since 1584.

CAMBRIDGE UNIVERSITY PRESS

Cambridge
New York   Port Chester   Melbourne   Sydney

PUBLISHED BY THE PRESS SYNDICATE OF THE UNIVERSITY OF CAMBRIDGE
The Pitt Building, Trumpington Street, Cambridge, United Kingdom

CAMBRIDGE UNIVERSITY PRESS
The Edinburgh Building, Cambridge CB2 2RU, UK
40 West 20th Street, New York NY 10011–4211, USA
477 Williamstown Road, Port Melbourne, VIC 3207, Australia
Ruiz de Alarcón 13, 28014 Madrid, Spain
Dock House, The Waterfront, Cape Town 8001, South Africa

http://www.cambridge.org

First published 1992
First paperback edition 2003

A catalogue record for this book is available from the British Library

Library of Congress cataloguing in publication data

Permo-Triassic events in the eastern Tethys : stratigraphy,
classification, and relations with the western Tethys / edited by
W.C. Sweet . . . [et al.] ; International Geological Correlation
Programme, Project 203: Permo-Triassic Events of east Tethys Region
and Their Intercontinental Correlation.
    p. cm.
Includes bibliographical references.
ISBN 0 521 38214 9 hardback
1. Geology, Stratigraphic—Permian. 2. Geology, Stratigraphic–
–Triassic. 3. Tethys (Paleography) I. Sweet, Walter C.
II. International Geological Correlation Programme. Project No.
203—"Permo-Triassic Events of the Tethys Region and Their
Intercontinental Correlation."
QE674.P474   1992
551.7′56—dc20   91-11010   CIP

ISBN 0 521 38214 9 hardback
ISBN 0 521 54573 0 paperback

Contents

Contents

Contents ix

List of contributors

Dr Bie Xianmei
Dept. Geology, China University of Geosciences, Wuhan, Hubei 430074, People's Republic of China

Dr Hamish J. Campbell
DSIR Geology and Geophysics, PO Box 30-368, Lower Hutt, New Zealand

Prof. Carmela Broglio Loriga
Dept. Scienze Geologiche e Paleontologiche, Universita di Ferrara, Corso Ercole 1 d'Este, 32, 44100 Ferrara, Italy

Prof. Giuseppe Cassinis
Dept. Scienze della Terra, Universita degli Studi, Strada Nuova 65, 27100 Pavia, Italy

Dr Cheng Zhengwu
Institute of Geology, Chinese Academy of Geological Sciences, Baiwanzhuang Road, Beijing 100037, People's Republic of China

Dr Chai Chifang
Institute of High Energy Physics, Academia Sinica, PO Box 2732, Beijing, People's Republic of China

Dr J. M. Dickins
Bureau of Mineral Resources, PO Box 378, Canberra A. C. T. 2601, Australia

Prof Ding Meihua
Dept. Geology, China University of Geosciences, Wuhan, Hubei 430074, People's Republic of China

Dr Yoram Eshet
Geological Survey of Israel, 30 Malkhei Israel, Jerusalem, Israel 95501

Dr Richard E. Grant
Dept. Paleobiology, National Museum of Natural History, Smithsonian Institution, Washington, DC 20560 USA

Prof H. J. Hansen
Institute of Historical Geology and Paleontology, University of Copenhagen, Oster Voldgade 10, DK-1350 Copenhagen, Denmark

Dr He Jingwen
Nanjing Institute of Geology and Paleontology, Academia Sinica, Chi-Ming-Ssu, Nanjing, People's Republic of China

Dr Hou Jingpeng
Institute of Geology, Chinese Academy of Geological Sciences, Baiwanzhuang Road, Beijing 100037, People's Republic of China

Dr Huang Siji
Dept. Geology, China University of Geosciences, Wuhan, Hubei 430074, People's Republic of China

Dr H. M. Kapoor
Geological Survey of India, River Bank Colony. Behind H Block, Lucknow 226018, India

Dr Kong Ping
Institute of High Energy Physics, Academia Sinica, PO Box 2732, Beijing, People's Republic of China

Dr Li Daiyun
Yunnan Institute of Geological Sciences, No. 33 Baita Road, Kunming, Yunnan, People's Republic of China

Dr Li Peixian
Institute of Geology, Chinese Academy of Geological Sciences, Baiwanzhuang Road, Beijing 100037, People's Republic of China

Dr Li Zishun
Institute of Geology, Chinese Academy of Geological Sciences, Baiwanzhuang Road, Beijing 100037, People's Republic of China

Dr Long Jiarong
Regional Geological Survey of Guizhou Province, Bagongli (Eight Kilometers), Guiyang, Guizhou 550011, People's Republic of China

Dr Ma Jianguo
Institute of High Energy Physics, Academia Sinica, PO Box 2732, Beijing, People's Republic of China

Dr Ma Shulan
Institute of High Energy Physics, Academia Sinica, PO Box 2732, Beijing, People's Republic of China

Dr Mao Xueying
Institute of High Energy Physics, Academia Sinica, PO Box 2732, Beijing, People's Republic of China

Dr Qu Lifan
Institute of Geology, Chinese Academy of Geological Sciences, Baiwanzhuang Road, Beijing 100037, People's Republic of China

Dr Sun Shuying
Institute of Geology, Chinese Academy of Geological Sciences, Baiwanzhuang Road, Beijing 100037, People's Republic of China

Prof Walter C. Sweet
Dept. Geological Sciences, The Ohio State University, 125 So. Oval Mall, Columbus, OH 43210 USA

Dr R. S. Tiwari
Birbal Sahni Institute of Palaeobotany, 53, University Road, GPO Box 106, Lucknow 226001, India

Dr Nadège Toutin-Morin
Universite de Nice, URA au CNRS 'Geodynamique', Parc Valrose, 06034 Nice Cedex, France

Dr Vijaya
Birbal Sahni Institute of Palaeobotany, 53, University Road, GPO Box 106, Lucknow 226001, India

Prof Carmina Virgili
Dept. de Estratigrafia, Facultad de Ciencias Geologicas, Univ. Complutense, Madrid 3, Spain

Dr Wu Shaozu
Institute of Geology, Geological and Mineral Bureau of Xinjiang, 16 Friend N. Road, Urumqi, Xinjiang, People's Republic of China

Prof Xu Guirong
Dept. Geology, China University of Geosciences, Wuhan, Hubei 430074, People's Republic of China

Dr Yang Fengqing
Dept. Geology, China University of Geosciences, Wuhan, Hubei 430074, People's Republic of China

Dr Yang Jiduan
Institute of Geology, Chinese Academy of Geological Sciences, Baiwanzhuang Road, Beijing 100037, People's Republic of China

Prof Yang Zunyi
China University of Geosciences (Beijing) Beijing 100083, People's Republic of China

Prof Yin Hongfu
Dept. Geology, China University of Geosciences, Wuhan, Hubei 430074, People's Republic of China

Dr Yuri D. Zakharov
Far Eastern Scientific Centre, USSR Academy of Sciences, 690022 Vladivostok, USSR

Dr Zhang Kexing
Dept. Geology, China University of Geosciences, Wuhan, Hubei 430074, People's Republic of China

Dr Zhou Huiqin
Intitute of Geology, Chinese Academy of Geological Sciences, Baiwanzhuang Road, Beijing 100037, People's Republic of China

Dr Zhou Tongshun
Institute of Geology, Chinese Academy of Geological Sciences, Baiwanzhuang Road, Beijing 100037, People's Republic of China

Dr Zhou Yaoqi
Institute of High Energy Physics, Academia Sinica, PO Box 2732, Beijing, People's Republic of China

Preface

The 16 reports that make up this volume constitute the final report of Project 203 of the International Geological Correlation Programme, which dealt with Permo-Triassic events of the East Tethys and their intercontinental correlation. During the five-year life of Project 203, participants met twice in Beijing, and once each in Columbus, Ohio (USA) and Brescia (Italy), to exchange views, consider new data, and examine pertinent sections in the field. These conferences have provided new insight into the stratigraphy, classification, and relations of strata within the Permo-Triassic boundary interval, and a wider appreciation of the problems involved in the correlation and interpretation of these rocks, which document an unusually significant period in Earth history.

It will be clear to the reader that the 40 contributors to this volume are in substantial agreement in their interpretation of many features of the Permo-Triassic boundary interval, but disagree, at least by implication, in their evaluation of others. Although as editors we have strived to achieve readability and uniformity in basic terminology, we have also attempted to avoid interference with strongly held individual views, even though they run contrary to those of other contributors. In brief, it would be inappropriate to suggest that in five short years participants in Project 203 solved all problems with respect to the Permo-Triassic boundary interval – for many of these are likely to persist as long as vigorous study continues. We do suggest, however, that data and ideas gathered and expressed during the five-year life of Project 203 and summarized in this volume are important contributions toward understanding and interpreting the rock and fossil record of the late Permian and early Triassic.

The co-leaders of Project 203 and Prof. Yin Hongfu have served as the editorial board for this volume. Prof. Sweet assisted authors with English versions of their manuscripts, coordinated reviews by the editorial board, arranged for revisions in figures, and served as principal contact with the publisher. All members of the editorial board, however, have seen and commented on every contribution, and in several cases have solicited reviews by other experts. This international division of labor has cost considerable time, but we believe it has resulted in a volume that is both authoritative in content and consistent internally.

YANG ZUNYI   J. M. DICKINS   WALTER C. SWEET
(Co-Leaders IGCP Project 203)

Acknowledgments

Preparation of the manuscript for this volume required a substantial investment by the Department of Geological Sciences of The Ohio State University in duplication, postage, and drafting assistance. This help is gratefully acknowledged.

Hari Kapoor, author of Chapter 3, thanks J. M. Dickins and W. C. Sweet for advice and help with his manuscript. His chapter is dedicated to the late Yuji Bando, Dept. Earth Sciences, Kagawa University, Japan, who studied most of the areas discussed in Chapter 3 and whose premature death has been a great loss to teams working on problems of the Permo-Triassic boundary.

Giuseppe Cassinis, senior author of Chapter 7, notes, with gratitude, that his work has been supported by grants from the MPI (40%) and the CNR.

Xu Guirong and Richard Grant, coauthors of Chapter 9, thank their many colleagues, particularly Yang Zunyi, Yin Hongfu, and Wu Shunbao, of the China University of Geosciences, for providing materials from South China and encouragement and instruction as the report was written. They also thank G. A. Cooper and R. A. Doerscher of the United States National Museum, and other colleagues who aided them in the study of brachiopods from Permo-Triassic boundary strata in South China.

Ding Meihua, author of Chapter 10, thanks members of the Permo-Triassic Boundary Working Group of the China University of Geosciences, for their friendly cooperation in the field and laboratory. She also expresses her gratitude to W. C. Sweet for reviewing her manuscript and making valuable suggestions, and to Xiao Siyu, who helped draw the figures.

W. C. Sweet, author of Chapter 11, thanks Gilbert Klapper and Brian Glenister, of the University of Iowa, for reviewing his manuscript and for their helpful comments on it. He also acknowledges the help of Karen Tyler, faculty draftsperson at The Ohio State University, for preparing the figures that accompany his report and the introductory chapter, and for making corrections, changes, and alterations to a number of the figures submitted by other authors for inclusion in this volume.

Yoram Eshet, author of Chapter 12, notes that his study was supported by grants from the City University of New York and the Ministry of Energy of Israel. He thanks D. Habib (CCNY), H. Cousminer (US Minerals Management, Los Angeles), and E. I. Robbins (US Geological Survey, Reston, VA) for their useful comments on his manuscript. H. Visscher and W. A. Brugman, of Utrecht State University, Netherlands, also contributed to his study by sharing their ideas on interpretation of the Permo-Triassic boundary in Europe; Y. Druckman and T. Weisbrod (Geological Survey Israel) provided useful information on Permo-Triassic lithostratigraphy, T. Beer processed samples; A. Pe'er drafted the figures; and B. Katz helped proofread and edit the text.

Yin Hongfu and coauthors of Chapter 13 express cordial appreciation to Yang Zunyi, Xu Guirong, and Wu Shunbao for their encouragement and help in preparing their contribution. They also thank Xu Daoyi, Chai Zhifang, and Li Zishun for helpful information, and Xiao Siyu for drafting the figures.

Chai Chifang and coauthors of Chapter 14 thank Xu Daoyi, Sun Yiying, He Xiling, and Yin Hongfu for providing some of the samples used in their work. They also thank the National Natural Science Foundation of China (NSFC) for financial support.

J. M. Dickins publishes with the approval of the Director, Bureau of Mineral Resources, Canberra, Australia.

1   Permo-Triassic events in the eastern Tethys – an overview

WALTER C. SWEET, YANG ZUNYI, J. M. DICKINS AND YIN HONGFU

Introduction

It has long been clear that waning stages of the Permian Period and initial phases of the Triassic make up an interval of time during which there were profound changes in the Earth's internal and surficial features, in its atmosphere and climate, and, conspicuously, in its biota. Unfortunately, however, rocks that record this fascinating segment of Earth history are greatly restricted in distribution by comparison with those from which we reconstruct events in earlier Paleozoic or later Mesozoic history. Furthermore, subsequent orogenic cycles have dismembered, effaced, and in other ways complicated the record. In combination, these factors make development of a stratigraphic scale for the Permo-Triassic boundary interval difficult, and this, in turn, serves to insure that the history of the latest Permian and earliest Triassic has been reconstructed thus far in only very general outline.

Project of 203 of the International Geological Correlation Programme (IGCP) was organized in 1984 with the aim of integrating modern studies of rocks and fossils in the Permo-Triassic boundary interval into an internally coherent and well-controlled body of information from which it might ultimately be possible to infer something meaningful about the history of late Paleozoic and early Mesozoic times. The attention of project participants naturally focused on Permian and Triassic strata in the eastern part of the Tethyan realm, in South China and India, for marine rock sequences of this age are most continuously developed there. However, close cooperation with students of the Permian and Triassic in Italy, Hungary, Yugoslavia, Israel, the Soviet Union, Australia, and New Zealand has enabled development of a data base that now includes information for essentially the entire length and breadth of the former Tethys seaway.

Several studies conducted as part of the overall mission of IGCP Project 203 were concluded before the Project ended in 1987 and have been published in other places, primarily in two important volumes (Yang Zunyi et al., 1987; Cassinis, 1988). In the following pages, repeated reference is made to these volumes, and data from contributions included in them are summarized in later parts of this chapter, as is information from other IGCP 203-related reports that are scattered through the literature.

Biostratigraphy

Although the Permian and Triassic systems are worldwide in their distribution, they were originally established in Europe for distinctive bodies of rock in the southern Urals and the Germanic Basin, respectively. In its type area, the uppermost part of the Permian System is nonmarine and, even though it has yielded fossils of a variety of terrestrial and lacustrine animals and plants, it has proved difficult to use those fossils to develop a biostratigraphic scale of much more than local utility. Added to this is the fact that the lower third of the typical Triassic, the Buntsandstein, records an interval of dominantly terrestrial deposition in a restricted central European basin. Thus, even if it could be established with precision that Permian rocks immediately beneath the Buntsandstein are equivalent to those atop the typical Permian, it would be very difficult to develop from the distribution of the fossils in such a composite section a biostratigraphic scale that would be very useful in assembling a worldwide record for historical interpretation.

Inadequacy of the typical Upper Permian and Lower Triassic as the bases for a biostratigraphy for marine rocks in the Permo-Triassic boundary interval was recognized nearly a century ago by Mojsisovics, Waagen & Diener (1895). However, the collective experience of these authors with Permian and Triassic rocks and fossils at various places along the length of the former Tethys seaway enabled them to recognize that the Werfen Formation of the Southern Alps is a close equivalent, in marine facies, of the basal Triassic Buntsandstein, and to postulate that the debut in Himalayan sections of a distinctive ceratitic ammonoid, *Otoceras woodwardi* Griesbach, marked a level approximately equivalent to the base of the Werfen. They thus proposed that the base of the Triassic in pelagic, marine facies be drawn at the base of the Zone of *Otoceras woodwardi* – and there it has remained ever since, at least conceptually.

Studies since 1895 have shown that *Otoceras woodwardi* is far from universal in its distribution, so it has been necessary to devise other means of differentiating between Permian and Triassic in the many places in the world where this species or its congeners are not represented. These supplementary procedures have resulted in a number of separate stratigraphies for the Permo-Triassic boundary interval (Fig. 1.1), and it is not easy to

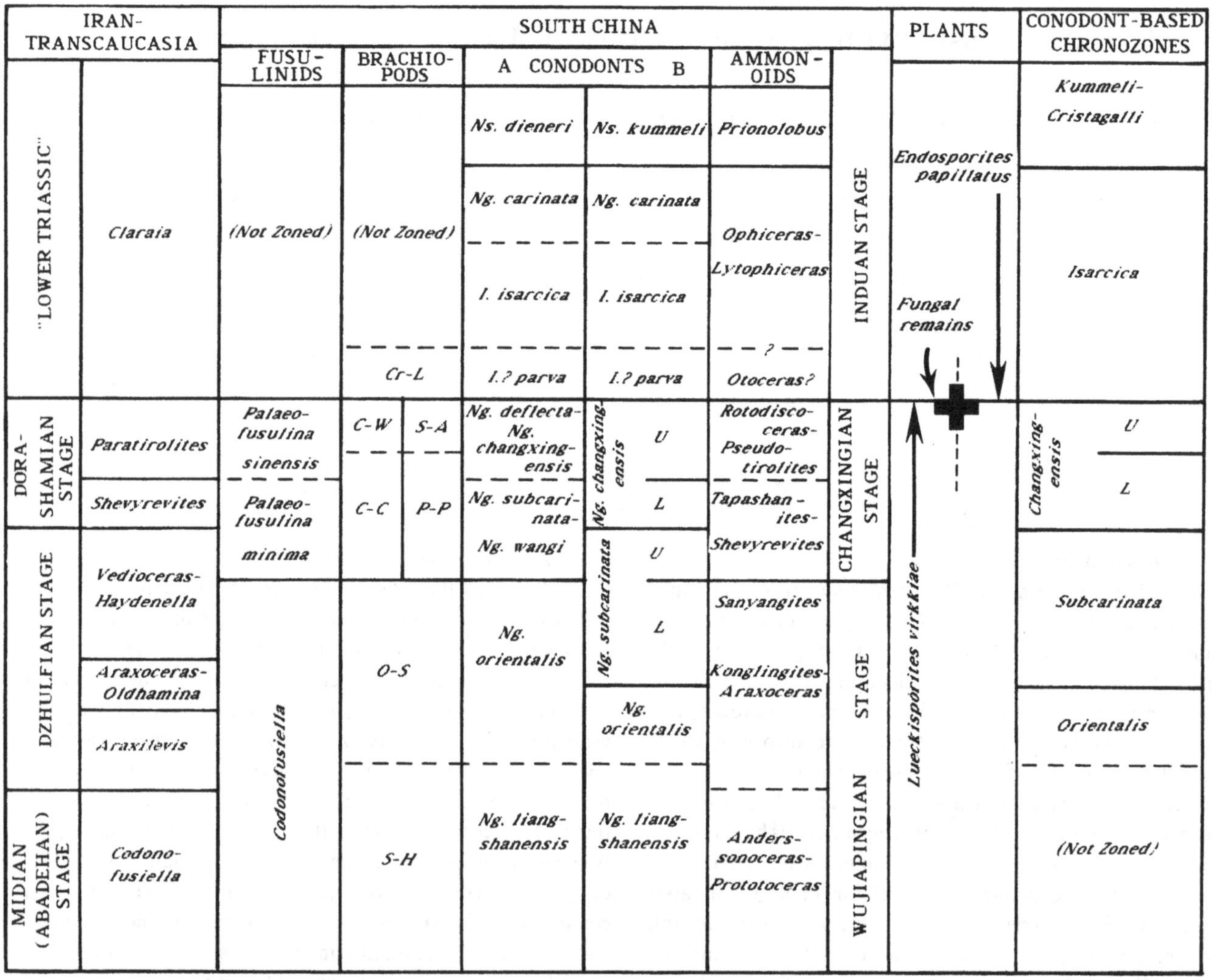

IRAN-TRANSCAUCASIA		SOUTH CHINA								PLANTS		CONODONT-BASED CHRONOZONES	
		FUSU-LINIDS	BRACHIO-PODS		A CONODONTS B			AMMON-OIDS					
"LOWER TRIASSIC"	Claraia	(Not Zoned)	(Not Zoned)		Ns. dieneri	Ns. kummeli		Prionolobus	INDUAN STAGE		Endosporites papillatus	Kummeli-Cristagalli	
					Ng. carinata	Ng. carinata		Ophiceras-Lytophiceras				Isarcica	
					I. isarcica	I. isarcica					Fungal remains		
			Cr-L		I.? parva	I.? parva		Otoceras?					
DORASHAMIAN STAGE	Paratirolites	Palaeo-fusulina sinensis	C-W	S-A	Ng. deflecta-Ng. changxingensis	Ns. changxingensis	U	Rotodisco-ceras-Pseudo-tirolites	CHANGXINGIAN STAGE	Luecki sporites virktiae		Changxing-ensis	U
	Shevyrevites	Palaeo-fusulina minima	C-C	P-P	Ng. subcari-nata		L	Tapashan-ites					L
					Ng. wangi	Ng. subcarinata	U	Shevyrevites					
DZHULFIAN STAGE	Vedioceras-Haydenella	Codonofusiella			Ng. orientalis		L	Sanyangites	WUJIAPINGIAN STAGE			Subcarinata	
	Araxoceras-Oldhamina		O-S			Ng. orientalis		Konglingites-Araxoceras				Orientalis	
	Araxilevis												
MIDIAN (ABADEHAN) STAGE	Codono-fusiella		S-H		Ng. liang-shanensis	Ng. liang-shanensis		Anders-sonoceras-Prototoceras				(Not Zoned)	

Fig. 1.1. Biostratigraphy of the Permo-Triassic boundary interval based on various groups of organisms. Assemblage zones in column headed 'Brachiopods' are abbreviated *Cr-L* (= *Crurithyris pusilla-Lingula subcircularis*); *C-W* (= *Cathaysia sinuata-Waagenites barusiensis*); *S-A* (= *Spirigerella discussella-Acosarina minuta*); *C-C* (= *Cathaysia chonetoides-Chonetinella substrophomenoides*); *P-P* (= *Peltichia zigzag-Prelissorhynchia triplicatioid*); *O-S* (= *Orthothetina ruber-Squamularia grandis*); *S-H* (= *Squamularia indica-Haydenella wenganensis*). Conodont zonal scheme A, with nomenclatural modifications, is that of Wang & Wang (1981) and Yang & Li (Chapter 2); conodont zonal scheme B is the one proposed by Ding (Chapter 10); in both, *Ns.* = *Neospathodus*, *I.* = *Isarcicella*; and *Ng.* = *Neogondolella*. Ranges in 'Plants' column are from Visscher & Brugman (1988) and Eshet (Chapter 12). Conodont-based chronozones are from Sweet (1988; Chapter 11). Note: The first occurrence in central Iran of *Shevyrevites shevyrevi*, index to the *Shevyrevites* Zone, is at a level equivalent to a point in the *Vedioceras–Haydenella* beds of Dorasham and at a level equivalent to the base of the Changxing Limestone of South China. It is for this reason we draw the base of the *Tapashanites–Shevyrevites* Zone of South China below the base of the *Shevyrevites* Zone of Iran–Transcaucasia.

combine them into a coherent, defensible, and cosmopolitan biostratigraphy for the Permo-Triassic boundary interval.

In recent years it has become increasingly obvious that Late Permian and Early Triassic time are most completely recorded in marine strata that accumulated at various places in the Tethyan realm outside the Himalayan chains. For example, in Chapter 8 of this volume Broglio Loriga and Cassinis conclude that, contrary to earlier opinions, the Bellerophonkalk-Werfen succession in the Southern Alps of northern Italy is a continuous record of Late Permian and Early Triassic time; Sweet, in Chapter 11, maintains – as do Iranian and Indian geologists (e.g. Kapoor, Chapter 3) – that highly fossiliferous Permo-Triassic sequences in Transcaucasia, Iran, and Kashmir are a complete sedimentary record through the boundary interval; and in the

last decade and a half Chinese geologists have described and interpreted apparently continuous sequences of Permian and Triassic rocks at more than 30 localities in South China. Although a few specimens from South China have been assigned tentatively to *Otoceras*, *O. woodwardi* itself is apparently restricted to Himalayan sections, only a few of which seem to be complete in the boundary interval. Hence, despite the generally very fossiliferous nature of the continuous Chinese and Iranian sections, it has been difficult to partition them into Permian and Triassic components that can be shown to be exact matches of those in sections with *O. woodwardi* or its congeners. Thus, several participants in IGCP Project 203 have worked to develop a high-resolution biostratigraphic framework within which to compile data on the occurrence and range of various fossils in continuous sections through the Permo-Triassic boundary interval, and to explore use of that framework as a means of correlating sequences without *O. woodwardi* with those that have yielded representatives of this historically important guide species. Biostratigraphic scales developed in these studies are compared in Fig. 1.1.

Brachiopods

In Chapter 9, IGCP 203 participants Xu Guirong and Richard Grant summarize information on the distribution of brachiopods in the Permo-Triassic boundary interval of South China. That distribution is used as the basis for recognizing a succession of assemblage zones and it is related where possible to the distribution of ammonoids, conodonts, and fusulinids. Xu and Grant compare the distribution of brachiopods in the Permo-Triassic boundary interval of South China with that recorded elsewhere and reach conclusions that are generally similar to those advocated by students of other groups of fossils. They note, however, that boundaries between assemblage zones based on brachiopods, conodonts, and ammonoids do not match exactly in the Changxingian, although zonal schemes based on these three fossil groups generally support one another.

Ammonoids

The ammonoid zonation outlined in Fig. 1.1 is essentially the one proposed by Zhao, Liang & Zheng (1978) and repeated by subsequent authors. The volume at hand contains no revision of this scheme, nor are we aware that one is required. We note only that the complete zonal succession is apparently not known from any one section but is based on materials collected at more than 11 widely separated localities in South China. Further, Yin Hongfu and his coauthors point out in Chapter 13 that ammonoids are confined for the most part to cherty facies of the Changxingian Stage and this limits their biostratigraphic utility in chert-free but coeval strata.

Conodonts

The ranges of conodonts in South China have been used in the last decade to define two rather distinct biozonal successions, and Sweet, in Chapter 11, makes graphic use of their distribution and that of several ammonoids to develop a composite standard section and several chronozones for the Permo-Triassic boundary interval. The relationship between these three conodont-based stratigraphic schemes is shown in Fig. 1.1.

In conodont-biozonal scheme A (Fig. 1.1), Wang & Wang (1981) established a succession of assemblage zones. Yang and Li, Chapter 2, relate these assemblage zones on a one-to-one basis to the succession of ammonoid-based assemblage zones established by Zhao *et al.* (1978), which are taken as standard but, as noted in a previous paragraph, are only very broadly controlled (cf. Zhao *et al.*, 1978, Fig. 1). Ding Meihua, in Chapter 10, notes that boundaries between the zones established by Wang & Wang (1981) are difficult to recognize and also that a good deal of new information on conodont distribution has come to light in the decade since the Wang & Wang scheme was proposed. Ding uses these new data as the basis for proposing the succession of conodont-based interval biozones identified in Fig. 1.1 as conodont scheme B.

In conodont zonal scheme B, Ding draws biozonal bases at the level of first occurrence of the nominal species. Thus, even though *Neogondolella subcarinata* is known from upper Wujiapingian rocks at only two localities, the base of the *N. subcarinata* Zone is drawn at a mid-Dzhulfian level rather than at the base of the Changxingian Stage, where the species first becomes dominant. In like manner, the base of the superjacent *N. changxingensis* Zone is at the level of first occurrence of *N. changxingensis*, even though representatives of that species are subordinate to those of *N. subcarinata* until somewhat higher in the Changxingian.

An unfortunate consequence of the proposal of two different biozonal schemes based on conodonts is the fact that the denominations *N. subcarinata* and *N. changxingensis* appear in the names of zones defined in two very different ways. Confusion may be avoided by using the full names of the assemblage zones proposed by Wang & Wang (e.g. *N. subcarinata–N. wangi* Zone, or *N. deflecta–N. changxingensis* Zone).

Sweet (Chapter 11) uses the distribution of conodonts in seven well-controlled sections to develop graphically a composite standard section for the Permo-Triassic boundary interval, which extends one developed previously for the Lower Triassic (Sweet, 1988). In the combined composite sections, first occurrences of conodont species with wide geographic distribution are used to define the bases of four chronozones, three of which, in the Dzhulfian-Dorashamian segment of the Permo-Triassic boundary interval, and in the Dienerian, correspond exactly to the *N. subcarinata*, *N. changxingensis*, and *Neospathodus kummeli* biozones of Ding's classification (Fig. 1.1). Between the top of the Changxingensis and the base of the Kummeli-Cristagalli chronozones, however, Sweet (1988) recognizes just a single unit, the Isarcica Chronozone, whereas Ding recognizes three biozones. The *Isarcicella? parva* Zone of her scheme is an interval zone bracketed between the first occurrences in South China of *I.? parva* and *I. isarcica*. The *I. isarcica* Zone, defined as the total range in South China of *I. isarcica*, is a local range zone. The *Neogondolella carinata* Zone, proposed by Sweet (1970) for an interval in the Lower Triassic of Pakistan, is used by Ding Meihua for an approximately correlative interval between the highest occurrence of *I. isarcica* in South China and the first occurrence there of *Neospathodus kummeli*, the nominal species of the next succeeding biozone. Sweet (1988) has abandoned the *Neogondolella carinata* Zone for purposes of international correlation and, although she includes it in her scheme, Ding also questions its utility.

It is particularly important to note that *I.? parva* (= *Anchignathodus* or *Hindeodus parvus* of authors) appears first in South China in rocks just above the Changxing Limestone and equivalent strata from which a few specifically indeterminate and generically questionable specimens of *Otoceras* have also been collected. *I.? parva* thus has special importance in China as an index to rocks that are regarded as lowermost Triassic and Yin *et al.* (1988) proposed that the base of the *I.? parva* Biozone, as defined in South China, be regarded as the Permo-Triassic boundary. In Kashmir and Tibet, however, conodonts identified as *I.? parva* first occur well below the top of the *Otoceras woodwardi* (or *O. latilobatum*) Zone, in rocks that also yield specimens of *Neogondolella subcarinata*, *N. changxingensis*, and *N. deflecta*, which are characteristic of the *N. changxingensis* biozone or chronozone. This suggests either that *I.? parva* made its debut earlier in Kashmir and Tibet than in South China or that *N. changxingensis* and *N. deflecta* lived longer in Tibet than elsewhere in the Tethyan realm. Also, illustrations indicate little agreement among different authors on the identity of *I.? parva*, which has a tangled nomenclatural and taxonomic history, and Sweet (Chapter 11) concludes from his graphic analysis that the oldest known occurrences of *I. isarcica*, in Kashmir and central Iran, are at the same level in his composite standard section as that of *I.? parva* in the two South Chinese sections he considers. Thus, in Sweet's opinion, *I.? parva* appeared in Kashmir and Iran before it did in South China. For this reason, Sweet uses the level of first occurrence of *I.? parva* in his composite standard section to mark the base of an informal 'upper member' of the Changxingensis Chronozone. Disagreement among authors is for the most part semantic, but it will need to be resolved before a final decision is made on the nomenclature of biostratigraphic and chronostratigraphic units in the Permo-Triassic boundary interval.

Conodonts are widespread and biostratigraphically very useful in the Permo-Triassic boundary interval and many of the species recognized are cosmopolitan in their distribution. It is clear, however, that species represent at least two major biofacies (a *Neogondolella* and an *Isarcicella–Hindeodus* biofacies), and this must be taken into account in evaluating conodont-based zonal schemes.

Plants

Visscher & Brugman (1988) and Eshet (Chapter 12) present valuable information on the distribution of plant microfossils in the Permo-Triassic boundary interval. Visscher & Brugman propose no formal zonation of this interval, but Eshet includes uppermost Permian rocks in a *Lueckisporites virkkiae* Zone and superjacent Lower Triassic ones in an *Endosporites papillatus* Zone. It is noted by both Vissher & Brugman and Eshet that fungal remains are especially prominent in the uppermost part of what Eshet defines as the *L. virkkiae* Zone, and that acritarchs become conspicuous components of the palynoflora in an interval at the base of the *E. papillatus* Zone.

From information supplied by Visscher & Brugman and by Eshet, it is clear that the boundary between the *L. virkkiae* and *E. papillatus* zones coincides with a major change in the flora and very closely to the one chosen by a majority of stratigraphers as the boundary between Permian and Triassic. Neither palynozone has great utility in detailed subdivision of the Permo-Triassic boundary interval, however, for both are based on rather long-ranging species.

Summary

In the following sections of this chapter we use the conodont-based biostratigraphic units defined by Ding (Chapter 10) and evaluated regionally by Sweet (Chapter 11) as the framework for summarizing other types of information from

and about strata in the Permo-Triassic boundary interval. As noted below, the South Chinese *Isarcicella? parva* Biozone, whose base is above that of the *Otoceras woodwardi* Biozone of Kashmir, is also distinguished in at least two Chinese sections by a distinctive paleomagnetic event; and in sections in China, Iran, Austria, and Italy by a prominent geochemical event. Italian participants in IGCP Project 203 conclude that this boundary, which coincides with the base of the Werfen Formation in the Southern Alps and thus approximates the base of the Buntsandstein of the Germanic Basin, is the one at which the Permo-Triassic boundary should be drawn, and it is the level at which Chinese geologists prefer to place that boundary.

Several participants in IGCP Project 203 are also members of a Permian–Triassic Boundary Working Group, which was charged several years ago with preparing a formal recommendation for the IUGS Commission on Stratigraphy on the biologic criteria that should be used in defining the base of the Triassic System and on a stratotype section for that boundary. At the time of this writing, no formal recommendation has been made. Thus, in this chapter we summarize our studies of uppermost Permian and lowermost Triassic strata in terms of the conodont-based chronozones outlined in Fig. 1.1.

Geochemical studies

In Fig. 1.2 we summarize studies made by Chen *et al.* (1984), Li *et al.* (1986), Holser & Magaritz (1987), Magaritz *et al.* (1988), and Baud, Magaritz, & Holser (1989) of the distribution of carbon and oxygen isotopes in several sections through the Permo-Triassic boundary interval. The framework within which these profiles are arrayed, as well as the vertical scale of Fig. 1.2, are the ones determined graphically by Sweet (1988; Chapter 11).

In the profiles of Fig. 1.2 values of $\delta^{13}C$ are generally high to a point near the top of the Changxingensis Chronozone, drop rapidly into the basal part of the Isarcica Chronozone, and are mostly below 0 to the top of that chronozone. Smoothness of the profiles suggested to Magaritz *et al.* (1988) that the sections for which they were determined are continuous or nearly so. For the rapid change in $\delta^{13}C$ values in the Changxingensis–Isarcica boundary interval Holser & Magaritz (1987) favored an explanation that ties oceanic depletion in ^{13}C to oxidation of primarily nonmarine organic carbon during an episode of intensive terrestrial erosion that was associated with widespread exposure of shelf areas during a general late Permian regression. Yin *et al.*, in Chapter 13, suggest that mass extinction of phytoplankton late in the time represented by the Changxingensis chronozone might have diminished the biomass, which would then have absorbed much less ^{12}C. This would then have reduced the relative abundance of ^{13}C in the seawater. Or, as proposed by Yin *et al.*, a decline in terrestrial vegetation as a result of the burn-out of forests suggested by soot layers in the boundary clays of South China, would have provided a great influx of ^{12}C through runoff and thus a decline in the relative abundance of ^{13}C in the oceanic reservoir. However, studies also summarized by Yin *et al.* suggest that carbon in the boundary clays is probably directly of volcanic origin and not from the burning of wood.

Magaritz *et al* (1988) and Holser *et al.* (1989) have also

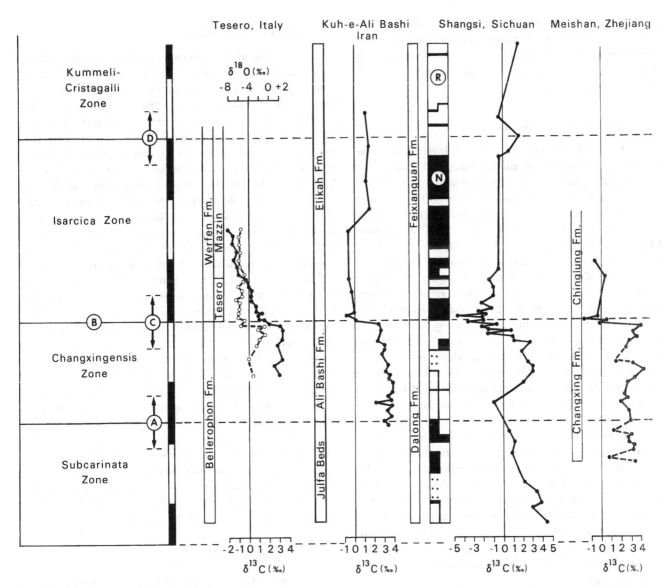

Fig. 1.2. Conodont-based chronozones, profiles of $\delta^{13}C$ (—•—) and $\delta^{18}O$ (—o—), and magnetostratigraphy for sections in the Permo-Triassic boundary interval. Geochemical data from Chen *et al.* (1984), Li *et al.* (1986), Magaritz *et al.* (1988), Holser & Magaritz (1987), and Baud *et al.* (1989); paleomagnetic profile for Shangsi section from Steiner *et al.* (1989). Profiles redrawn at vertical scale indicated by results of graphic correlation (Sweet, 1988; Chapter 11); units in ladder scale to left of Tesero profile are standard time units (STUs) and represent the shortest divisions of time whose records can be recognized at the 95% confidence level in all the sections shown. Each STU probably represents about 110 000 years. Letters on the left identify levels and intervals at, or within which various stratigraphers have proposed siting the Permo-Triassic boundary.

documented a pronounced decline in $\delta^{18}O$ values in the uppermost part of the Changxingensis and the lower part of the succeeding Isarcica Chronozone at two localities in the western part of the Tethyan realm. Holser *et al.* (1989) suggest that depletion of ^{18}O in this interval may indicate a temperature rise of perhaps 5 °C. This suggestion is consistent with data summarized in Chapter 15, contributed by Dickins, which indicate that the climate of late Permian and early Triassic times became very hot by present-day standards.

In their detailed study of the Gartnerkofel core, drilled near Nassfeld in the Carnic Alps of Austria, Holser *et al.* (1989) record a sharp increase in the relative abundance of sulphur, iron, rare-earth elements, and iridium in the interval of low and declining abundance of $\delta^{13}C$ in the lower part of the *Isarcica* Chronozone. Holser *et al.* (1989) suggest that these increases

may indicate that, at least in this part of the western Tethys, regression was followed by an interval of marine anoxia during which metals dissolved from exposed volcanics or underlying redbeds were redeposited either on contact with anoxic waters or at the oxidation/reduction interface. Accumulation of metals, they note, may also have been aided by algae. These observations are generally supported by other geochemical data, such as those documented by Chai *et al.* in Chapter 14. Clark *et al.* (1986) also noted an increase in the relative abundance of K and rare-earth elements at the boundary between the Changxingensis and Isarcica chronozones in South China, as do Chai *et al.* in Chapter 14. Clark *et al.* (1986) suggest, however, that the suite of elements just mentioned may have been volcanic in origin and the fact that their increased abundance is recorded primarily in black shales may be related to the carbon included in those rocks. Chai *et al.* agree that this is possible, but they have also identified in the South Chinese boundary clays minor quantities of sedimentary and metamorphic-rock debris, which suggest to them that bolide impact may also have contributed to the complex boundary 'event'.

Paleomagnetic investigations

Heller *et al.* (1988) and Steiner *et al.* (1989) have developed a paleomagnetic record from studies of sections in Sichuan that are also parts of the biostratigraphic framework established by Ding and Sweet in Chapters 10 and 11. Although the two studies cited differ in the spacing of samples and in areal coverage, results of both indicate (Fig. 1.2) that the Changxingensis Chronozone is characterized by dominantly reversed polarity, whereas the following Isarcica Chronozone is almost continuously normal. Comparably detailed paleomagnetic studies of sections elsewhere in the Tethyan realm have yet to be published, so it can not be determined if data from the two recent Chinese studies will be generally useful stratigraphically. In this regard it may be significant that Klootwijk *et al.* (1983) report that virtually all samples from Permian and Triassic strata in Kashmir well below, in, and well above the interval shown in Fig. 1.2 are normal. However, conodonts indicate that Permian and Triassic rocks in the Kashmir syncline have been heated by prolonged deep burial to at least 300°C, and, although this is well below the Curie point of either magnetite or hematite, temperature might have affected the paleomagnetic record adversely or in some other way modified it. We also note that based on studies in central Europe, Menning (1990) draws the Permo-Triassic boundary between a reversed and a normal interval within a much more extensive 'Permo-Triassic mixed megazone'. It is not possible now to determine if this is the same level at which Heller *et al.* (1988) or Steiner *et al.* (1989) recognized a similar reversal in polarity in South China, but the coincidence is interesting and potentially significant.

Volcanism

As Yin *et al.* and Dickins note in Chapters 13 and 15, respectively, volcanism was unusually active and widespread in its influence through the Late Permian. Yin *et al.* demonstrate that the 'boundary clay' at some 17 localities in South China is tuffaceous in origin and note that at an additional 18 localities there are clays of volcanic origin within the Changxingensis Chronozone. They relate the high chert content of rocks with the Changxingensis conodont fauna to pervasive volcanism, and Dickins suggests that cherts in the famous Upper Permian Phosphoria Formation of western United States may also be modified volcanogenic rocks. Dickins and Yin *et al.* regard widespread Late Permian volcanism as a primary contributor to fundamental changes in the biota, through the ways in which volcanic deposits may have affected the chemistry of seawater, triggered forest fires on land, and inhibited photosynthesis by clouding the atmosphere and thus reducing the amount of sunlight available to both aquatic and terrestrial plants. CO may also have contributed to the climatic warming suggested by oxygen-isotopic studies, although its relationship to Late Permian and Early Triassic volcanism has not been established.

Causes of biotic turnover

A majority of those who have contributed to this volume agree that the distribution of strata in the Permo-Triassic boundary interval records a major marine regression, and that the strata themselves, and the fossils they contain, document the acme in a time of widespread intermediate and acid volcanism, with which it is reasonable to associate basic changes in the composition of seawater and the atmosphere. Although Chai *et al.* (Chapter 14) identify sedimentary and metamorphic rock fragments in the boundary clays of South China, which they conclude originated by spattering from a bolide impact, most other authors seem to agree that bolide impact is not required by any of the geochemical, sedimentologic, or paleontologic data assembled in the course of this study. That is, although accumulation of rare-earth and siderophile elements to quantities above background is a feature of rocks in the lower part of the Isarcica Chronozone, such anomalous accumulations are readily accounted for through precipitation in the anoxic environments documented by the carbon-rich black shales in which the anomalies occur.

In brief, the biotic turnover that took place in the Permo-Triassic boundary interval may be regarded as a natural response by already stressed animal groups to prolonged marine regression, during which shallow shelves dwindled in area, and to long-continued acid and intermediate volcanism, which may have blanketed large land areas with ash, caused great forest fires, and contributed materials to both the atmosphere and the ocean basins that altered their composition and thus directly affected the biosphere. There is no compelling evidence of extraterrestrial interference in this earthbound process, although bolide impact can not be excluded.

The Permo-Triassic boundary

As noted in an earlier part of this chapter, the Permian–Triassic Boundary Working Group has not yet made a formal recommendation to the Commission on Stratigraphy concerning either the biologic criteria that should be employed in defining the base of the Triassic or a stratotype for that boundary. However, contributors to this volume supply information that should be very useful to PTBWG members as they ponder a recommendation. For that reason we summarize features of the three positions within the Permo-Triassic boundary interval that have been most commonly proposed as base of the Triassic.

In Chapter 11, Sweet uses the ranges of conodonts, several cephalopods, and representatives of the bivalve *Claraia* to effect correlation of important sections of Permian and Triassic strata in South China, Kashmir, Pakistan, Iran, and the southern USSR. From the exercise described in Chapter 11, Sweet concludes that, relative to the other sections considered in Fig. 1.1, a level equivalent to the base of the *Otoceras woodwardi* Zone in a section at Guryul Ravine, Kashmir, is somewhere within interval A of Fig. 1.2 and that interval A also includes the base of the Dorasham Beds in the Nakhichevan ASSR, the stratotype of the Dorashamian Stage of Rostovtsev & Azaryan (1973). He also concludes that a level equivalent to the base of the *O. woodwardi* Zone of Kashmir is to be found somewhere in the lower third of the Changxing Limestone in its type section in Zhejiang Province, China. In Fig. 1.2, interval A is also shown to include the base of the *Neogondolella changxingensis* Biozone of Ding and that of Sweet's Changxingensis Chronozone.

Most stratigraphers who have considered the correlation of rocks in the Permo-Triassic boundary interval agree, at least in general, with the correlations indicated in Fig. 1.1, but few have seriously considered the possibility suggested by Yin *et al.* (1988) or Sweet (this volume) that some part of the Changxingian Stage of South China and its equivalents elsewhere might be equivalent temporally to the *O. woodwardi* Zone of the Himalaya. In fact, most recent correlation charts (e.g. Tozer, 1979; Yang & Li, this volume; Kapoor, this volume) show the *O. woodwardi* Zone superimposed directly on the Changxingian and Dorashamian stages, even though direct evidence of such superposition is lacking. In short, most stratigraphers would probably conclude that level B of Fig. 1.2, which is also the base of Sweet's (Chapter 11) Isarcica Chronozone, is approximately equivalent to the base of the *O. woodwardi* Zone of Kashmir and thus the traditional level of the Permo-Triassic boundary.

In Sweet's opinion (Chapter 11), level B, tops of the Changxingian and Dorashamian stages, and the top of the *O. woodwardi* Zone as it is developed in the Guryul Ravine section, Kashmir, all fall somewhere within interval C of Fig. 1.2. In numerous South Chinese sections level B is marked by clays of volcanic origin, which Chinese geologists use to mark the Permo-Triassic boundary. In earlier parts of this chapter we have pointed out that in numerous sections from South China on the east to the Southern Alps of Austria and Italy on the west, level B also coincides with the acme of a major decline in $\delta^{13}C$ values, which have higher, more typically Upper Paleozoic values in the subjacent Changxingensis Chronozone. We also note that there is an abrupt shift at level B from dominantly reversed to dominantly normal polarity in the magnetostratigraphic profiles developed for three South Chinese sections by Heller *et al.* (1988) and Steiner *et al.* (1989). Finally, we note that correlations based on conodonts indicate to Sweet (1988) that level B, which falls somewhere in interval C of Fig. 1.2, corresponds to the base of the Werfen Formation in the Southern Alps, which Mojsisovics *et al.* (1895) and other authors regarded as the approximate equivalent, in marine facies, of the base of the Buntsandstein, the lowest unit in the classical Triassic. Yin *et al.* (1988) proposed that the Permo-Triassic boundary be drawn at the base of the *Isarcicella? parva* Biozone, which, in their view, is level B in Fig. 1.2. Broglio Loriga and Cassinis also conclude, in Chapter 8, that the base of the Werfen is the level in the Southern Alps and Hungary at which the major biotic turnover is recorded and for that reason they opt for Level B as the base of the Triassic. Kozur (1989) also recommends Level B as Triassic base, but bases his recommendation on correlations that differ in detail from those expressed by contributors to this volume.

Interval D of Fig. 1.2 includes the base of the Kummeli–Cristagalli Chronozone and probably also the base of the Dienerian (or Nammalian) Stage. These levels are readily recognized in Kashmir, Pakistan, Iran, western United States, and South China by the first occurrences of various species of the conodont *Neospathodus*, and in other sections by the appearance of gyronitid ammonoids. In the section at Shangsi, Sichuan, interval D is just above the top of the consistently

normal magnetostratigraphic interval that distinguishes the Isarcica Chronozone there, and it is also the interval there within which values of $\delta^{13}C$ again become positive. These factors in combination may make boundaries within interval D at least as widely recognizable as level B and it may be that interval D approximates the time at which highly stressed environmental conditions began to ameliorate. The base of the Dienerian (or Nammalian) Stage, which is somewhere within interval D, has long been supported as base of the Triassic by Newell (e.g. 1973, 1988), who argues that it represents the beginning of the radiation of typically Triassic groups of ammonoids and conodonts. Budurov *et al.* (1988) also draw the base of the Triassic within interval D.

Summary

The Permo-Triassic boundary interval, probably not much more than 500,000 years long, was marked in the Tethyan realm and worldwide by the culmination of a major marine regression, by widespread acid and intermediate volcanism, by a climate that was probably much warmer than now, and by a substantial reduction in biotic diversity, which was probably the result of the influence on already stressed groups of organisms of a complex of climatic factors on land, chemical factors in the sea and greatly reduced areas of shallow-water marine habitats. The work of participants in Project 203 on various aspects of the Permo-Triassic boundary interval permits the conclusion that not all animal groups were affected in the same way or at the same time by the interaction of these factors. In short, the concept of a mass extinction at or near the end of the Permian may be a generalization based on a record too grossly organized to show the patterns of extinction recognized in various chapters in this book.

Conflicting ideas on the most appropriate position in the Permo-Triassic boundary interval at which to site the base of the Triassic seem to center on whether it is most desirable to draw the boundary at the beginning (interval A), in the middle (level B, interval C), or at the end (interval D) of the boundary interval. Practical recognition of boundaries in any of these positions will depend on rather different sets of criteria, and it is unlikely that any boundary level can ever be recognized worldwide by a single criterion or by the same set of criteria. Participants in Project 203, however, provide in this volume and in other published reports information on a wide variety of features, both organic and inorganic, in the Permo-Triassic boundary interval and a biostratigraphic framework within which to evaluate these features historically. For these reasons we suggest that the history of this fascinating segment of the past can now be written with far greater meaning and significance than previously.

References

Baud, A., Magaritz, M. & Holser, W. T. (1989). Permian-Triassic of the Tethys: carbon isotope studies. *Geol. Rundschau*, **78**(2): 649–77.

Budurov, K. J., Gupta, V. J., Kachroo, R. K. & Sudar, M. N. (1988). Problems of the Lower Triassic conodont stratigraphy and the Permian-Triassic boundary. *Mem. Soc. Geol. Ital.*, **34**(1986): 321–8.

Cassinis, G. (Ed.) (1988). Permian and Permian-Triassic boundary in the South-Alpine segment of the western Tethys, and additional regional reports. *Mem. Soc. Geol. Ital.*, **34**(1986): 1–366.

Chen Jinchi, Shao Maorong, Huo Weiguo & Yao Yuyuan (1984). Carbon isotope of carbonate strata at Permian-Triassic boundary in Changxing, Zhejiang. *Scientia Geol. Sin.*, 1984(1): 88–93 (in Chinese, with English abstract).

Clark, D. L., Wang Cheng-yuan, Orth, C. J. & Gilmore, J. S. (1986). Conodont survival and low iridium abundances across the Permian-Triassic boundary in South China. *Science*, **233**: 984–6.

Heller, F., Lowrie, W., Li Huamei & Wang Junda (1988). Magnetostratigraphy of the Permo-Triassic boundary section at Shangsi (Guangyuan, Sichuan Province, China). *Earth and Planet. Sci. Lett.*, **88**(1988): 348–56.

Holser, W. T. & Magaritz, M. (1987). Events near the Permian-Triassic boundary. *Modern Geol.*, **11**: 155–80.

Holser, W. T., Schönlaub, H.-P., Attrep, M. Jr., Boeckelmann, K., Klein, P., Magaritz, M., Orth, C. J., Fenninger, A., Jenny, C., Kralik, M., Mauritsch, H., Pak, E., Schramm, J.-M., Stattegger, K., & Schmöller, R. (1989). A unique geochemical record at the Permian/Triassic boundary. *Nature*, **337**(6202): 39–44.

Klootwijk, C. T., Shah, S. K., Gergan, J., Sharma, M. L., Tirkey, B. & Gupta, B. K. (1983). A paleomagnetic reconnaissance of Kashmir, northwestern Himalaya, India. *Earth Planet. Sci, Lett.*, **63**: 305–24.

Kozur, H. (1989). The Permian-Triassic boundary in marine and continental sediments. *Zbl. Geol. Paläont., Teil I*, 1988(1): 1245–77.

Li Zishun, Zhan Lipei, Zhu Xiufang, Xie Longchun, Liu Guifang, Zhang Jinghua, Jin Ruogu, Huang Hengquan, Dai Jinye & Sheng Huabien (1986). Mass extinction and geological events between Palaeozoic and Mesozoic era. *Acta Geol. Sin.*, **60**(1): 1–15.

Magaritz, M., Bär, R., Baud, A. & Holser, W. T. (1988). The carbon-isotope shift at the Permian/Triassic boundary in the southern Alps is gradual. *Nature*, **331**(6154): 337–9.

Menning, M. (1990). A new scheme for the Permian and Triassic succession of central Europe. *Permophiles*, **16**: 14.

Mojsisovics, E. von, Waagen, W. & Diener, C. (1895). Entwurf einer Gliederung der pelagischen Sedimente des Trias-Systems. *Sitz.-Ber. Akad. Wiss. Wien*, **104**(1): 1271–302.

Newell, N. D. (1973). The very last moment of the Paleozoic Era. *Canadian Soc. Petrol. Geol. Mem.*, **2**: 1–10.

Newell, N. D. (1988). The Paleozoic/Mesozoic erathem boundary. *Mem. Soc. Geol. Ital.*, **34**(1986): 303–11.

Rostovtsev, K. O. & Azaryan, N. R. (1973). The Permian-Triassic boundary in Transcaucasia. *Canadian Soc. Petrol. Geol. Mem.*, **2**: 89–99.

Steiner, M., Ogg, J., Zhang, Z. & Sun, S. (1989). The Late Permian/Early Triassic magnetic polarity time scale and plate motions of South China. *J. Geophys. Res.*, **94**(B6): 7343–63.

Sweet, W. C. (1970). Uppermost Permian and Lower Triassic conodonts of the Salt Range and Trans-Indus ranges, West Pakistan. In *Stratigraphic Boundary Problems: Permian and Triassic of West Pakistan*, ed. Kummel, B. & Teichert, C., pp. 207–75. (Dept. Geology Spec. Pub. 4). Lawrence: Univ. Kansas Press.

Sweet, W. C. (1988). A quantitative conodont biostratigraphy for the Lower Triassic. *Senckenberg. leth.*, **69**(3/4): 253–73.

Tozer, E. T. (1979). The significance of the ammonoids *Paratirolites* and *Otoceras* in correlating the Permian-Triassic boundary beds of Iran and the People's Republic of China. *Canadian J. Earth Sci.*, **16**(7): 1524–32.

Visscher, H. & Brugman, W. A. (1988). The Permian-Triassic boundary in the Southern Alps: A palynological approach. *Mem. Soc. Geol. Ital.*, **34**: 121–8.

Wang Cheng-yuan & Wang Zhi-hao (1981). Permian conodont biostratigraphy of China. *Geol. Soc. Am. Spec. Paper*, **187**: 227–36.

Yang Zunyi, Yin Hongfu, Wu Shunbao, Yang Fengqing, Ding Meihua, Xu Guirong, *et al.* (1987). Permian-Triassic boundary stratigraphy and fauna of South China. *Ministry of Geol. and Min. Res., Geol. Mem., ser. 2*, **6**: 1–379, 37 pl., charts.

Yin Hongfu, Yang Fengqing, Zhang Kexing & Yang Weiping (1988). A proposal to the biostratigraphic criterion of Permian/Triassic boundary. *Mem. Soc. Geol. Ital.*, **34**: 329–44.

Zhao Jinko, Liang Xiluo & Zheng Zhuoguan (1978). Late Permian cephalopods of South China. *Palaeontologia Sinica, n. ser. B*, **154**(12): 194 pp.

2 Permo-Triassic boundary relations in South China

YANG ZUNYI AND LI ZISHUN

Introduction

Study of the Permo-Triassic boundary is important in that the boundary marks a great event in geologic history and the transition from the Paleozoic Era to the Mesozoic Era. In many parts of the world, gaps representing varying amounts of time separate Permian and Triassic rocks, but in South China as many as 33 fairly continuous marine Permo-Triassic sections have been studied in detail (Fig. 2.1) (Yang *et al.*, 1987). Four of the better of these (Figs. 2.2, 2.3) represent different environments and give a satisfactory record from which problems

concerned with Permo-Triassic relations can be discussed and conclusions drawn.

Review of typical Permo-Triassic sections

The Meishan section in Changxing, Zhejiang Province

This well-known section (Fig. 2.1, locality 28; Fig. 2.2), which consists of platformal slope deposits, was carefully worked out by Zhao and his colleagues of the Nanjing Institute of Geology and Paleontology (Zhao *et al.*, 1981) and Sheng *et al.* (1984), and has been restudied by Wu Shunbao and his colleagues (in Yang *et al.*, 1987), whose results (Fig. 2.2) are partially as follows:

Yinkeng Formation (part), Lower Triassic . . . more than 12 m

(32) Greenish grey marl alternating with fine silty montmorillonite-hydromica clay, yielding bivalves (*Claraia fukianensis* Chen, *C. griesbachi* (Bittner), *Claraia* sp., *Pseudoclaraia wangi* (Patte)), ammonoids (*Lytophiceras* sp., *Ophiceras* sp., *Prionolobus* sp.) and conodonts (*Hindeodus* sp.) . . . . **10.46 m**

(31) Greenish grey, iron-cemented, fine-grained quartz siltstone, bearing ammonoids (*Ophiceras* sp.), bivalves (*Claraia griesbachi* (Bittner), *Claraia* sp., *Pseudoclaraia wangi* (Patte)) . . . . **1.00 m**

(30) Grey, medium-bedded marl with bluish grey clayrock (4 cm) at top, with bivalves (*Pseudoclaraia wangi* (Patte), *C. griesbachi* (Bittner)) . . . . **0.52 m**

(29) Grey, medium-bedded argillaceous limestone, yielding brachiopods (*Paryphella orbicularia* (Liao)), conodonts (*Hindeodus* sp., *Neogondolella* sp., *Xaniognathus* sp.), bivalves (*Pseudoclaraia wangi* (Patte)), and ammonoids (*Ophiceras* sp.) . . . . **0.26 m**

Transitional beds

(28) Orange-yellow clayrock . . . . **0.04 m**

(27) Greyish white, medium-bedded, silty limestone, yielding brachiopods (*Acosarina* sp. cf. *A. minuta* (Abich), *Crurithyris flabelliformis* Liao, *Fusichonetes pigmaea* Liao, *Neochonetes* sp. (?), *Waagenites* sp., *W. barusiensis* (Davidson), *Paryphella orbicularis* (Liao), *P. triquetra* Liao) and conodonts (*Isarcicella*? *parva* (Kozur & Pjatakova), *Hindeodus typicalis* (Sweet)) . . . . **0.15 m**

(26) Dark brown calcareous mudstone, yielding brachiopods

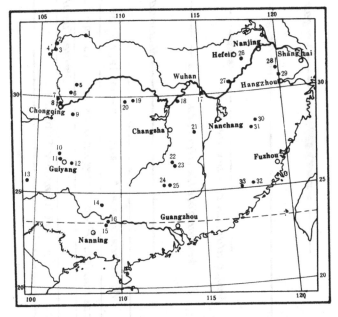

Fig. 2.1. Location of Permo-Triassic sections in South China. 1, Xixiang, Shaanxi; 2, Chaotian, Guangyuan, Sichuan; 3, Xindianzi, Guangyuan, Sichuan; 4, Shangsi, Guangyuan, Sichuan; 5, Guangan, Sichuan; 6, Linshui, Sichuan; 7, Hechuan, Sichuan; 8, Liangfengya, Chongqing, Sichuan; 9, Qijiang, Sichuan; 10, Qingzhen, Guizhou; 11, Wengjing, Guiyang, Guizhou; 12, Xiaochehe, Guiyang, Guizhou; 13, Panxian, Guizhou; 14, Yishan, Guangxi; 15, Paoshui, Laibin, Guangxi; 16, Hongshuihe, Laibin, Guangxi; 17, Daye, Hubei; 18, Puchi, Hubei; 19, Sangzhi, Hunan; 20, Cili, Hunan; 21, Yichun, Jiangxi; 22, Laiyang, Hunan; 23, Yongxing, Hunan; 24, Jiahe, Hunan; 25, Yizhang, Hunan; 26, Chaoxian, Anhui; 27, Huaining, Anhui; 28, Changxing, Zhejiang; 29, Wuxing, Zhejiang; 30, Guangfeng, Jiangxi; 31, Shangrao, Jiangxi; 32, Zhangping, Fujian; 33, Longyan, Fujian.

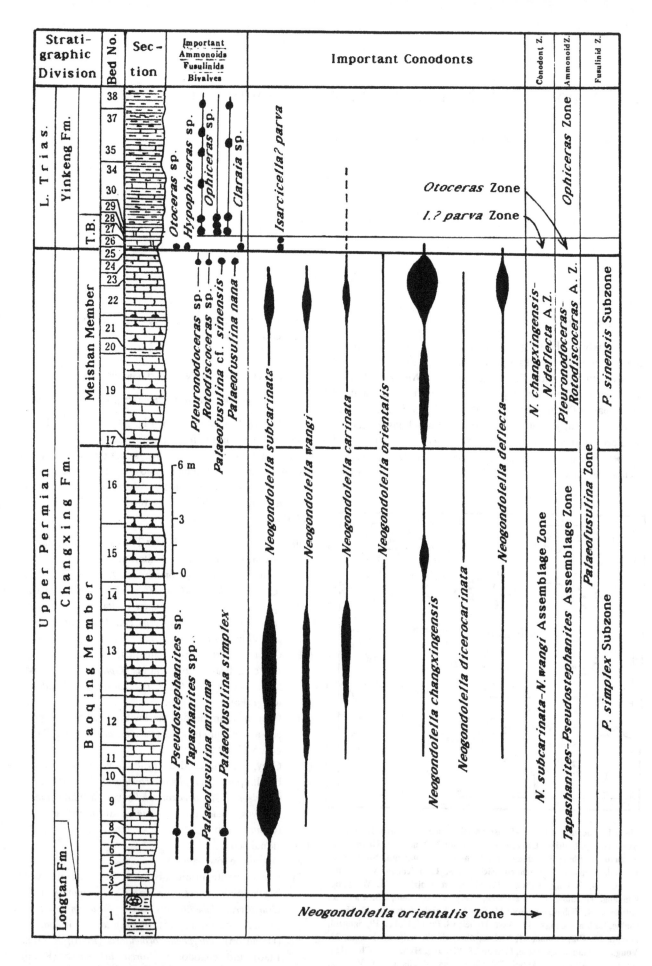

Fig. 2.2. The section at Meishan, Changxing Co., Zhejiang Province, with ranges (in m) of species and other taxa useful in correlation. T.B. = Transitional Beds. (After Li *et al.*, 1991; and Yin *et al.*, 1988.)

(*Cathaysia chonetoides* (Chao), *Crurithyris flabelliformis* Liao, *Neochonetes convexa* Liao, *Paryphella orbicularis* (Liao), *P. triquetra* Liao, *Uncinunellina* sp., *Waagenites barusiensis* (Davidson), *W.* sp. cf. *W. soochowensis* (Chao), *W. wongiana* (Chao)), bivalves (*Claraia* sp.), and conodonts (*Neogondolella changxingensis* Wang & Wang, *N. deflecta* Wang & Wang). . . . . **0.07 m**

– – – – conformity – – – –

Changxing Formation, Upper Permian. Divided into Meishan Member (units 17–25) and Baoqing Member (units 2–16).

(25) White montmorillonite clayrock, bearing conodonts (*Neogondolella changxingensis* Wang & Wang, *N. deflecta*

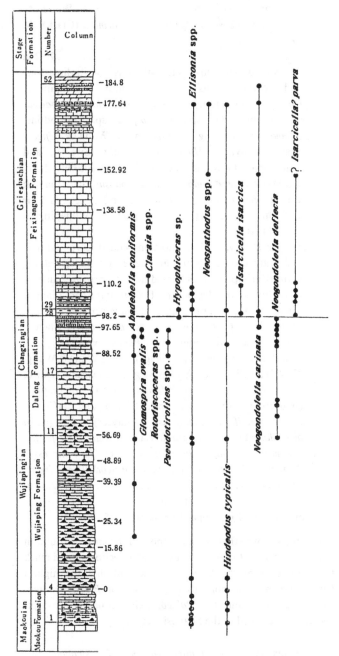

Fig. 2.3. The section at Shangxi, Guangyuan Co., Sichuan Province, with ranges (in m) of species and other taxa useful in correlation. 'Number' column includes numbers of beds just above horizontal lines. (Modified from Yin *et al.*, 1988.)

Wang & Wang, *N. orientalis* (Barskov & Koroleva)) . . . .**0.04 m**

(24) Grey medium-bedded, micritic limestone, containing fusulinids (*Palaeofusulina* sp.), ammonoids (*Rotodiscoceras* sp.), conodonts (*Neogondolella changxingensis* Wang & Wang, *N. deflecta* Wang & Wang). . . . .**0.20 m**

(23) Black medium-bedded, bioclastic, mud-banded, micritic limestone, bearing conodonts (*Xaniognathus?* sp., *Neogondolella changxingensis* Wang & Wang) . . .**0.22 m**

(17–22) Argillaceous, dolomitized, bioclastic, siliceous-banded, and micritic carbonate rocks, with a rich conodont fauna characterized by *Xaniognathus?* sp., *Neogondolella carinata* (Clark), *N. changxingensis* Wang & Wang, etc.

. . . .**1.30 m**

(2–16) Bioclastic limestone, in part siliceous, argillaceous, carbonaceous, or asphaltic, or intercalated with siltstone; containing about the same conodonts as above, and the brachiopods *Crurithyris* sp., *Leptodus.* sp., *Meekella* sp., *Orthotetina* sp., *Araxathyris* sp., *Cathaysia chonetoides*, *Chonetes chonetoides* (Chao). Bed 6 (5.65 m thick) is exceptionally rich in conodonts (same as in beds 17–22), ammonoids (*Lopingoceras quangdense* Zhao, Liang & Zheng, *Longmenshanoceras changxingense* Yang & Huo, *Mingyuexiaceras changxingense* Zhao, Liang & Zheng, *M. radiatus* Zhao, Liang & Zheng, *Parametaceras* sp., *Pseudogastrioceras gigantum* Zhao, Liang & Zheng, *Pseudostephanites meishanensis* Yang & Huo, *Sinoceltites costatus* Zhao, Liang & Zheng, *S. opimus* Zhao, Liang & Zheng, *S. sichuanensis*, *Tapashanites chaotianensis* Zhao, Liang & Zheng, *T. compressus* Zhao, Liang & Zheng, *T. curvoplicatus* Zhao, Liang & Zheng) and brachiopods (*Cathaysia chonetoides* (Chao), *Squamularia* sp.).

– – – – conformity – – – –

Longtan Formation (Upper Permian)

In the Meishan section, the Permo-Triassic boundary was formerly set between beds 24 and 25 (Zhao *et al.*, 1981; Sheng *et al.*, 1984), but recent study by Zhang (1987) shows that it should be placed between beds 25 and 26 because of the presence in bed 25 (the 'boundary clay') of Permian-type conodonts (*Neogondolella changxingensis*, *N. deflecta*, and *N. orientalis* of the *N. changxingensis*–*N. deflecta* assemblage zone). Bed 24 is characterized by fusulinids (*Palaeofusulina* sp.), ammonoids, and conodonts of late Changxingian age.

The Shangsi section, in Guangyuan, Sichuan Province

The Shangsi section (Fig. 2.1, locality 4; Fig. 2.3), which consists of basinal deposits, was studied as early as 1961 by Zhao Jinke, and has been restudied by members of IGCP Project 106 and the Chinese Working Group of IGCP Project 203, especially Yang Fengqing (in Yang *et al.*, 1987) and Li Zishun (Li *et al.*, 1986). The detailed section given by Li (submitted) is partially reproduced as follows:

Feixianguan Formation (Lower Triassic), lower member

(36–51) Mainly micritic limestone rich in conodonts *Hindeodus typicalis* (Sweet), *Isarcicella? parva* (Kozur &

Pjatakova)).......................**65.86 m**

(35) Grey-brown, medium-bedded micritic limestone intercalated with thin-plated calcareous shale having stromatolitic structure; a few forminifer and bivalve fragments; top part with conodonts..........**6.57 m**

(34) Grey, thin-bedded, micritic limestone, intercalated with thin-plated shale.......................**1.20 m**

(33) Grey, medium-bedded, pebbly, micritic limestone; lower part with conodonts (*Isarcicella isarcica* (Huckriede), *Chirodella* sp., *Hindeodus typicalis* (Sweet)).....**2.68 m**

(32) Grey, laminated, algal-sheeted limestone, medium- to thin-bedded, with some organic matter; well-developed, horizontal bedding; middle part with conodonts (*Isarcicella? parva* (Kozur & Pjatakova), *Xaniognathus? sp.*).....................................**0.96 m**

(31) Yellowish-green, thin-bedded, argillaceous, micritic limestone, alternating with greyish, laminated, algal, dolomitic limestone; top part with 'birdseye' structure; basal part yielding conodonts (*I.? parva* (Kozur & Pjatakova))............................**2.69 m**

(30) Yellowish-green, argillaceous limestone, intercalated with grey argillaceous, micritic limestone; laminated, horizontally bedded; yielding conodonts (*I.? parva* (K. & P.), *Hindeodus typicalis* (Sweet) and, in basal part, the palynomorphs *Chordasporites* sp., *Abietineaspollenites* sp., *Pinuspollenites* sp., *Catonispollenites* sp., and the bivalve *Claraia* sp........................**1.80 m**

(29) Yellowish-green illite clayrock, intercalated with calcareous argillite and argillaceous limestone containing conodonts (*Hindeodus typicalis* (Sweet), *Ellisonia triassica* Müller, *Neospathodus* sp.), bivalves (*Claraia griesbachi* (Bittner), *Pseudoclaraia wangi* (Patte), *Towapteria scythica* (Wirth)), and sporopollen (*Verrucosisporites* sp., *Krauselisporites?* sp., *Alisporites* sp., *Pinuspollenites* sp., *Podocaroidites* sp., *Cedripites* sp., *Taeniaesporites* sp., *Chadrosporite* sp., *Jugasporites* sp.)..........**1.91 m**

(28) Yellowish-green, silty, illite clayrock, intercalated with small amount of calcareous argillite and argillaceous limestone; conodonts (*Hindeodus typicalis* (Sweet)) in top part; *Eumorphotis venetiana* (Hauer) 45 cm from base; *Promyalina* sp. 10 cm from base; and, within 16 cm of base, the conodonts *Neogondolella changxingensis* Wang & Wang, *N. subcarinata* Sweet, the sporopollen *Circumpolis* sp., *Cycadopites* sp., and the ammonoids *Hypophiceras* sp., *Metophiceras* sp., *Tompophiceras* sp., and *Pseudogastrioceras* sp....................**1.78 m**

– – – – conformity – – – –

Dalong Formation. Upper Permian, partly Changxingian (beds 17–27c).

(27c) Greyish black, irregularly mixed montmorillonite or illite clayrock, yielding ammonoids (?*Huananoceras* sp.), bivalves (*Claraia guanyuanensis* Li), and conodonts (*Neogondolella subcarinata* Sweet, *N. changxingensis* Wang & Wang)......................**0.04 m**

(27b) Greyish white, irregularly mixed, montmorillonite–illite clayrock, with volcanic shards..............**0.06 m**

(27a) Greyish black microlaminated montmorillonite-illite clayrock, containing some organic substance but with only rare terrestrial clastics. Yields ammonoids

(*Pseudotirolites asiaticus* (Jaeckel)), brachiopods (*Crurithyris changjianggouensis* Zhan, *C. pigmaea* (Liao)), and conodonts (*Neogondolella changxingensis*, *N. deflecta* Wang & Wang, *N. subcarinata* Sweet, *N. tulongensis* Wang & Wang)......................**0.05 m**

(20–26) Mainly grey or dark grey, thin-bedded, siliceous limestone or shale, including two beds (23, 25) of illite-montmorillonite clayrock; all richly fossiliferous. Ammonoids (*Pseudotirolites disconnectens* Zhao, Liang & Zheng, *P. regularis* Zhao, Liang & Zheng, *Rotodiscoceras* sp., *Pleuronodoceras* sp. cf. *P. dushanense* Zhao, Liang & Zheng, *Pseudogastrioceras* sp. cf. *P. gigantum* Zhao, Liang & Zheng); foraminifers (*Glomospira ovalis* Malachova, *G. regularis* Lipina, *Nodosaria* sp.) and conodonts (*Neogondolella deflecta* Wang & Wang, *N. changxingensis* Wang & Wang, *N. carinata* (Clark), *N. subcarinata* Sweet)
....**7.92 m**

(17–19) Medium- to thin-bedded, siliceous, micritic limestone and carbonaceous shale, yielding foraminifers (*Shangsilites shangsiensis* Sheng, *Nodosaria* sp., *Neodiscus* sp., *Hemigordius* sp., *Glomospira* sp.), conodonts (*Neogondolella changxingensis* Wang & Wang, *N. deflecta* Wang & Wang, *Hindeodus typicalis* (Sweet)), and ammonoids (*Tapashanites* sp. cf. *T. floriformis* Zhao, Liang & Zheng, *Pseudostephanites nodosus* Zhao, Liang & Zheng, *Shevyrevites* sp., *Sinoceltites* sp.)......**11.69 m**

Dalong Formation. Upper Permian, partially Wujiapingian (beds 11–16).

(11–16) Dark grey, medium- to thin-bedded, carbonaceous and argillaceous, micritic limestone and shale, yielding various groups of fossils at different horizons, such as ammonoids (*Konglingites* sp. cf. *K. striatus* Zhao, Liang & Zheng, *Prototoceras* sp., *Pseudogastrioceras* sp.), fusulinids (*Codonofusiella kueichowensis* Sheng); nonfusulinid foraminifers (*Palaeotextularia* sp.), conodonts (*Neogondolella changxingensis* Wang & Wang, *N. deflecta* Wang & Wang, *N. guangyuanensis* (Dai & Zhang)), corals (*Tachylasma* sp.), brachiopods (*Acosarina* sp., *Waagenites wongiana* (Chao), *Spinomarginifera lopingensis* (Kayser), *Paryphella orbicularis* Liao, *P. elegantula* Zhan, *Cathaysia subpusilla* (Licharew), *Compressoproductus* sp. cf. *P. compressus* (Waagen), *Leptodus* sp., *Permophricodothyris indica* (Waagen))........................**22.37 m**

– – – – conformity – – – –

Wujiaping Formation. Upper Permian (beds 5–10).

It is to be noted that the ammonoid *Hypophiceras* found in association with '*Otoceras*' in the Meishan section should mark the basal part of the Lower Triassic and that the highest Changxingian ammonoid zone, the zone of *Pseudotirolites–Pleuronodoceras*, and the *Palaeofusulina* zone, are conspicuously represented in the Shangsi section.

Beifengjing section, Liangfenya, Chonqing, Sichuan Province

This sequence of Permian and Triassic rocks (Fig. 2.1, locality 8) has been carefully worked out by Chinese members of

IGCP Project 203 (Yin, 1985; Wu, in Yang *et al.*, 1987). It is briefly reproduced here, for reference:

Feixianguan Formation, first member (part). Lower Triassic

(7) Alternating beds of grey, thin-bedded, argillaceous limestone, marl, and calcareous argillite, containing bivalves (*Claraia griesbachi* (Bittner), *C. painkhandana* (Bittner)) . . . . **1.00 m**

(6) Alternating beds of yellowish-green, thin-bedded marl and yellow calcareous argillite, yielding bivalves (*Claraia griesbachi* (Bittner), *C. hunanica* (Hsu), *C. stachei* (Bittner)), and brachiopods (*Lingula tenuissima* Wirth *sinensis*) . . . . **1.40 m**

Transitional beds

(5) Alternating beds of yellowish-green marl and yellow, laminated calcareous argillite, with bivalves (*Claraia* sp, *Eumorphotis inaequicostata* (Benecke), *Leptochondria minima* (Kiparisova)), brachiopods (*Crurithyris speciosa* Wang, *Lingula subcircularis* Wirth) and conodonts (*Isarcicella? parva* (Kozur & Pjatakova), *Hindeodus typicalis* (Sweet), *Isarcicella sichuanensis* Ding) . . **0.56 m**

(4) Earthy yellow clayrock bearing calcite crystals . . **0.04 m**

– – – – conformity – – – –

Changxing Formation. Upper Permian

(3) Greyish argillaceous limestone with pyrite crystals, yielding brachiopods (*Crurithyris* sp.) and conodonts (*Hindeodus typicalis* (Sweet)) . . . . **0.15 m**

(2) Yellowish-white calcareous argillite and clayrock **0.06 m**

(1) Greyish-green argillaceous limestone with pyrite crystals, yielding fusulinids (*Palaeofusulina* sp., *Reichelina* sp. cf. *R. tenuisssima* M.-Maclay), brachiopods (*Cathaysia parvula?* Chan, *Crurithyris speciosa* Wang, *C. pusilla* Chan, *Paracrurithyris pigmaea* (Liao), and nonfusulinid foraminifers (*Geinitzina spandeli* Tscherd, *Geinitzina* sp. cf. *G. caucasica* M.-Maclay, *Nodosaria netchajewi subquadrata* Lipina, *Robuloides acutulus* M.-Maclay) . . . . **0.19 m**

Huaying section, Lingshui, Sichuan Province

This section is locality 6, Fig. 2.1. We show only the lower three beds, 2.89 m thick, of the 140-m thick first member of platformal deposits of the Daye Fm.

Daye Formation. Lower Triassic

(25) Grey, thin-bedded marl, intercalated with calcareous argillite, containing bivalves (*Eumorphotis multiformis* (Bittner), *Leptochondria* sp., *Pseudoclaraia wangi* (Patte), *Claraia griesbachi* (Bittner), *C. hubeiensis* (Chen)) and brachiopods (*Lingula* sp.). . . . . **1.71 m**

(24b4) Greyish-green marl, containing brachiopods (*Lingula* sp., *L. subelliptica* Mansuy) and cephalopods (*Ophiceras* sp.) . . . . **0.62 m**

(24b3) Greyish-green marl, with brachiopods (*Lingula* sp.), bivalves (*Pseudoclaraia wangi* (Patte), *Claraia griesbachi* (Bittner), *C. dieneri* (Nakazawa), *Eumorphotis* sp.) **0.56 m**

Transitional beds

(24b2) Greyish-green, thin-bedded marl with brachiopods (*Lingula* sp., *Crurithyris pusilla* (Chan), *Acosarina indica* (Waagen)), bivalves (*Eumorphotis* sp., *Leptochondria* sp.,

Pteria ussurica variabilis Chen & Lan, *Towapteria scythica* (Wirth), *Bakevellia* sp.). **0.04 m**

(24b1) Grey, thin-bedded argillaceous limestone, with calcareous argillite in lowermost 0.03 m. Foraminifers (*Nodosaria netchajewi* Lipina) and brachiopods (*Crurithyris speciosa* Wang) . . . . **0.26 m**

– – – – conformity – – – –

Changxing Formation, Upper Permian. Only uppermost three beds, representing 2.99 m of the entire 125.51 m thickness, are described.

(24a) Dark grey, thin-bedded limestone, with trilobites (*Pseudophillipsia chongingensis* (Lu)), rugose corals (*Paracaninia sinensis* (Chi)), brachiopods (*Neoplicatifera multispinosa* Ni, *Waagenites barusiensis* (Davidson), *W. soochowensis* (Chan), *Crurithyris pusilla* Chan, *Plicochonetes dissulcata* Liao, *Acosarina minuta* (Abich), *A. indica* (Waagen), *Paracrurithyris pigmaea* (Liao)), nonfusulinid foraminifers (*Nodosaria meitienensis* Hao & Lin, *N. mirabilis caucasica* M.-Maclay, *Geinitzina* sp., *Glomospirella* sp., *Pseudoglandulina* sp.). . . . . **0.42 m**

(23) Dark grey calcareous shale. . . . . **0.21 m**

(22) Grey, medium- to thin-bedded chert-nodular limestone, yielding conodonts (*Hindeodus typicalis* (Sweet), *Neogondolella carinata* (Clark), *N. changxingensis* Wang & Wang). . . . . **2.36 m**

It is worth noting that the trilobites, rugose corals, and brachiopods in bed 24a, and especially the fusulinid, *Palaeofusulina*, in bed 15 (not included in the above section), indicate a Late Permian Changxingian age for this part of the section, whereas the superposed marls (beds 24b1–4, 25) are marked predominantly by bivalves such as *Leptochondria, Eumorphotis, Towapteria*, and *Claraia*, which indicate a new stage of faunal development. For that reason beds 24b1 and above are referred to the Triassic.

Biostratigraphic zonation and correlation

Faunal analyses of the four sections just described, and of other sections in South China, show that the distribution of ammonoids, conodonts, and fusulinids permits establishment of the following zones (Table 2.1).

Changxingian ammonoid zones

Five zones were originally established by Zhao *et al.* (1981), but these were later reduced to two by Yang F. Q. (in Yang *et al.*, 1987). These two zones are widely recognized as standard for the youngest known marine Permian formations, particularly in the siliceous and clastic Dalong facies. They are (1) the *Tapashanites–Shevyrevites* Zone, including *Paratirolites*, and (2) the *Rotodiscoceras–Pseudotirolites* Zone, with *Pleuronodoceras*. These two zones represent, respectively, the early and later parts of the Changxingian Stage.

The *Tapashanites–Shevyrevites* Zone is confined to the lower Changxingian and is recognized at seven localities spread over six provinces (Shaanxi, Sichuan, Guangxi, Hunan, Zhejiang, Guizhou). The fauna of the zone includes *Tapashanites, Paratirolites, Shevyrevites, Sinoceltites, Pseudostephanites, Mingyuex-*

Table 2.1. *Zonation of Permo-Triassic boundary interval in South China on the basis of ammonoids, conodonts, fusulinids, and bivalves*

Series	Stage		Ammonoid Zones	Conodont Zones	Fusulinid Zones		Bivalve Zones
Lower Triassic	Griesbachian	Upper	*Ophiceras*	*Isarcicella isarcica*	– – –		*Pseudoclaraia wangi*
		Lower	*Otoceras?/ Hypophiceras*	*Isarcicella? parva*	– – –		– – –
Upper Permian	Changxingian		*Rotodiscoceras- Pseudotirolites Pleuronodoceras*	*Neogondolella deflecta- N. changxingensis*	*Palaeofusulina*	*P. sinensis*	– – –
			Tapashanites- Shevyrevites- Pseudostephanites (Paratirolites)	*N. subcarinata – N. wangi*		*P. simplex*	– – –

iaceras, *Changhsingoceras*, *Liuchengoceras*, *Laibinoceras*, *Penglaites*, *Pseudogastrioceras*, and a minor contingent of primitive *Pseudotirolites*. Among these, only *Paratirolites*, *Shevyrevites*, *Pseudogastrioceras*, and doubtful *Pseudostephanites* have been reported from localities outside China.

Shevyrevites occurs in the lower Changxingian in Anshun, Guizhou; Xixian, Shaanxi; and Laibin, Guangxi. It has also been reported from the Dorashamian of Transcaucasia, USSR; from the Ali Bashi Formation of northwest Iran; and from the Hambast Formation, in the Abadeh region, central Iran.

Paratirolites is known from the same regions as *Shevyrevites*. In terms of species and individuals, it is an important faunal element in the Dorashamian of Transcaucasia, and also in Anshun, Guizhou.

The *Rotodiscoceras–Pseudotirolites* Zone is extensively distributed in South China and its fauna includes many genera and species. Among genera may be mentioned: *Pleuronodoceras*, *Pachydiscoceras*, *Dushanoceras*, *Qianjiangoceras*, *Xenodiscus*, *Chaotianoceras*, *Huananoceras*, *Pseudogastrioceras*, and *Pernodoceras*. The great majority of these are confined to the upper Changxingian of South China. *Rotodiscoceras* has been reported from the Upper Permian of the South Kitakami Massif, Japan, but only in Iran have fossils resembling *Pseudotirolites* and *Pleuronodoceras* been found.

Pseudogastrioceras and *Huananoceras* range through the entire Upper Permian. *Pseudotirolites* and *Pleuronodoceras* are each represented by a dozen or so species, and representatives are encountered commonly in the middle and upper parts of the Dalong Formation. *Rotodiscoceras* is widely distributed in Sichuan, Guizhou, Guangxi, Zhejiang, Anhui, Shaanxi, and Hunan provinces, and is always associated with *Pseudotirolites* and *Pleuronodoceras*.

Lower Triassic ammonoid zones

Except at two localities in India (Shalshal Cliff and Spiti) where *Ophiceras* and *Otoceras* occur in the same bed, representatives of the Ophiceratidae are found elsewhere to be abundant immediately above the highest *Otoceras* or its associate, *Hypophiceras*, as in sections near Mt. Qomolangma, at Meishan in Zhejiang, in Kashmir, in Arctic Canada, and in Greenland. So far as Chinese materials go, there are two distinct Lower Triassic zones (see Table 2.1), a lower *Otoceras*, or better *Hypophiceras* Zone and, above it, an *Ophiceras* Zone. In their correlation of the lowermost Triassic of South China, Dagys & Dagys (1988) omitted the *Otoceras* or *Hypophiceras* Zone, leaving a glaring gap between the *Ophiceras* Zone and the *Rotodiscoceras–Pseudotirolites–Pleuronodoceras* Zone. Such a correlation can not be accepted.

Changxingian conodont zones

Two conodont zones have been proposed by Wang & Wang (1981) and advocated by Zhang (1987). The *Neogondolella subcarinata–N. wangi* Zone, in the lower member of the Changxing Formation (Fig. 2.2; Table 2.1) is an acme zone, which corresponds to the *Tapashanites–Shevyrevites* ammonoid zone (including *Pseudostephanites* and *Paratirolites*) and to the *Palaeofusulina simplex* fusulinid subzone. This conodont zone is widely distributed in South China and has been recognized at more than a dozen localities in Jiangsu, Zhejiang, Anhui, Sichuan, Guizhou, and Guangxi provinces.

Neogondolella subcarinata Sweet occurs in the *Phisonites–Paratirolites* beds (lower part of Dorasham Formation) of Transcaucasia (Kozur, 1978), in the Ali Bashi Formation of northwest Iran (Teichert *et al.*, 1973), and in the upper part of the Hambast Formation in the Abadeh region of central Iran (Iranian–Japanese Research Group, 1981). Consequently, the beds mentioned have been correlated with the lower Changxing Formation. It should be noted, however, that *N. subcarinata* is also represented in the Lower Triassic Dinwoody Formation, at a locality in the Terrace Mountains of Utah, in western United States (Paull, 1982). Hence *N. subcarinata* should be used with caution as an indicator of an early Changxingian age for strata in which it is represented.

The *Neogondolella changxingensis–N. deflecta* Zone is also an acme zone, which is confined to the upper Changxingian and corresponds to the *Rotodiscoceras–Pseudotirolites–Pleuronodoceras* ammonoid zone and to the *Palaeofusulina sinensis* fusulinid subzone. According to available data, *N. changxingensis* has been identified at more than 21 localities in 10 provinces of South China. *N. deflecta*, on the other hand, has been observed at only 16 localities in seven provinces. The fauna of this conodont zone is abundantly represented in the upper member of the Changxing and Dalong formations, and is associated at eight localities with ammonoids of the upper Changxingian *Rotodiscoceras–Pseudotirolites–Pleuronodoceras* ammonoid zone. At present, the zone is known primarily from South China. However, *N. changxingensis* and *N. deflecta* have both been reported from beds with *Otoceras latilobatum* in the Selong section of Tibet (Yao & Li, 1987); *N. changxingensis* ranges through much of the Dorashamian Ali Bashi Formation of northwest Iran (Wang & Wang, 1981; Sweet, this volume); and Sweet (this volume) reports *N. deflecta* from a bed at the top of the *Otoceras woodwardi* Zone in the section at Guryul Ravine, Kashmir.

Lower Triassic conodont zones

The *Isarcicella? parva* Zone is represented at 13 localities in six provinces of South China. It is also known from beds in the *Otoceras woodwardi* Zone of Guryul Ravine, Kashmir; from the *Claraia* beds at Dorasham, in Transcaucasia; and from the basal part of Unit A in the Abadeh region of central Iran. *I.? parva* has also been reported from the main dolomite of the Kathwai member of the Salt Range, Pakistan, and it first appears near the base of the Werfen Formation in the Southern Alps of Italy.

It is to be noted that six species of *Neogondolella* that flourished in latest Changxingian times disappeared altogether at the end of the Changxingian. Only two species, *N. subcarinata* and *N. carinata*, continued into the Early Triassic.

The Changxingian, especially in its carbonate facies (the Changxing Formation) is characterized by the *Palaeofusulina* Zone, which is divisible into the *P. simplex* and *P. sinensis* subzones.

The Permo-Triassic boundary

Previous parts of our report follow the traditional practice of placing the Permo-Triassic boundary below the *Otoceras* Zone, which is at least partially equivalent to the *Hypophiceras* or *Isarcicella? parva* zones (Yin *et al*, 1988). This boundary is favored for the following reasons:

1 The major alteration of the marine invertebrate fauna took place most conspicuously between the Permian and the Triassic, that is, at the turn of the Paleozoic into the Mesozoic (Fig. 2.4). In Fig. 2.4, the boundary we favor is at E (for event) 1, at which level a mass extinction of Paleozoic life involved the total disappearance of an entire class (Trilobita), two orders (Rugosa, Cryptostomata), and a suborder or superfamily (Fusulinida), although some other groups dropped out at various times in the late Permian. Also, at this level Changxingian ammonoids such as *Rotodiscoceras*, *Pseudotirolites*, and *Pleuronodoceras*, which flourished in numbers of species

and individuals, were replaced by the easily distinguishable *Hypophiceras* fauna, whose key members, *Otoceras* and *Hypophiceras* were allied more to the Permian *Julfotoceras* and Xenodiscus, respectively, than to any Triassic forms. Finally, just above the boundary clayrock the Ophiceratidae begin to expand gradually and proliferation of bivalves (*Towapteria*, *Claraia*, and *Eumorphotis*) represents another wave of pectinoid development that has generally been considered a hallmark of early Triassic (Werfenian) age.

2 Placement of the Permo-Triassic boundary between the Griesbachian and Dienerian stages, as Newell (1988) does, has the advantage of marking 'the beginning of a new cycle of diversification and radiation of ammonoids (meekoceratids) and conodonts (*Neospathodus*). It is a biostratigraphic boundary that is easier to recognize and far more widespread than the Dorashamian/ Griesbachian boundary.' This is the boundary marked E 3 in Fig. 2.4. However, this level is much less important in magnitude of faunal change than level E1, which was marked by mass extinctions of orders or even a class of Paleozoic invertebrates.

3 Event 2 is marked by the appearance of numerous ophiceratids in many sections in Canada, Greenland, Siberia, India (Spiti), and Kashmir. In these sections beds with *Ophiceras* overlie strata with *Otoceras* (Tozer, 1978), or at least *Ophiceras* ranges higher than *Otoceras*. Kozur (1973) proposed to include the *Otoceras* Zone in the Permian, thus drawing the Permo-Triassic boundary between the *Ophiceras* and *Otoceras* zones. Kozur maintains the same view in subsequent papers (Kozur, 1978, 1980). Bhatt & Arora (1984) also chose to place the Permo-Triassic boundary at the base of the *Ophiceras* beds in Spiti sections on the basis of such evidence as the extinction of the Otoceratidae, the Productidae, and several species of *Neogondolella*, as well as the proliferation at this level of the Triassic pelecypod *Claraia*. In our opinion, however, event 2 is much less significant than event 1.

4 The 'boundary clayrock' of South China is tuffaceous for it contains volcanic glass. A similar clayrock occurs at nine localities in Guizhou, Guangxi and Guangdong provinces. Li *et al*. (1986) discovered submarine eruptives (tuffs), now altered into montmorillonite-illite clay. Geochemically there is an abrupt shift in carbon-isotopic composition that corresponds precisely to a Permo-Triassic boundary based on faunal evidence in the Changxing Formation (Chen *et al*., 1984). Analyses by Zhang *et al*. (1983) indicate that the content of Cs, Hf, Ta, Th, U, and Zr is higher in the boundary clayrock than in those above or below it. The Th/U ratio shows a higher anomaly, but that of Zr/Hf is lower at both Changxing and Guangyuan. Abundance patterns of 29 elements from Changxing samples studied by Chai *et al*. (1986) show that Ir, Os, Pt, Au, Re, and other siderophile and chalcophile elements are more or less enriched at the boundary. Further, the discovery by Sun *et al*. (1984) of an iridium anomaly in the boundary clayrock at Changxing, and a study by Xu *et al*. (1985) of the variation

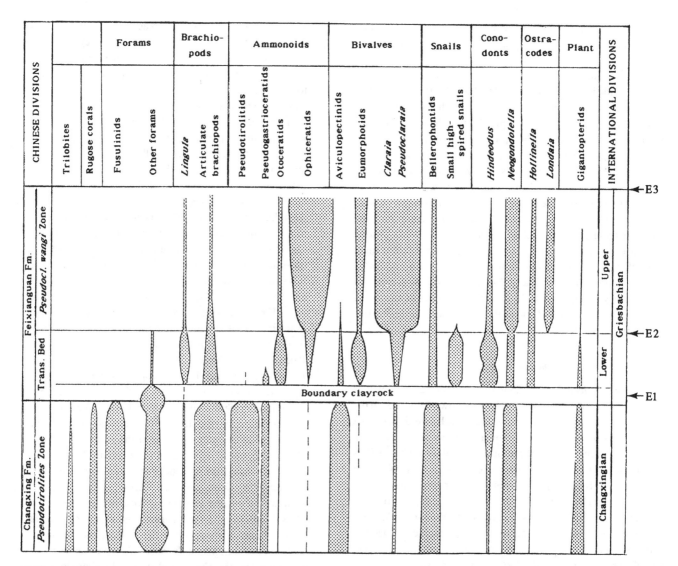

Fig. 2.4. Schematic chart of alterations in Permo-Triassic biota in South China. (After Yin Hongfu, 1985; E1, E2, and E3 added; names of conodont genera up-dated.)

in abundance of iridium and trace elements at the Permian-Triassic boundary at Shangsi have brought to light the presence of some great nonbiotic event at the Permo-Triassic boundary. No similar geochemical anomalies have been detected at the other boundary levels (E2, E3).

Permo-Triassic transitional beds (with mixed faunas)

We use the term 'transitional beds' for beds in a Permo-Triassic sequence, a few dozen centimeters to several meters thick, of fairly homogeneous or only slightly variable lithology, and with a mixture of Permian survivors (dwarfed chonetids, reticulariids, etc.) and a small number of 'Triassic-type' forms (*Isarcicella? parva*). Such beds have been found at some 20 localities in 11 provinces (Liao, 1979, 1980; Yin, 1985), and they form a special unit that lies conformably on rocks of Changxingian age and is succeeded by the conventional basal Triassic

Pseudoclaraia wangi Zone, rich in *Claraia* and *Ophiceras* (Table 2.2).

Based on the presence in them of key fossils such as *Isarcicella? parva* the transitional beds are regarded as forming the basal unit of the Triassic in China (Yin, 1985). *I.? parva* is widely distributed in the Tethyan region (Table 2.3), but *Otoceras* is known in China only from the Mt. Qomolangma area of Tibet, and possibly from Meishan, Zhejiang. Both *I.? parva* and *Otoceras* indicate an early, most probably a late early Griesbachian age for the transitional beds.

Permian-type brachiopods in the transitional beds are treated as holdovers because (1) by comparison with their Permian predecessors, they are impoverished in terms of genera and species and (2) they are minute, or dwarfed, and fragile, hence were most probably pseudoplanktonic in habit and might thus have survived the terminal Permian mass extinction.

Once the boundary is placed at the 'boundary clay', then Permian-type elements are naturally treated as holdovers.

Table 2.2. *Upper Permian and Lower Triassic stages and zones in South China*

Age	Rock Units	International Chronostratigraphic Units	Traditional Chinese Chronostratigraphic Units	Ammonoid Zones	Conodont Zones	Bivalve Zones	Brachiopod assemblages and Assemblage Zones	Fusulinid Zones	Non-fusulinid Foraminiferal assemblages
Lower Triassic	Jialingjiang Fm	Spathian	Olenekian — Columbitan	*Subcolumbites*	*Neospathodus timorensis*	*Pteria* cf. *murchisoni*			
		Smithian	Columbitan	*Tirolites-Columbites*	*N. homeri - N. anhuiensis* / *N. collinsoni*	*Eumorphotis inaequicostata*			
			Owenitan	*Anasibirites*	*N. waageni*				
				Owenites		*Eumorphotis multiformis*			
	Daye Fm	Dienerian	Induan — Gyronitan	*Flemingites*	*N. pakistanensis*	*Claraia aurita*			
				Prionolobus	*N. cristagalli*	*Claraia stachei*			
		Ellesmerian / Gangetian (Griesbachian)	Otoceratan	*Ophiceras*	*Neogondolella carinata* / *N. dieneri*	*Pseudoclaraia wangi*			
				Lytophiceras	*Isarcicella isarcica*				
				Otoceras?	*Isarcicella ? parva*	*Pteria ussurica variabilis-* / *Towapteria scythica*	*Crurithyris speciosa-* / *Lingula subcircularis* assemblage		
Upper Permian	Changxing Fm	Dorashamian (?)	Changxingian — Upper Substage	*Rotodiscoceras-* / *Pseudotirolites*	*Neogondolella deflecta-* / *Neogondolella changxingensis*	*Hunanopecten exilis*	*Waagenites barussiensis-* / *Crurithyris pusilla* Zone	*P. sinensis-* *Reichelina changhsiangensis* Subzone — Palaeofusulina Zone	non-*Colaniella* mixed assemblage / *Colaniella* assemblage / *Glomospira*
			Lower Substage	*Tapashanites-* / *Shevyrevites*	*Neogondolella subcarinata-* / *Neogondolella wangi*		*Enteletina zigzag-* / *Neowellerella pseudoutah* Zone	*P. simplex-* *Gallowayinella meitienensis* Subzone — Palaeofusulina Zone	
	Wujiaping Fm	Dzhulfian	Wujiapingian	*Sanyangites* *Araxoceras-* *Konglingites*	*Neogondolella orientalis*	*Guizhoupecten regularis*	*Orthotetina ruber-* *Squamularia grandis* Zone	*Codonofusiella schubertelloides*	
				Anderssonoceras- *Prototoceras*	*Neogondolella liangshanensis*		*Squamularia indica* *Haydenella wenganensis* Zone		

Source: (After Yang Zunyi *et al.,* 1987.)

Table 2.3. *International correlation of Permo-Triassic strata*

South China (this paper)	Mt. Qomolangma area (Wang & He, 1976)	KASHMIR (Nakazawa & al. 1975)	Salt Range (Kummel & Teichert, 1970; Pakistan–Jap. Res. Gp. 1981)	Dzhulfa USSR & Ali Bashi, Iran (Rostovtsev & al., 1973; Teichert & al., 1973)	Abadeh, Iran (Iran–Jap. Res. Gp., 1981)	NE Japan	Maizuru Zone	SW Japan	Greenland (Tozer, 1967; Teichert & Kummel, 1976)	No. America Summarized	USSR Summarized
Subcolumbites Tirolites Columbies (Jialingjiang Fm.)	*Procarnites–Anasibirites*	H	Narmia Fm.	*Owenites–Eumorphotis venetiana* Bed	*Neospath-odus dieneri* (Unit A)	Inai Gp.	Yakuno Gp.	Kamura Fm.		Spathian	Olenekian
Anasibirites										Smithian	
Owenites	*Owenites*	G									
Flemingites		F	Mittiwali Mbr.	*U. Claraia* Bed; *C. aurita*; *C. stachei*; *Gyronites*	*Claraia– Ophiceras*; *Isarcicella isarcica*				*Ophiceras commune*	Dienerian	Induan
Prionolobus	*Gyronites psilogyrus*										
Ophiceras– Pseudoclaraia wangi (Feixianguan (Daye) Fm.)	*Lytophiceras sakuntala*	E3 (Khunamuh Fm.)	Ls. Bed (Kathwai Mbr.)	*L. Claraia* Bed; *Lytophiceras*; *Pseudoclaraia wangi*	*Isarcicella? parva*		*(Glypto-ophiceras)*		*Otoceras boreale*	Griesbachian / Gangetian / Ellesmerian	
(Transitional bed)		E2	Main Dolomite Bed	*Isarcicella? parva*		Toyama Fm.	Gujo Fm.	Mitai Fm.	*O. con-cavum* / *Hyp-ophi-ceras*		
Isarcicella? parva, Otoceras?	*Otoceras latilobatum*		L. Dolomite Bed								
Rotodiscoceras– Pseudotirolites (Changxing Fm.)		E1 (Zewan Fm.)	Chhidru Fm.	Dorasham Fm.	? / *Para-tirolites* / *Shevyrevites* (Hambast Fm.)	Kanokura Fm.	L. Mai-zuru Gp. (M. & U. Maizuru Gp.)	Kuman Fm.	Foldvik Creek Fm.	? / Ochoan	? / Tatarian
Tapashanites– Shevyrevites (Paratirolites)	?	D		Julfa Fm.	*Vedioceras*						
Sanyangites Araxoceras Konglingites (Wujiaping Fm.)		C	Kalabagh Fm.	*Codonofusiella*; *Reichelina*; *Codonofusiella*; *Chusenella* (Khachik Fm.)	*Araxilevis* (Abadeh Fm.)		L. Mai-Akasa-kan Fm.			Guada-lupian / Capi-tanian / Word-ian	Kazanian
Anderssonoceras– Protootoceras		B	Wargal Fm.								
(Yengiao Fm.)	Jielong Gp.	A / Panjal Fm.									

Source: (After Yang Zunyi *et al.*, 1987.)

Should the boundary suggested by Newell (1988) or Budurov *et al.* (1988) be accepted, however, the transitional beds with mixed faunas would disappear altogether. For the time being, this unit is considered valid in our analysis.

Correlation of Permo-Triassic boundary strata

Based primarily on biostratigraphic zonation and faunal succession, strata of the Permo-Triassic boundary interval (Changxing/Feixianguan, or Daye formations) in South China are correlated in Table 2.3 with equivalent units in ten other areas (Mt. Qomolangma, Kashmir, Salt Range, the Dzhulfa region of Transcaucasia, northwest and central Iran, three localities in Japan, Greenland, North America, and the USSR). Table 2.3 is self-explanatory.

Conclusions

From the foregoing discussion, the following conclusions may be drawn:

1 The great majority of Paleozoic marine invertebrates disappeared at, or somewhat below the top of the Changxingian, thus marking the end of the Paleozoic Era.

2 Among the three Permo-Triassic boundaries proposed by different workers (Newell, 1978, 1988; Tozer, 1979, 1988; Waterhouse, 1976; Budurov *et al.*, 1988) the traditional one is preferred in the present study for it is marked by mass extinction of the highest order and is thus equivalent to Gould's (1985) third level of organism evolutionary events.

3 The ammonoid *Hyophiceras* and the conodont *Isarcicella.? parva* may be used to replace *Otoceras* in defining the basal zone of the Triassic.

References

Bhatt, D. K. & Arora, R. K. (1984). *Otoceras* beds of Himalaya and Permian–Triassic boundary – assessment and elucidation with conodont data. *J. Geol. Soc. India*, **25**(11): 720–7.

Budurov, K. J., Gupta, V. J., Kachroo, R. K. & Sudar, M. M. (1988). Problems of the Lower Triassic conodont stratigraphy and the Permian-Triassic boundary. *Mem. Geol. Soc. Ital.*, **34**(1986): 321–8.

Chai, C. F., Ma, S. L., Mao, X. Y., Sun, Y. Y., Xu, D. Y., Zhang, Q. W. & Yang, Z. Z. (1986). Elemental geochemical characters at the Permian-Triassic boundary section in Changxing, Zhejiang, China. *Acta Geol. Sin.*, **60**(2): 139–50 (in Chinese with English abstract).

Chen, J. S., Shao, M. R., Huo, W. G. & Yao, Y. Y. (1984). Carbon isotopes of carbonate strata at Permian-Triassic boundary in Changxing, Zhejiang. *Scientia Geol. Sin.*, 1984(1): 88–93 (in Chinese with English abstract).

Dagys, A. S. & Dagys, A. A. (1988). Biostratigraphy of the lowermost Triassic and the boundary between Paleozoic and Mesozoic. *Mem. Soc. Geol. Ital.*, **34**(1986): 313–20.

Gould, S. J. (1985). The paradox of the first tier: An agenda for paleobiology. *Paleobiol.*, **11**(1): 2–12.

Iranian–Japanese Research Group (1981). The Permian and Lower Triassic systems in Abadeh region, central Iran. *Mem. Fac. Sci. Kyoto Univ., ser. Geol. Min.*, **47**(2): 60–132.

Kozur, H. (1973). Beiträge zur Stratigraphie und Paläontologie der Trias. *Paläont. Mitt. Innsbruck*, **3**(1): 1–30.

Kozur, H. (1978). Beiträge zur Stratigraphie des Perms: Teil II, Die Conodonten-chronologie des Perms. *Freiberger Forschungshefte*, C334: 85–161.

Kozur, H. (1980). Die Faunenänderungen an der Basis und innerhalb des Rhäts und möglichen Ursachen für die Faunenänderungen nahe der Perm/Trias- und Trias/Jura-Grenze. *Freiberger Forschungshefte*, **C357**: 111–34.

Li, Z. S., Zhan, L. P., Zhu, X. F., Zhang, J. H., Jin, R. G., Liu, G. F., Sheng, H. B., Shen, G. M., Dai, J. Y., Huang, H. Q., Xie, L. C., & Yan, Z. (1986). Mass extinction and geological events between Paleozoic and Mesozoic eras. *Acta Geol. Sin.*, **60**(1): 1–17 (in Chinese with English abstract).

Liao, Z. T. (1979). Brachiopod assemblage zones of Changxingian stage and brachiopods of the mixed fauna in S. China. *J. Strat.*, **3**(3): 200–7 (in Chinese).

Liao, Z. T. (1980). Brachiopod assemblages from the Upper Permian and Permian-Triassic boundary beds, S. China. *Canadian J. Earth Sci.*, **17**(2): 289–95.

Newell, N. D. (1978). The search for a Palaeozoic-Mesozoic boundary stratotype. *Schr. Erdwiss. Komm. Österr. Akad. Wiss.*, **4**: 9–20.

Newell, N. D. (1988). The Paleozoic/Mesozoic erathem boundary. *Mem. Soc. Geol. Ital.*, **34** (1986): 303–11.

Paull, R. K. (1982). Conodont biostratigraphy of Lower Triassic rocks, Terrace Mountains, northwestern Utah. *Utah Geol. Assoc. Pub.*, **10**: 235–49.

Sheng, J. Z. Chen, C. C., Wang, Y. G., Rui, L., Liao, Z. T., Bando, Y., Ishii, K., Nakazawa, K. & Nakamura, K. (1984). Permian-Triassic boundary in middle and eastern Tethys. *J. Fac. Sci. Hokkaido Univ., Ser. 4*, **21**(1): 133–81.

Sun, Y. Y., Chai, Z. F., Ma, S. L., Mao, Z. Y., Xu, D. Y., Yang, Z. Z., Sheng, J. Z., Chen, C. Z., Rui, L., Liang, X. L., Zhao, J. M. & He, J. W. (1984). The discovery of Iridium anomaly in the Permian-Triassic boundary clay in Changxing, Zhejiang, China and its significance. *Dev. Geosci. Contr. 27th Int. Geol. Congr. (Moscow) 1984*, p. 235–45. Beijing: Science Press.

Teichert, C., Kummel, B. & Sweet, W. (1973). Permian-Triassic strata, Kuh-e-Ali-Bashi, northwestern Iran. *Bull. Mus. Comp. Zool.*, **145**(8): 359–472.

Tozer, E. T. (1978). Review of the Lower Triassic ammonoid succession and its bearing on chronostratigraphic nomenclature. In *Beiträge zur Biostratigraphie der Tethys-Trias*, ed. H. Zapfe, pp. 21–36. Vienna: Österr. Akad. Wiss., Schriftenreihe d. Erdwiss. Komm. 4.

Tozer, E. T. (1979). The significance of the ammonoids *Paratirolites* and *Otoceras* in correlating the Permian-Triassic boundary beds of Iran and the People's Republic of China. *Canadian J. Earth Sci.*, **16**: 1524–32.

Tozer, E. T. (1988). Definition of the Permian-Triassic (P-T) boundary: the question of the age of the *Otoceras* beds. *Mem. Soc. Geol. Ital.*, **34**(1986): 291–301.

Wang, C. Y. & Wang, Z. H. (1981). Permian conodont biostratigraphy of China. *Geol. Soc. America, Spec. Paper*, **187**: 227–36.

Waterhouse, J. B. (1976). World correlations for marine Permian faunas. *Queensland Univ. Dept. Geol., Papers*, **7**(2): 1–232.

Xu, D. Y., Ma, S. L., Chai, Z. F., Mao, Z. Y., Sun, Y. Y., Zhang, Q. W. & Yang, Z. Z. (1985). Abundance variation of iridium and trace elements at the Permian-Triassic boundary at Shangsi in China. *Nature*, **314**: 154–6.

Yang, Z. Y., Yin, H. F., Wu, S. B., Yang, F. Q, Ding, M. H. & Zu, G. R., et al. (1987). Permian-Triassic boundary stratigraphy and fauna of South China. *Ministry of Geol. and Min. Res., Geol. Mem., ser. 2*, **61**: 379 pp. Beijing: Geological Publishing House.

Yao, J. & Li, Z. (1987). Permian-Triassic conodont faunas and the Permian-Triassic boundary at the Selong section in Nyalam county, Xizang, China. *Kexue Tongbao*, **32**(22): 1555–60.

Yin, H. F. (1985). On the transitional bed and the Permian-Triassic boundary in South China. *Newsl. Strat.*, **15**(1): 13–27.

Yin, H. F., Yang, F. Q., Zhang, K. X. & Yang, W. P. (1988). A proposal to the biostratigraphic criterion of the Permian/Triassic boundary. *Mem. Soc. Geol. Ital.*, **34**(1986): 329–43.

Zhang, J. H., Zhang, Y. J., Wang, Y. Q., Chang, B. R. & Sun, J. X. (1983). Features of rare earth elements of Permian-Triassic boundary clay rocks in South China with their stratigraphical significance. *Acta Petrol. Min. Analyt.*, **2**(2): 81–6.

Zhang, Kexin (1987). The Permo-Triassic conodont fauna in Changxing area, Zhejiang province and its significance. *Earth Sci. J. Wuhan College of Geol.*, **12**(2): 193–200.

Zhao, J. K., Sheng, J. Z., Yao, Z. Q., Liang, X. L., Chen, C. C., Rui, L. & Liao, Z. T. (1981). The Changhsingian and Permian-Triassic boundary of South China. *Bull. Nanjing Inst. Geol. Paleont., Acad. Sinica*, **2**: 1–95.

3 Permo-Triassic boundary of the Indian subcontinent and its intercontinental correlation

HARI M. KAPOOR

Introduction

Of the various factors involved in determining placement and correlation of the Permo-Triassic boundary, the biological is the most difficult one upon which to decide. This is due mainly to the fact that Permian sequences show too much provincialism and their faunas have too many endemic characters.

The Indian subcontinent has both marine and terrestrial deposits of Late Permian and Early Triassic age. Marine sequences are confined to the extrapeninsular region (Fig. 3.1) and represent the Perigondwana Province of the Tethys. Terrestrial deposits represent Gondwana. In the body of this chapter only the extrapeninsular region is considered. Gondwana deposits are discussed in the next chapter, prepared at my request by Drs R. S. Tewari and Vijaya of the Birbal Sahni Institute of Palaeobotany.

In his monograph on the Triassic of the Himalaya, Diener (1912) placed the boundary between the Permian and Triassic systems at the base of the *Otoceras woodwardi* Zone because the zone was thought to be equivalent to the lower part of the Lower Triassic sequence of the Alps and because, in the Himalaya, the beds in which it is represented are intimately related lithologically to beds above. Diener's analysis was based on Painkhanda sections, with support from sections in the Spiti district. This traditional boundary has been reconsidered in the last 20 years, however, in the light of new fossil and stratigraphic data that have been described from areas such as Iran and China.

There are many features of the Permo-Triassic boundary

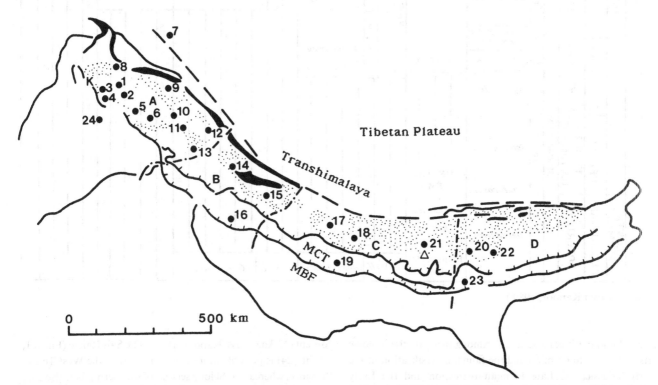

Fig. 3.1. Index map. 1–6: Kashmir localities (1, Guryul Ravine; 2, Pastannah; 3, Pir Panjal; 4, Thanamandi–Pira; 5, Bhallesh–Chamba; 6, Kalhel–Salooni); 7, northern Ladakh (Karakoram); 8, southern Ladakh; 9, Zanskar; 10, Chharap; 11, Tandi; 12, Kinnaur; 13, Lilang (Spiti); 14, western Kumaon; 15, eastern Kumaon; 16, Garhwal; 17, Dolpo; 18, Thakkhola; 19, Tansen; 20, Chumbi Valley (Sikkim); 21, northern slope of Mt. Everest; 22, Bhutan; 23, Daling Basin; 24, Salt Range (Pakistan). A, Punjab Himalaya; B, Kumaon Himalaya; C, Nepal Himalaya; D, eastern Himalaya. K, Kashmir; MCT, main central thrust; MBF, main boundary fault. Stippled: Phanerozoic Tethyan sediments; dark-shaded areas: ophiolitic melange belts. (Redrawn from Kapoor & Tokuoka, 1985.)

Table 3.1. *Correlation of Upper Permian and lowermost Triassic rocks in the East, West, and Perigondwana provinces of the Tethyan realm*

	URALS	TETHYS	WEST TETHYS		CENTRAL (PERIGONDWANA) TETHYS			
			ABADEH	KARA-KORAM	SALT RANGE	KASHMIR	BHALLESH	CHAMBA
TRIASSIC		GRIESBACHIAN	L. TRIAS. Unit a — *Ophiceras* / *Claraia stachei*	MORGO FM.	MIANWALI FM. Kathwai Member: U *Ophiceras*; M *Eumorphotis* "Perm brachs"; L "Perm Brachs"	KHUNAMUH FM. Mbr. E: E3 *Ophiceras*; E2 *Otoceras*; E1 *Claraia bioni* "Perm brachs"	BISHOT FM. *Ophiceras* / *Otoceras* / "Mixed fauna"	KALHEL FM. *Ophiceras*
UPPER PERMIAN	TATARIAN	DORASHAMIAN	HAMBAST FM. Unit 7 — *Paratirolites* / *Shevyrevites*			Mbr. D *Xenodiscus*	TALAI FM. *Xenodiscus*	
		DZHULFIAN	Unit 6 — *Vedioceras* / *Araxoceras* / *Araxilevis*	AQTASH FM. / *Marginifera*	CHHIDRU FM.: 4 *Oldhamina*; 3 *Cyclolobus*; 2 *Colaniella*; 1	ZEWAN FM. Mbr. C: Fauna III; Mbr. B: Fauna II *Marginifera*	*Marginifera*	SALOONI FM. *Marginifera*
		MIDIAN	ABADEH FM. 5 *Codonofus.*; Unit 4 *Sphaerulina*		WARGAL FM. Kala: 5 *Colaniella*; 4 *Codonofus.*; 3	Mbr. A: Fauna I *Colaniella* *Abadehella*	*Polypora*	*Polypora*
	KAZANIAN	MURGABIAN	3 *Orientoschw. abichi*; 2 *Neoschwag. margaritae*; *Eopolydiex.*	*Neoschwag.*	2 *Neoschwag.*; 1 *Chusenella*			
	UFIMIAN	KUBER-GANDINIAN	SURMAQ FM. Unit 1 *Schwagerina*	? ?				
LOWER PERMIAN	KUNGURIAN	BOLORIAN	*Chalarosch.* / *Darvasites*	CHONGTASH FM. Volcanics and HARPATSO FM.	AMB U *Glossopteris*	MAMAL *Glossopteris*		

Source: (From Kapoor, 1989.)

interval that merit consideration. Among these are the biologic crisis of the Late Permian and the faunal diversification of the Early Triassic; the Late Permian regression and the Early Triassic transgression; and the terminal effects of the end-phase of the Hercynian movements. It will be difficult to deal with all these matters here.

The Perigondwana Province of the Tethys is distinct from the East and West Tethys provinces in both lithologic and fossil content (Nakazawa & Kapoor, 1977). The Salt Range (Fig. 3.1, locality 24) is possibly more closely related to the West Tethys Province, whereas the Selong section of southern Tibet, also part of the Perigondwana Province, is related to the East Tethys Province. The Karakoram region (Fig. 3.1, locality 7), however, has some characteristics of the West Tethys Province, but too few data are available to provide answers to boundary and related questions.

Table 3.1. (*cont.*)

CENTRAL (PERIGONDWANA) TETHYS							EAST TETHYS	
SPITI	PAIN-KHANDA	BYANS	NEPAL	SIKKIM	BHUTAN	S. TIBET	N. TIBET	S. CHINA
TAMBAKURKUR — *Ophiceras* / *Otoceras*	RAMBAKOT FM. — *Ophiceras* / *Otoceras*	CHOCOLATE — *Ophiceras*	PANGJANG FM. — *Ophiceras* / *Otoceras* "Perm brachs"	TRIASSIC ?	LINGSHI FM	TULONG GP. KANSHARE FM. — *Ophiceras* / *Otoceras*	KANGLU FM — *Ophiceras* / *Claraia stachei* / ?	LR. CHINGLUNG Mbr. I – Mbr. III — *Otoceras* "Perm brachs"
			SENJA FM. — *Marginalosia* / *Krotovia* / *Pyramus*				RAGYORCAKA FM. — *Giganto-pteris* / *Lobatannul.* / *Palaeo-fusulina*	CHANGXING FM. U. Mbr. — *Rotodis-coceras* / *Pseudo-tirolites* / *Palaeofus.*; Lr. Mbr. — *Paratir.* / *Shevyrev.* / *Schwager.*
KULING FM. Gungri Mbr. — *Cyclolobus*	KULING FM. — *Cyclolobus*	KULING FM. — *Cyclo-lobus*	*Cyclo-lobus*			CHUBUJEKA FM. — *Paracruri thyris* / *Araxa thyris* / *Chonetes nasuta*	?	WUCHIAPING FM. — *Pseudostaph.* / *Araxoceras* / *Codono fusiella* / *Tapashanites*
Marginifera	*Marginifera*	*Marginifera*	NANJUNG — *Margini-fera*	LACHHI FM *Margini-fera*	TANGCHU FM. *Margini-fera*	SELONG GP.	?	MAOKOU FM. — *Lepidolina-Yabeina* / *Neoschwag. margaritae* / *Neoschwag. cratic.* / *Cancellina*
						CHUBUK FM. *Glossopteris*	?	CHIHSIA FM. — *Misellina* / *Pamirina* / *Schwagerina tsch.*

The literature identifies as 'Himalayan' the sequence in Kashmir, through Nepal, to Bhutan. The Pira-Thanamandi parautochthone belt connects Kashmir with the Salt Range, and the Bhallesh-Chamba and Zanskar basins connect it with Spiti (Fig. 3.1, locality 13). Spiti, on the other hand, is connected with Kumaon through Kinnaur and south Tibet, and the development of the Permo-Triassic sequence of Kumaon-Nepal can be traced into south Tibet (Table 3.1). There is also thought to be a development of this system in Sikkim and Bhutan, but biostratigraphic details are still unknown.

The Permian and Triassic formations of the Salt Range, Kashmir, Spiti, Kumaon and Nepal have been worked out in fine detail and are dealt with here. However, emphasis in this chapter is on strata above the *Cyclolobus* Zone (Dzhulfian Stage), which is persistently traceable throughout the province, easily correlated with other provinces, and significant in that it is followed by the beginning of the regressive phase in the Himalaya.

Upper Permian

Salt Range

The Upper Permian Chhidru Formation of the Salt Range (Figs 3.2, 3.7, Table 3.1) is customarily divided into four members, of which only the upper two are considered here. Unit 3 of the Chhidru Formation, composed of sandstone and calcareous sandstone, contains representatives of *Cyclolobus* (Pakistani–Japanese Research Group, 1985) in association with *Xenodiscus* (Furnish & Glenister, 1970). In addition, it has yielded a few fusulinids, such as *Codonofusiella*, *Reichelina*, *Nanlingella*, and *Nankinella*, and the brachiopod *Megasteges nepalensis*. The latter is also known from the Senja Formation of Nepal, which was referred to the Dorashamian by Waterhouse (1978). Fusulinids represented in Unit 3 are absent in the Himalayan areas of the Perigondwana Province. Their presence in Salt Range sections may be a result of the intermixing of waters of the West Tethyan and Perigondwanan provinces, which were in close proximity there.

Unit 4 of the Chhidru Formation is composed of sandstone, which is white in its uppermost part and varies in thickness from 0 to 3 m as a result of subaerial erosion (Kummel & Teichert, 1970). Unit 4 contains smaller foraminifers, brachiopods such as *Oldhamina decipiens* and *Crurithyris* sp., and bivalves representing several species of *Permophorus*, *Schizodus*, and *Cyrtorostra*. Sweet (1970b, 1979) reported *Hindeodus typicalis* and *Neogondolella carinata* from Unit 4.

The cycle of Late Permian, post-*Cyclolobus* Zone, sedimentation was therefore brief in the Salt Range, and the sea regressed from the area much earlier than elsewhere, perhaps in the Late Dzhulfian or early Dorashamian (Fig. 3.7, Table 3.1; see also Sweet, Chapter 11).

Kashmir

The middle part of member C of the Zewan Formation (Figs 3.2, 3.3), which is characterized by rhythmic alternation of calcareous sandstone and arenaceous shale, has yielded specimens of *Cyclolobus walkeri* and *Xenaspis* (Nakazawa *et al.*, 1975). *Cyclolobus* is also represented in a 10 cm dark shale at the contact with Unit D in the Pahalgam section, and this may represent its uppermost limit in Kashmir. Member C also contains foraminifers, bryozoans, brachiopods, gastropods (*Bellerophon*) and conodonts (*Hindeodus typicalis* and *Neogondolella carinata*).

Member D of the Zewan Formation (Fig. 3.3) is mostly thick-bedded sandy limestone, with minor sandy shale, argillaceous sandstone, and calcareous sandstone. The latter two rock types are dominant in the upper part of the unit. Bivalves, gastropods, and brachiopods are represented at several horizons and a few smaller foraminifers and bryozoans have also been observed. Member D includes specimens of the conodonts *Neogondolella carinata*, *Hindeodus typicalis*, and *Ellisonia triassica*. Representatives of *Neogondolella subcarinata* have also been reported (Murata, 1981) from Member D, but the identification is questioned by Matsuda (1981b).

The fauna of units C and D was included in Division III by

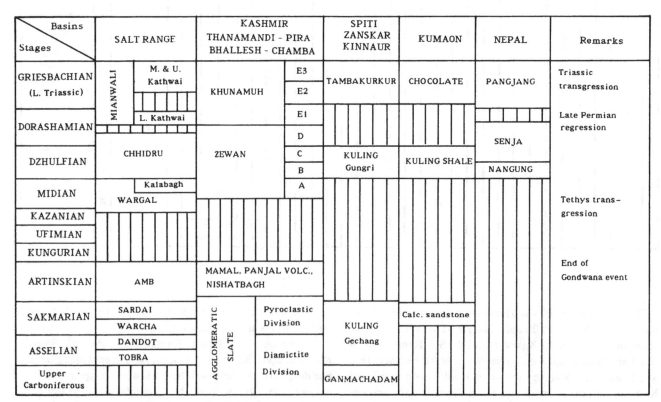

Fig. 3.2. Permian and lowermost Triassic stratigraphic units of the Tethyan Himalaya.

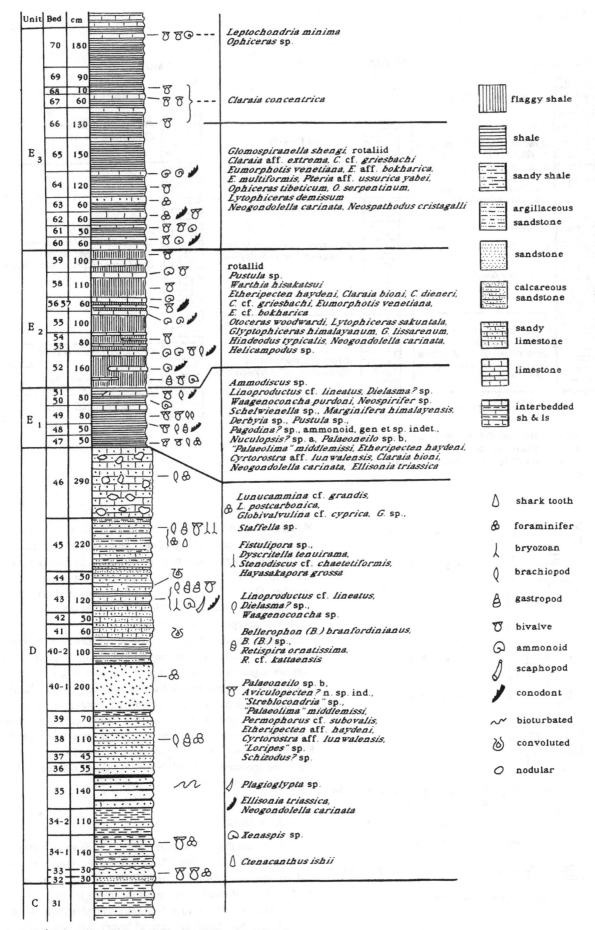

Fig. 3.3. The section at Guryul Ravine, Kashmir. Units C and D make up the upper part of the Zewan Formation; units E1, E2, and E3 compose the lowermost part of the Khunamuh Formation. (Redrawn, with slight emendation, from Nakazawa *et al.*, 1975.)

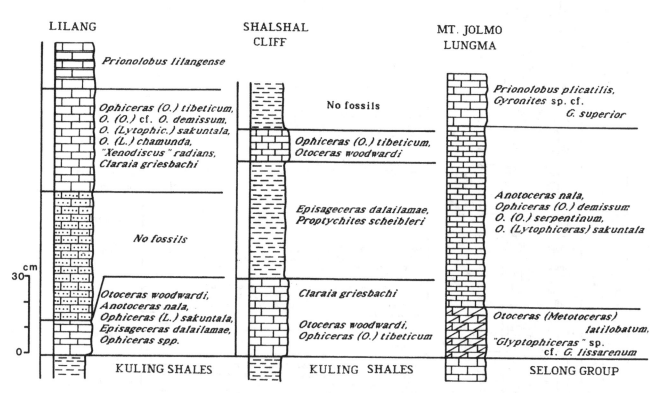

LILANG

Prionolobus lilangense

Ophiceras (O.) tibeticum,
O. (O.) cf. O. demissum,
O. (Lytophic.) sakuntala,
O. (L.) chamunda,
"Xenodiscus" radians,
Claraia griesbachi

No fossils

cm
30

Otoceras woodwardi,
Anotoceras nala,
Ophiceras (L.) sakuntala,
Episageceras dalailamae,
Ophiceras spp.

0

KULING SHALES

SHALSHAL
CLIFF

No fossils

Ophiceras (O.) tibeticum,
Otoceras woodwardi

Episageceras dalailamae,
Proptychites scheibleri

Claraia griesbachi

Otoceras woodwardi,
Ophiceras (O.) tibeticum

KULING SHALES

MT. JOLMO
LUNGMA

Prionolobus plicatilis,
Gyronites sp. cf.
        G. superior

Anotoceras nala,
Ophiceras (O.) demissum,
O. (O.) serpentinum,
O. (Lytophiceras) sakuntala

Otoceras (Metotoceras)
        latilobatum,
"Glyptophiceras" sp.
        cf. *G. lissarenum*

SELONG GROUP

Fig. 3.4. Columnar sections and distribution of significant fossils in the Lilang and Shalshal Cliff sections of the Spiti district, and in the Mt. Jolmo Lungma section of southern Tibet. (Redrafted from Nakazawa, Bando & Matsuda, 1980.)

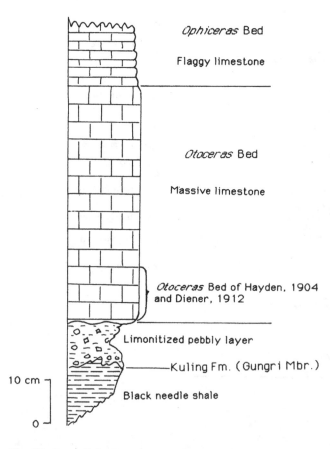

Ophiceras Bed

Flaggy limestone

Otoceras Bed

Massive limestone

Otoceras Bed of Hayden, 1904
and Diener, 1912

Limonitized pebbly layer

Kuling Fm. (Gungri Mbr.)

10 cm

Black needle shale

0

Fig. 3.5. Permo-Triassic section at Lalung, Spiti district, Himalaya. (Redrawn from figure in Bhatt, Joshi & Arora, 1981.)

Nakazawa *et al.* (1975). The fauna of Unit C suggests an infraneritic to shallow bathyal environment, and later restriction of the environment may be indicated by the fact that brachiopods of Unit D show a tendency toward dwarfing. Member D, the uppermost unit of the Zewan Formation, suggests uplift of the land and shallowing of the sea, but there is no sign of emergence above sea level.

In Pahalgam, the uppermost part of the Zewan Formation contains an interval of bioherms and shale intercalations, and suggests a gradational contact with the overlying formation. *Xenodiscus* has been reported from Member D at several localities, but in Pahalgam some of the hollows contain molds, which become brittle and are different from 'xenodiscids'. One such mold was suspected to represent '*Stacheoceras*' when Dr B. Kummel and Dr C. Teichert visited Kashmir in 1968. Unfortunately the specimen could not be preserved and none like it has subsequently been found.

Spiti–Kumaon

The Late Permian Gungri Member (Srikantia, 1981) of the Kuling Formation (Figs 3.2, 3.4, 3.5; Table 3.1) is characterized by the dominance of dark shales, which contain thin, irregular beds of calcareous sandstone in the lower part. The lower part also contains a few brachiopods and bryozoans, but in the upper part there are several layers with specimens of *Cyclolobus*, particularly *C. walkeri*. The highest bed with *Cyclolobus* is only 20 cm from the top of the Gungri Member and here and there also contains phosphatic nodules. In Spiti (Bhatt,

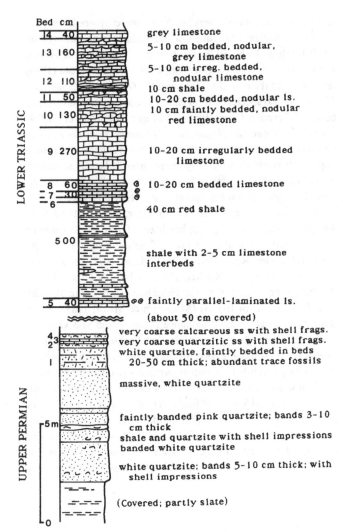

Fig. 3.6. Schematic columnar section of Permian and Triassic rocks at Thakkhola, Nepal (Loc. Th-A of Tokuoka, 1985). (From Tokuoka, 1985.)

Joshi & Arora 1981; Fig. 3.5), a pebbly limonitic band of variable thickness suggests subaerial exposure following deposition of the Kuling Formation and indicates complete withdrawal of the sea during or soon after depositon of rocks with *Cyclolobus*.

Nepal

Late Permian and Early Triassic successions have been worked out in three areas, Dolpo (Karnali), Thakkhola (Gandaki River), and Nyi Shang (Fig. 3.1, localities 17, 18). The Senja Formation (Waterhouse, 1977, 1978, 1979; Fig. 3.2), which represents the upper part of the Upper Permian in Nepal, is divisible into the Popa, Pija, Nisal, Nambda, Luri, and Kuwa members. The Popa member is quartzite, whereas other members are dominantly shale and siltstone with varying proportions of sandstone (Fig. 3.6). The lithofacies do not indicate much difference from adjoining Kumaon, although they are considered to be younger and to include the *Marginalosia kalikoti* and *Atomodesma variable* zones of the Dorashamian. Fuchs (1977) reported a thin limonitic layer and reworked dark shale in the Kuwa Member at the top of the Senja Formation in

the Dolpo region, and these features suggest regression (Fig. 3.2; Table 3.1).

Lower Triassic

Salt Range

The Lower Triassic Mianwali Formation (Figs 3.2, 3.7) is a markedly different sedimentary unit from the underlying Upper Permian Chhidru Formation, from which it is separated by a paraconformity (Kummel & Teichert, 1970; Pakistani–Japanese Research Group, 1985). The Mianwali Formation is divisible, in ascending order, into Kathwai, Mittiwali, and Narmia members. The Kathwai Member, sometimes referred to as the Permo-Triassic transition beds, is further divisible into lower, middle, and upper parts. The lower part, composed of dolostone, dolomitic limestone, and minor calcareous sandstone, contains several coquinoid layers. A number of brachiopods (Grant, 1970), reworked smaller foraminifers, bryozoans, and syndepositional specimens of *Reichelina* are known from the lower part, as are representatives of the conodont *Hindeodus typicalis*. The middle unit of the Kathwai Member is bedded dolostone that contains chonetids, representatives of *Crurithyris*, the bivalves *Eumorphotis* and *Entolium*, and the conodonts *H. typicalis*, *Isarcicella? parva*, and *I. isarcica*. The upper unit consists of bedded limestone, with shale and dolostone interbeds, and also includes coquinoid layers. Schindewolf (1954) reported *Ophiceras connectens* from this unit, which also yields rhynchonellids, arenaceous foraminfers, and the conodont *Neogondolella carinata*. The absence of *Otoceras* suggests a time gap between the lower and middle units.

The Mittiwali Member, which overlies the Kathwai Member, is characterized by limestone and contains at its base the fauna of the *Gyronites* Zone. A succession of conodont zones based on samples from the Mittiwali and Narmia members (Sweet, 1970b) has come to be a standard for the Lower Triassic throughout the Tethyan realm and in western North America.

Transgression began in the Salt Range in the latest Permian, with deposition of the lower member of the Kathwai Member, which records a very shallow marine environment and a wide continental shelf (Pakistani–Japanese Research Group, 1985). The lower part of the Mittiwali Member represents essentially the same environment, but overlying beds suggest deposition in an outer-shelf basin, which deepened rapidly along with concurrent uplift of source areas (Fig. 3.7; Table 3.1).

Kashmir

The Khunamuh Formation (Figs 3.2, 3.3), composed mostly of interbedded limestone and shale, is divisible, in ascending order, into members E through J. Division is based on the relative thicknesses of limestone. The formation includes several Lower Triassic ammonoid zones, but attention is focused here on Member E, which is divisible into units E1, E2, and E3.

Unit E1, 2.6 m thick, is composed of black shale with thin, discontinuous intercalations of limestone. Unit E2, 6.1 m thick, consists of flaggy shale with thin limestone beds, and unit E3, 9.9 m thick, is composed of an alternation of shale and subordinate limestone. Detailed petrographic studies of Zewan and E1 sediments show a gradational change. Thus, unit E1 may be regarded as a Permo-Triassic transition bed, for it contains specimens of a Triassic-type bivalve, *Claraia bioni*, together with dwarfed Permian-type brachiopods and additional bivalves such as *Etheripecten haydeni* and '*Paleolima*' *middlemissi* (Nakazawa et al., 1975; Nakazawa, 1981). Conodonts are rare in unit E1 and have been referred to *Hindeodus typicalis*, *Neogondolella carinata*, and *Ellisonia triassica* (Sweet, 1970a; Nakazawa et al., 1975; Murata, 1981; Matsuda, 1981a).

Dickins (1987) has pointed out that some of the brachiopods from Unit E1 retain matrix of Zewan type. This matter has been reviewed by examining the specimens and through petrologic studies, particularly of the specimen of *Costiferina* sp. (GSI No. 19346) (= *Dictyoclostus* sp. ind. in Nakazawa et al., 1970, pl. 29, fig. 9, from bed 47b; Nakazawa & Kapoor, 1981, pl. 7, fig. 16, from bed 48; these are both the same specimen). The specimens were found to have come from Member B of the Zewan Formation and to have been mislabeled in the field. Brachiopods of this size are not found in Unit E1 and no matrix of Member B-type occurs in unit E1.

Unit E2 is characterized by the association of *Otoceras woodwardi*, *Glyptophiceras himalayanum*, '*G.*' *lissarenum*, and *Lytophiceras sakuntala* with many bivalves, in particular *Eumorphotis venetiana*, *E. bokharica*, *Claraia bioni*, *C. dieneri*, *C. extrema*, and *C. griesbachi*. In addition, conodonts abruptly become very numerous and represent *Hindeodus typicalis*, *Isarcicella? parva*, *I. isarcica*, and *Neogondolella carinata*.

Unit E3 yields abundant representatives of several species of *Ophiceras* (viz. *tibeticum*, *serpentinum*, *subdemissum*), several species of the bivalves *Eumorphotis* and *Claraia*, as well as *Leptochondria minima*. Conodonts are referrable to *Neogondolella carinata*, *Neospathodus kummeli*, *N. dieneri*, and *N. cristagalli* (Sweet, 1970a; Matsuda, 1981; Nakazawa, Bando & Matsuda, 1980).

Higher members of the Khunamuh Formation include the *Paranorites–Vishnuites*, *Prionites–Koninckites*, and *Owenites* zones.

In Kashmir, marine transgression was marked by a rapid deepening of the sedimentary basin. According to Kapoor & Tokuoka (1985) this is indicated by '. . . the common occurrence of typical limestone turbidite alternating with black shale, especially in Unit E3 and later stages.'

Spiti–Kumaon

The Spiti region includes the Zanskar, Lahul, and Kinnaur areas. Kumaon encompasses Painkhanda in the west and Byans in the east. In the Spiti belt, the Lower Triassic-Anisian succession is included in the Tambakurkur Formation (Fig. 3.2; Table 3.1), as designated by Srikantia (1981) and revised by Bhargava & Gadhoke (1988), to the Mikin Formation. Lower Triassic strata in the lower part of the sequence (Figs. 3.4, 3.5) vary in thickness from 18 to 30 m and include, at the base, the *Otoceras* and *Ophiceras* beds, with younger beds above, including the *Hedenstroemia* beds. The Lower Triassic in this region, although much condensed, is similar lithologically to that in Kashmir. No major biostratigraphic revisions have resulted

from recent studies of megafossils. However, there have been numerous studies of microfossils, particularly conodonts (Goel, 1977; Bhatt & Joshi, 1978; Bhatt et al., 1981; Bhatt & Arora, 1984; Nicora, Gaetani & Garzanti, 1984). Bhatt & Arora (1984) reported that the conodont fauna of the *Otoceras* beds includes *Hindeodus* and species of *Neogondolella* considered elsewhere to be latest Permian (Dorashamian) in age. Nicora et al. (1984) regard the *Neogondolella carinata–Hindeodus typicalis* Zone to be Lower Griesbachian. Sweet (Chapter 11) suggests that these apparently contradictory assignments may be largely semantic. Succeeding conodont zones, based on *Neospathodus dieneri*, *N. novaehollandae*, and other species are certainly Triassic.

The Chocolate Shale (Heim & Gansser, 1939) in the Lower Triassic of Kumaon (Fig. 3.2) has been referred to variously as the Chocolate Formation (Kumar, Singh & Singh, 1977), the Chocolate Series (Kumar, Mehdi & Prakash, 1971), the Pari Tal Formation (Jamwal & Kacker, 1989; Malviya & Pande, 1989), the Rambakot Formation (in Painkhanda), and the Pankha Gad Formation (in Byans) (Mamgain & Misra, 1989). As in Spiti, the Lower Triassic sequence is condensed and varies in thickness from 18 to 25 m. The biostratigraphy erected by Diener (1912) still holds good for the region, although there have been a large number of papers written on microfossils from random samples and detailed studies have recently been begun by the Geological Survey of India. It is of interest to note that in a recent collection from Painkhanda there is a sample that contains both *Otoceras* and *Ophiceras* – a co-occurrence noted long ago by Diener (1912). In Byans the *Otoceras* bed is not known and the sequence begins with a bed that yields *Ophiceras serpentinum*.

Nepal

The Lower Triassic Pangjang Formation (Waterhouse, 1977; Fig. 3.2), which has been studied in detail by Tokuoka (1985), is characterized by dolomite, limestone, and interbedded limestone and shale. The formation is divisible into two parts, the lower dominated by orange-colored dolomite, and the upper consisting of limestone and interbedded limestone and shale (Fig. 3.6). Bassoullet & Colchen (1977) reported *Otoceras woodwardi* from the upper part of the formation at Thini Khola, and *Ophiceras* in higher layers. In the Dolpo region, Waterhouse (1979) noted the co-occurrence of *Otoceras concavum* and productid brachiopods in a dense purple carbonate band 2 to 4 cm thick. Fuchs (1977) also mentions the occurrence of *Otoceras woodwardi* and *Ophiceras* in the lower 10 cm of the basal bed of the formation. Conodonts have been reported from strata in the Permo-Triassic boundary interval of Nepal in numerous papers (Kozur & Mostler, 1973; Clark & Hatleberg, 1983; Hatleberg & Clark, 1984; Matsuda, in Tokuoka, 1985). Most specimens are referable to *Hindeodus typicalis*, *Neogondolella carinata*, and *Ellisonia triassica*, all of which have long ranges in the boundary interval. However, Matsuda (in Tokuoka, 1985) has identified *Isarcicella? parva* (as *Hindeodus parvus*) and *Neospathodus kummeli* in the upper part of the Pangjang Formation in the Thakkhola area. The former first appears just above the base of the Werfen Formation in the Southern Alps, is also known from the upper part of the *Otoceras* beds in Kashmir and Tibet, and occurs in numerous South Chinese sections immediately above rocks of Changxingian age. The first occurrence of *N. kummeli*, on the other hand, corresponds regionally to the base of the Dienerian Stage.

Tokuoka (1985) concludes that an abrupt deepening of the depositional basin is indicated by changes in lithology and fauna between the middle and upper parts of the Pangjang Formation.

Karakoram (Permian and Triassic)

Northern Ladakh, in the Karakoram region (Fig. 3.1, locality 7), represents another area that may provide a link between deposition in central Asia, to the north, and the Perigondwana Himalaya south of the Indus suture. In northern Ladakh, the Chongtash, Aqtash, and Morgo formations represent the Permian and Triassic systems.

The Permian Chongtash Formation, 800 m thick, is composed of siltstone, sandstone, and black shale, interbedded with thick volcanics. A number of intertrappean beds of sandy slate and calcareous shale contain specimens of *Parafusulina*, *Pseudofusulina*, *Schwagerina*, and corals.

The Aqtash Formation, 200 m thick, is composed of volcanics, conglomerate, and grey, massive limestone and shale. Intertrappean beds of limestone and shale yield brachiopods and fusulinids. The Triassic Morgo Formation, about 300 m thick, is composed of limestone and shale (Gergan & Pant, 1983).

Biostratigraphic details are few, but the northern Ladakh area has played an important role in developing a geological history for the Permian of the Perigondwana Province.

Correlation of Perigondwana Province with East and West Tethys provinces

Correlation of Permian and Triassic rocks of the West Tethys Province with those of the Perigondwana and East Tethys provinces are shown schematically in Fig. 3.7 and Table 3.1. Note in that figure that the East and West Tethys provinces differed from the Perigondwana Province during the Permian in having a dominantly fusulinid fauna throughout. The fact that Late Permian rocks in the Salt Range, Afghanistan, and the Karakoram have some of the faunal characteristics of the East and West Tethys provinces is attributable to the paleogeographic position of these areas between the East and West Tethys provinces and possibly to a mixing of waters between these provinces. Thus, even though they lack fusulinids, Upper Permian rocks of the Perigondwana (or Himalayan) Province can be compared with those of other regions by way of the Salt Range. Lower Triassic rocks, on the other hand, are readily comparable because they show a more or less uniform development throughout the region studied.

Iran

The Hambast Formation of the Abadeh region of central Iran (Fig. 3.8), considered to be of latest Permian age, consists of thin-bedded limestone. The lower part yields fusulinids, and ammonoids occur throughout. The formation was deposited under calm, possibly lagoonal conditions (Iranian–Japanese Research Group, 1981), and following its deposition, the area emerged to form a flat coastal plain. After a short interval of emergence, the area was rapidly submerged and Early Triassic sediments were deposited under infraneritic to littoral conditions.

Faunal elements in common between the Abadeh region and the Perigondwana Province include *Abadehella coniformis*, *Leptodus nobilis*, *Xenodiscus carbonarius*, *Cyclolobus*, *Stenopora carbonarius*, *Hindeodus typicalis*, and *Neogondolella carinata*. Dorashamian ammonoids such as *Shevyrevites* and *Paratirolites* have not been found in the Perigondwana Province, although they do occur in Transcaucasia (Rostovtsev & Azaryan, 1973; Ruzhentsev & Sarycheva, 1965). In rocks above the Hambast Formation there are ammonoids referable to *Acanthophiceras*, *Lytophiceras*, and *Vishnuites*, and conodonts such as *Isarcicella? parva* and *I. isarcica*, which aid in correlation with strata assigned to the Lower Griesbachian Stage in the Perigondwana Province.

Afghanistan

The lower part of the Permian Khingi Formation exposed at Kohe-Safe (Fig. 3.7) is conglomerate, with argillaceous and sandy interbeds; the mid-portion of the formation is calcareous and argillaceous and contains Middle Permian fusulinids; and the upper part, some 200 m thick, consists of limestone with *Waagenophyllum*, *Costiferina*, *Oldhamina* and *Lyttonia*, all known from the Upper Permian of the Perigondwana region. Lower Triassic rocks are yellowish brown limestone, which contrast distinctly with argillaceous, uppermost Permian limestones (Fig. 3.9).

At the beginning of the Triassic, Upper Permian carbonates were exposed to subaerial denudation and dried quickly in an intertidal environment. At this time the tidal-flat surface was penetrated by vertical burrows and now records the maximum of the Late Permian regressive phase. The Early Triassic transgression began as a marginal sea that changed soon to a subtidal environment. As a result, Lower Triassic rocks show rhythmic alternations of limestone and shale, which represent repeated oscillatory fluctuations of the sea (Ishii, Fischer & Bando, 1971). The fauna of these Lower Triassic rocks includes *Ophiceras*, *Kymatites*, *Gyronites*, and other genera.

South Tibet

The Selong section of South Tibet (Fig. 3.4) is another representative of Himalayan type and, although part of the Perigondwana Province, is situated in near proximity to the East Tethys province. Sheng *et al.* (1984a, 1984b) and Wang, Chen & Rui (1988) described the Permian Selong Group and the Lower Triassic Kangshare Formation.

Units 1 to 3 in the lower part of the Selong Group are lithologically similar to members A to C of the Zewan Formation of Kashmir and contain many of the same fossils, particularly brachiopods. The latter correspond to those in faunal division II of Kashmir. Unit 4 of the Selong Group, however, includes brachiopods that are not known elsewhere in the

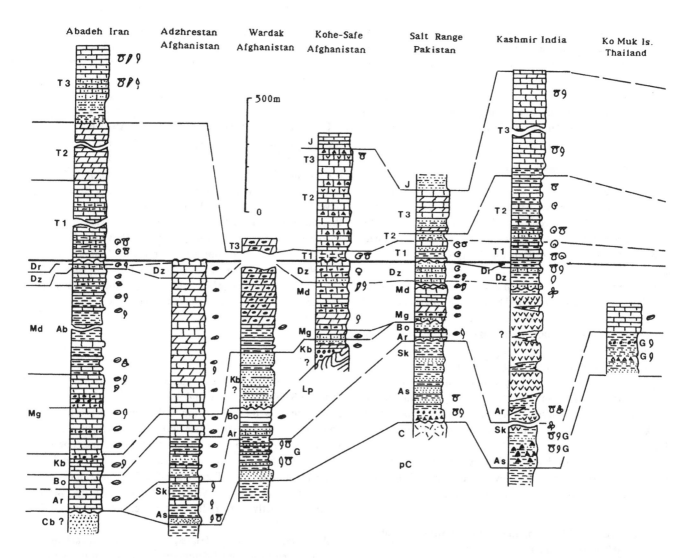

Fig. 3.7. Correlation of Permo-Triassic sequences between the West Tethys Province (Abadeh, Adzhrestan, Wardak, Kohe-Safe) on left and East Tethys Province (e.g., Yangtze Block) on right, by way of the Perigondwana Province (central sections). Lithic symbols: a, sandstone; b, conglomerate; c, mudstone; d, siltstone; e, dolostone; f, limestone; g, sandy limestone; h, marlstone; i, siliceous nodules; j, calcareous nodules; k, basic volcanic rocks; 1, tuffaceous sandstone; C, coal. Fossil symbols: m, bivalve; n, ammonoid; o, brachiopod; p, fusulinid; q, coral; r, bellerophontid; s, plant; t, trilobite; u, gastropod; G, Gondwana fauna or flora. Symbols for stages or series: pC, Precambrian; C, Cambrian; Cb, Carboniferous; As, Asselian; Sk, Sakmarian; Ar, Artinskian; Bo, Bolorian; Kb, Kubergandinian; Mg, Murgabian; Md, Midian; Ab, Abadehian; Dz, Dzhulfian; Dr, Dorashamian; Tl, Lower Triassic, T2, Middle Triassic; T3, Upper Triassic; J, Jurassic; Jl, Lower Jurassic; Cs, Chihsian; Mk, Maokouan; Wu, Wuchiapingian; Ch, Changxingian; Lp, Lower Paleozoic. (Slightly modified from Kapoor & Tokuoka, 1985.)

Perigondwana Province. Unit 4 also differs in lithology from underlying members of the Selong Group.

The Kangshare Formation represents a development of the Lower Triassic like that in Kumaon and Spiti and is readily correlated with strata in those districts. In ascending order, the lowermost 1.5 m of the formation yields *Otoceras latilobatum*, *O. woodwardi*, and *Ophiceras*, and beds with these fossils are followed upward by ones with *Gyronites*, *Koninckites*, and other ammonoids. According to Yao & Li (1987) the lower beds with *Otoceras* also yield the conodonts *Neogondolella changxingensis*, *N. deflecta*, *N. carinata*, and *Hindeodus typicalis*. *Isarcicella? parva* appears in the upper part of the *Otoceras* beds, just as it does in Kashmir, and representatives of *Neospathodus dieneri* and *N. cristagalli* have been collected from higher units with *Gyronites*, which almost certainly represent the Dienerian Stage.

Wang *et al.* (1988) report that a hiatus between Permian and Triassic deposits may be indicated by the presence on the irregular upper surface of Permian strata of reworked brachiopods, corals, and crinoids.

South China

Permian and Triassic rocks of South China (Fig. 3.7) represent the East Tethys Province and are thoroughly described by other contributors to this volume. Here it is sufficient to note that the Upper Permian Lungtan–Changxing or Wuchiaping–Talung successions are faunally quite different from presumably equivalent successions in the Perigondwana Province, primarily in yielding numerous fusulinids and in containing ammonoids that are mostly unknown at Perigondwana localities. On the other hand, Lower Triassic rocks, which overlie Permian strata without a break, yield fossil faunas that are similar in their

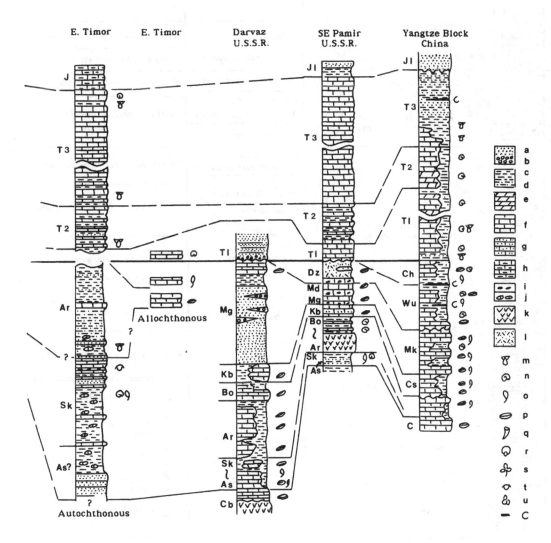

Fig. 3.7. (*cont*)

Correlation of Himalayan and Salt Range sections

Upper Permian strata in Himalayan and Salt Range sections differ from one another primarily in lithofacies, and this is an expression of differences in depth of water and sediment supply in the various basins in which the strata accumulated. Extent of Permian deposits in different areas, as determined biostratigraphically, is related to the times at which the Late Permian regression and the Early Triassic transgression affected various basins of deposition. Relations across the Tethyan realm are shown diagrammatically in Table 3.1 and Fig. 3.7.

Permo-Triassic strata in Kashmir and the Salt Range are mostly shallow-marine deposits, whereas those in the Spiti–Kumaon belt formed in slightly deeper water. In Nepal a shallow-water depositional regime is again indicated.

The Late Permian transgression is recorded first in the Salt Range (Table 3.1; Figs 3.2, 3.10), possibly in the Kazanian.

development to those in Himalayan sections. *Otoceras* has been reported from the base of the South Chinese Lower Triassic at several localities, but doubts have been expressed as to identification of the fossils.

Transgressing Late Permian seas reached Kashmir a bit later, in the Midian, and arrived even later in areas farther to the east. Because deposits in the Spiti–Kumaon belt suggest a depositional environment that was deeper than ones farther west, some pre-Late Permian uplift is indicated. And, because this is also the belt in which the Late Permian regression began the earliest, it is also assumed that the belt was remobilized following a brief incursion of Late Permian seas and that the depositional basin again assumed a high relief. Although we are still in the dark as to details of these movements, continued high relief of the depositional basin in the Early Triassic may be inferred from the fact that deposits of that age are greatly condensed.

In the Salt Range, the last phase of the Late Permian regression is marked by the upper beds of Unit 4 of the Chhidru Formation (Table 3.1). In Nepal this latest phase is recorded in the upper part of the Senja Formation (Table 3.1). In Kashmir sedimentological studies indicate that complete regression did not take place. Subsequent transgression reached the Salt Range and Nepal in the latest Permian and more eastern districts of the Perigondwana Province in earliest Triassic times (Fig. 3.10).

As noted in an earlier part of this chapter, correlation of Permo-Triassic rocks above the *Cyclolobus* Zone is difficult in

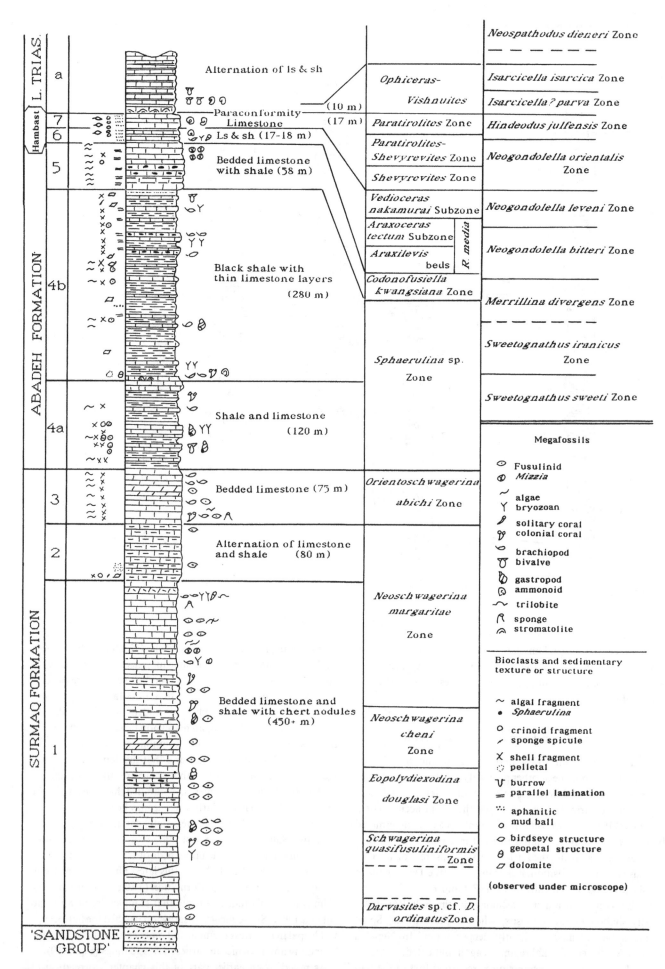

Fig. 3.8. Schematic columnar section of Permian and Lower Triassic rocks in the Abadeh region of central Iran. Zones based on fusulinids, brachiopods, conodonts, and ammonoids, are those recognized by the Iranian–Japanese Research Group (1981). (Redrafted from Sheng *et al.*, 1984, with corrections and emendations of several conodont names.)

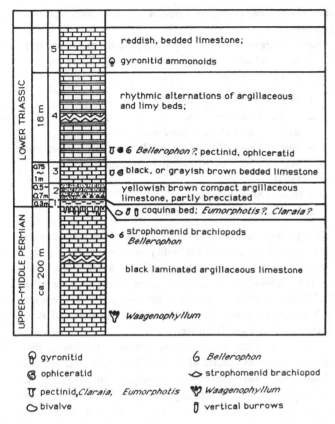

Fig. 3.9. Schematic column of Permian and Lower Triassic rocks at Khoja Ghare Wali, Afghanistan. (From Ishii, Fischer, & Bando, 1971.)

the absence of Dorashamian ammonoids, fusulinids or brachiopods, which are almost entirely absent from Himalayan sections and only sparsely represented at Salt Range localities. This suggests that the Dorashamian fauna developed in an environment that was quite different from the rapidly shallowing one that accompanied Late Permian regression in the Salt Range and Himalaya. Regression may have been initiated by volcanism in the Karakoram region, which was near the Perigondwana Province in the Late Permian and may have changed configuration of the Tethyan basin. This may also have affected development of the Kashmir Basin during deposition of Zewan Member D, for that basin apparently grew in isolation as a remnant of Late Permian Tethys after withdrawal of the sea from the Salt Range on one side and from Spiti–Kumaon on the other. Organisms with wide environmental tolerance survived, but many others disappeared, at least from the Perigondwana Province. Thus, correlation of the latest Permian of Kashmir with Dorashamian or Changxingian strata is very difficult, even though they may all represent the same interval of time.

Correlation of Lower Triassic strata is relatively simple through use of ammonoids, bivalves, and conodonts. In the Salt Range, transgression began in the latest Permian (Table 3.1; Fig. 3.10) and, except for a possible interruption during deposition of *Otoceras*-bearing beds elsewhere, continued through the Early Triassic. In Kashmir, deposition was continuous and the basin merely deepened slightly in the earliest Triassic. Thus unit E1 of the Khunamuh Formation in Kashmir, and the lowermost Pangjang Formation in Nepal with dwarfed brachiopods, which both occur below beds with *Otoceras*, as well as the lower part of the Kathwai Member of the Salt Range, represent almost the same chronostratigraphic level. It is not certain if these beds belong in the Permian or the Triassic.

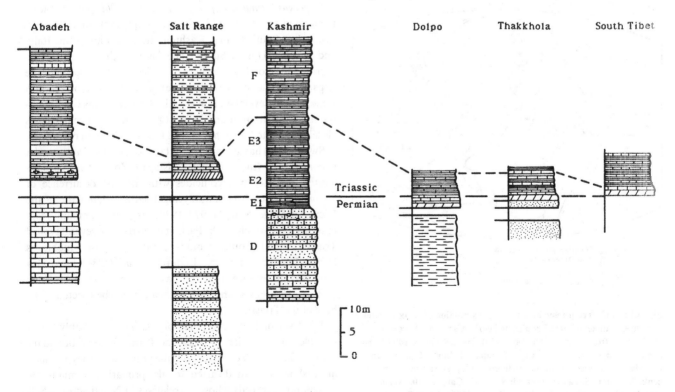

Fig. 3.10. Lithic changes across the Permo-Triassic boundary at various places in the Tethyan realm. (From Kapoor & Tokuoka, 1985.)

In support of a Triassic age for the units just mentioned we note the presence in unit E1 of Kashmir of *Claraia bioni*, thought by some to be a primitive species of the genus, but referred by some Chinese workers to the Triassic *Peribositra* on the basis of similarity in valve outline and rounded hinge margin. On the other hand, *Reichelina* and brachiopods of 'Permian' type occur in the lower part of the Kathwai Member of the Salt Range, and this suggests a Permian age, even though stragglers of several dominantly 'Permian' lineages (including *Otoceras* itself) are known in and above the *Otoceras* Zone in Kashmir and China. Reworking of these fossils is not supported by field evidence.

The contact between the regressive and transgressive parts of the Permo-Triassic transitional sequence might, under ideal conditions, be the most suitable level at which to place the Permo-Triassic boundary. However, it is clear that this level is not the same in all parts of the region considered here, even though it is probably the one that can be recognized most readily by the field geologist. Lower Triassic zones are easy to correlate throughout the Himalaya, and are given in Table 3.1. In connection with the zones listed in Table 3.1, it is of interest that a recent find in the Vardebukta Formation of West Spitzbergen (Nakazawa *et al.*, 1987; Nakamura *et al.*, 1990) of *Otoceras boreale* Spath together with *Claraia stachei*, an index to the Upper Griesbachian, suggests that *Otoceras* may have lingered longer in the Boreal region than elsewhere.

Permo-Triassic boundary

Problems associated with defining a boundary between the Permian and Triassic systems are numerous and have generated extensive discussion in the recent literature. Here only the biochronostratigraphic aspects of the discussion are considered.

Several suggestions have been made as to the most appropriate level at which to situate the boundary. These may be summarized as follows:

1. Base of *Otoceras woodwardi* Zone
2. Base of *Ophiceras* Zone
3. Base of *Isarcicella isarcica* Zone
4. Base of *Neospathodus kummeli* Zone (= base of Dienerian Stage or base of Kummeli–Cristagalli Zone of Sweet, 1988)
5. Base of Smithian Stage
6. First appearance of *Claraia*

The principal objection to fixing the base of the Triassic at the base of the *Otoceras* Zone has to do primarily with the fact that *Otoceras*, the youngest member of a Permian family (Bando *et al.*, 1980), is nowhere found in succession with its ancestors (Fig. 3.11). The latter are not known in the Perigondwana Province, although they are well represented in both the East and West Tethys provinces and elsewhere.

If the base of the *Ophiceras* Zone is fixed at the level of first occurrence of any species of this highly variable, exceedingly plastic stock it will presumably be coincident with the base of the *Otoceras* Zone. That is, representatives of both *Otoceras woodwardi* and *Ophiceras* (*Lytophiceras*) *sakuntala* have been reported from the same level in the famous sections at Guryul Ravine, Kashmir (Nakazawa *et al.*, 1975) and Painkhanda (Diener, 1912).

Sweet (1988) has shown that the level of first occurrence of the distinctive conodont species *Isarcicella isarcica* (Huckriede) (in which he includes conodonts variously described as *Anchignathodus parvus*, *Hindeodus parvus*, and *Isarcicella? parva*) can be recognized at stratigraphically equivalent levels in sections in Primor'ye, South China, Kashmir, Iran, Nakhichevan, Italy, and western United States. The base of the Isarcica Zone corresponds closely to the base of the Werfen Formation of the Alps and to the top (not the base) of the *Otoceras woodwardi* Zone, as it is developed in Kashmir. *Isarcicella* was cosmopolitan in its distribution and in most places its first occurrence is closely associated with the acme of the Late Permian regression.

Sweet (1988) has also shown that the base of the *Neospathodus kummeli* Zone (the Kummeli–Cristagalli Zone of his chronostratigraphic scheme) coincides with the first occurrence of *Gyronites*, which is commonly regarded as a hallmark of the Dienerian Stage. Newell (1973, 1988) has argued for many years that this level, which marks the first occurrence of several typical Triassic faunal groups, would be the one most suitable as Permo-Triassic boundary. Budurov *et al.* (1988) have also proposed, in essence, that the base of the *N. kummeli* Zone (= base of Sweet's Kummeli–Cristagalli Zone) be selected as the base of the Triassic.

The level of first occurrence of *Claraia* is probably not a suitable one at which to draw the Permo-Triassic boundary because the solitary occurrence of the primitive species, *C. bioni*, in Kashmir is with dwarfed fossils, primarily Permian-type brachiopods, in beds whose correlatives in Nepal and the Salt Range lack any Triassic forms.

The suitability of regarding the *Otoceras* Zone as Triassic has

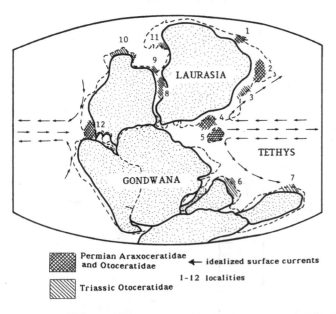

Fig. 3.11. Paleogeographic map showing distribution of Araxoceratidae and Otoceratidae in latest Permian and earliest Triassic. Reconstruction of Pangaea from Dietz & Holden (1970). Dashed line is 1000-fathom continental margin. Localities are: 1, Japan; 2, South China; 3 and 6, Himalaya; 4, Transcaucasia and northern Iran; 5, central Iran microcontinent; 7, Timor; 8, east Greenland; 9, Arctic Canada; 10, Alaska; 11, Siberia; 12, Mexico. Modified from Bando (1980). Note: Recent paleomagnetic evidence suggests that South China may have been much closer to the equator than shown in this reconstruction.

been discussed in detail by Bando (1971, 1973, 1979, 1980), Kummel (1972), Nakazawa *et al.* (1980), the Iranian–Japanese Research Group (1981), and Tozer (1988) and their views are endorsed here. The base of this zone is apparently the level at which a maximum change in biota is recorded and, although the zone is restricted in its geographic distribution, it is traceable throughout the Himalaya. Accordingly, the Permo-Triassic boundary is at the base of the middle unit of the Kathwai Member in the Salt Range, at the base of unit E2 in Kashmir, and at the base of the *Otoceras* bed in other parts of the Himalaya. Only in Kashmir, however, do rocks of Permian age grade without a break into strata of Triassic age.

The Guryul Ravine section, in Kashmir, with support from the Pahalgam section, is one of the best reference sections for the transition beds. The Guryul Ravine section has the disadvantage of lacking Dorashamian fossils, but it has an excellent development of Lower Triassic strata. Sections in South China, on the other hand, include a well-developed Late Permian biota, but the Permo-Triassic transition is not as clear as in Kashmir. The differences are related to original paleogeographic position of the two areas, which developed independently.

References

Bando, Y. (1971). On the Otoceratidae, Triassic ammonoids and its stratigraphical significance. *Mem. Fac. Educ. Kagawa Univ.*, Pt. II, no. 203: 1–11 (in Japanese).

Bando, Y. (1973). On the Otoceratidae and Ophiceratidae. *Sci. Rep. Tohoku Univ.*, 2d ser. (Geol.), Spec. vol. **6**: 337–51.

Bando, Y. (1979). Upper Permian and Lower Triassic ammonoids from Abadeh, central Iran. *Mem. Fac. Educ. Kagawa Univ.*, Pt. II, **29**(2): 103–38.

Bando, Y. (1980). On the Otoceratacean ammonoids in the central Tethys with a note on their evolution and migration. *Mem. Fac. Educ. Kagawa Univ.*, Pt. II, **30**(1): 23–49.

Bando, Y., Bhatt, D. K., Gupta, V. J., Hayashi, S., Kozur, H., Nakazawa, K. & Wang, Z. (1980). Some remarks on the conodont zonation and stratigraphy of the Permian. *Rec. Res. in Geol*, **8**: 1–53.

Bassoullet, J. P. & Colchen, M. (1977). La limite Permien-Trias dans le domain tibetian de l'Himalaya du Nepal (Annapurnas-Ganesh Himal.). In: *Himalaya; Sciences de la Terre*, ed. Jest, C., pp. 41–52. Colloq. Int. no. 268 (Ecologie et Geologie de l'Himalaya). Paris: C. N. R. S.

Bhargava, O. N. & Gadhoke, S. K. (1988). Triassic microfauna of the Lilang Group with special reference to Scythian-Anisian conodonts, Spiti Valley, Himachal Himalaya. *J. Geol. Soc. India*, **32**(6): 494–505.

Bhatt, D. K. & Arora, R. K. (1984). *Otoceras* bed of Himalaya and Permian-Triassic boundary – Assessment and elucidation with conodonts data. *J. Geol. Soc. India*, **25**(11): 720–7.

Bhatt, D. K. & Joshi, V. K. (1978). Early Lower Triassic conodonts from Spiti River section. *Curr. Sci.*, **47**(4): 118–20.

Bhatt, D. K., Joshi, V. K. & Arora, R. K. (1981). *Neospathodus praekummeli* – A new species of conodont from Lower Triassic of Spiti. *J. Geol. Soc. India*, **22**(9): 444–7.

Budurov, K. J., Gupta, V. J., Kachroo, R. K. & Sudar, M. N. (1988). Problems of the Lower Triassic conodonts stratigraphy and the Permian-Triassic boundary. *Mem. Soc. Geol. Ital.*, **34**: 321–8.

Clark, D. L. & Hatleberg, E. W. (1983). Paleoenvironmental factors and the distribution of conodonts in the Lower Triassic of Svalbard and Nepal. *Fossils Strata.*, **15**: 171–5.

Dickins, J. M. (1987). Correlation charts of Upper Permian with notes. *Permophiles*, no. 12: 8–11.

Diener, C. (1912). The Trias of the Himalayas. *Mem. Geol. Surv. India*, **36**(2): 202–347.

Fuchs, G. (1977). The geology of the Karnali and Dolpo region, western Nepal. *Jb. Geol. B. A.*, **120**(2): 165–217.

Furnish, W. M. & Glenister, B. F. (1970). Permian ammonoid *Cyclolobus* from the Salt Range, West Pakistan. In: *Stratigraphic Boundary Problems: Permian and Triassic of West Pakistan*, ed. Kummel, B. & Teichert, C., pp. 153–75. (Dept. Geol., Univ. Kansas, Spec. Publ. 4). Lawrence: Univ. Press of Kansas.

Gergan, J. T. & Pant, P. C. (1983). Geology and stratigraphy of eastern Karakoram, Ladakh. In: *Geology of Indus Suture Zone of Ladakh*, ed. Thakur, V. C. & Sharma, K. K., pp. 99–106.

Goel, R. K. (1977). Triassic conodonts from Spiti (Himachal Pradesh), India. *J. Paleont.*, **51**(6): 1085–101.

Grant, R. E. (1970). Brachiopods from Permian-Triassic boundary beds and age of the Chhidru Formation, West Pakistan. In *Stratigraphic Boundary Problems: Permian and Triassic of West Pakistan*, ed. Kummel, B. & Teichert, C., pp. 117–52. (Dept. Geol., Univ. Kansas, Spec. Publ. 4). Lawrence: Univ. Press of Kansas.

Hatleberg, E. W. & Clark, D. L. (1984). Lower Triassic conodonts and biofacies interpretations: Nepal and Svalbard. *Geologica Palaeont.*, **18**: 101–25.

Heim, A. & Gansser, A. (1939). Central Himalaya; geological observations of the Swiss expedition 1936. *Denkschr. Schweizer. Naturf. Ges.*, **73**(1): ix + 245 pp.

Iranian–Japanese Research Group (1981). The Permian and Lower Triassic systems in Abadeh region, central Iran. *Mem. Fac. Sci. Kyoto Univ., ser. Geol. Min.*, **47**(2): 61–133.

Ishii, K., Fischer, J. & Bando, Y. (1971). Notes on the Permo-Triassic boundary in eastern Afghanistan. *J. Geosci. Osaka City Univ.*, **14**(1): 1–18.

Jamwal, J. S. & Kacker, A. K. (1989). Lithostratigraphy and regional correlation of the Tethyan sediments of Gori Valley, Pithoragarh District, Uttar Pradesh. *Rec. Geol. Surv. India*, **122**(8): 281–4.

Kapoor, H. M. (1989). Report of Indian Working Committee on Upper Permian correlation and standard scale. *Permophiles*, **15**: 4–8.

Kapoor, H. M. & Tokuoka, T. (1985). Sedimentary facies of the Permian and Triassic of the Himalaya. In *The Tethys, Her Paleogeography and Paleobiogeography from Paleozoic to Mesozoic*, ed. K. Nakazawa, & J. M. Dickins, pp. 23–58. Tokyo: Tokai Univ. Press.

Kozur, H. & Mostler, H. (1973). Beiträge zur Mikrofauna permotriadischer Schichtfolgen. Teil I: Conodonten aus der Tibetzone des Niederen Himalaya (Dolpogebiet, Westnepal). *Geol. Paläont. Mitt. Innsbruck*, **3**(9): 1–23.

Kumar, G., Mehdi, S. H. & Prakash, G. (1971). A review of stratigraphy of parts of Uttar Pradesh, Tethys Himalaya. *J. Palaeont. Soc. India*, **15**: 36–98.

Kumar, S., Singh, I. B. & Singh, S. K. (1977). Lithostratigraphy, structure, depositional environment, palaeocurrent and trace fossils of the Tethyan sediments of Malla Johar area, Pithoragarh-Chamoli districts, Uttar Pradesh, India. *J. Palaeont. Soc. India*, **20**: 396–435.

Kummel, B. (1972). The Lower Triassic (Scythian) ammonoid *Otoceras*. *Bull. Mus. Comp. Zool.*, **143**(6): 365–418.

Kummel, B. & Teichert, C. (1970). Stratigraphy and paleontology of Permian-Triassic boundary beds, Salt Range and Trans-Indus ranges, West Pakistan. In *Stratigraphic Boundary Problems: Permian and Triassic of West Pakistan*, ed. Kummel, B., & Teichert, C., pp. 1–110. (Dept. Geol., Univ. Kansas, Spec. Publ. 4). Lawrence: Univ. Press of Kansas.

Malviya, A. K. & Pande, A. C. (1989). Lithostratigraphy and regional correlation of the Tethyan sediments of the Kali and Dhauliganga valleys, Pithoragarh District. *Rec. Geol. Surv. India*, **122**(8): 285–8.

Mamgain, V. D. & Misra, R. S. (1989). Biostratigraphical studies of the Palaeozoic and Mesozoic sediments of the Tethyan facies in U. P. Himalaya. *Rec. Geol. Surv. India*, **122**(8): 296–8.

Matsuda, T. (1981a). Early Triassic conodonts from Kashmir, India. Pt. 1. *Hindeodus* and *Isarcicella*. *J. Geosci. Osaka City Univ.*, **24**(3): 75–108.

Matsuda, T. (1981b). Appendix to conodonts of Guryul Ravine. *Palaeont. Indica*, n. ser., **46**: 187–8.

Murata, M. (1981). Late Permian and Early Triassic conodonts from Guryul Ravine. *Palaeont. Indica*, n. ser., **46**: 179–86.

Nakamura, K., Kimura, G., Winsnes, T. S. & Lauritzen, O. (1990). Permian and Permian-Triassic bounary in central Spitzbergen. In *The Japanese Scientific Expeditions to Svalbard 1983–1988*, ed. T. Tatsumi. Kyoikusha, Japan.

Nakazawa, K. (1981). Permian and Triassic bivalves of Kashmir. *Palaeont. Indica*, n. ser., **46**: 87–122.

Nakazawa, K., Bando, Y. & Matsuda, T. (1980). The *Otoceras woodwardi* Zone and the time gap at Permian-Triassic boundary in east Asia. *Geol. Palaeont. S. E. Asia*, **21**: 75–90.

Nakazawa, K. & Kapoor, H. M. (1977). Correlation of the marine Permian in the Tethys and Gondwana. *IV Internat. Gondwana Symp. Papers*, **2**: 409–19.

Nakazawa, K. & Kapoor, H. M., (Eds) (1981). The Upper Permian and Lower Triassic faunas of Kashmir. *Palaeont. Indica*, no. ser., **46**: 204 pp.

Nakazawa, K., Kapoor, H. M., Ishii, K., Bando, Y., Maegoya, T., Shimizu, D., Nogami, Y., Tokuoka, T. & Nohda, S. (1970). Preliminary report on the Permo-Trias of Kashmir, India. *Mem. Fac. Sci. Kyoto Univ., Ser. Geol. Min.*, **37**(2): 163–172.

Nakazawa, K., Kapoor, H. M., Ishii, K., Bando, Y., Okimura, Y. & Tokuoka, T. (1975). The Upper Permian and the Lower Triassic in Kashmir, India. *Mem. Fac. Sci. Kyoto Univ., Ser. Geol. Min.*, **42**(1): 1–106.

Nakazawa, K., Nakamura, K., & Kimura, G. (1987). Discovery of *Otoceras boreale* Spath from West Spitzbergen. *Proc. Japanese Acad.*, ser. B, **63**: 171, 174.

Newell, N. D. (1973). The very last moment of the Paleozoic Era. *Can. Soc. Petrol. Geol. Mem.*, **2**: 1–10.

Newell, N. D. (1988). The Paleozoic/Mesozoic erathem boundary. *Mem. Soc. Geol. Ital.*, **34**(1986): 303–11.

Nicora, A., Gaetani, M. & Garzanti, E. (1984). Late Permian to Anisian in Zanskar (Ladakh, Himalaya). *Rend. Soc. Geol. Ital.*, **7**: 27–30.

Pakistani–Japanese Research Group (1985). Permian and Triassic systems in the Salt Range and Surghar Range, Pakistan. In *The Tethys, Her Paleogeography and Paleobiogeography from Paleozoic to Mesozoic*, ed. Nakazawa, K. & Dickins, J. M., pp. 219–312. Tokyo: Tokai Univ. Press.

Rostovtsev, K. O. & Azaryan, N. R. (1973). The Permian-Triassic boundary in Transcaucasia. *Mem. Can. Soc. Petrol. Geol.*, **2**: 89–99.

Ruzhentsev, V. E. & Sarycheva, T. G. (Eds) (1965). Evolution and change in the marine organisms at the boundary between Paleozoic and Mesozoic. *Trudy Paleont. Inst.*, **119**: 273 pp. (in Russian).

Schindewolf, O. H. (1954). Über die Faunenwende vom Paläozoikum zum Mesozoikum. *Zt. Deutsche Geol. Ges.*, **105**: 154–83.

Sheng Jin-zhang, Chen Chu-zhen, Wang Yi-gang, Rui Lin, & Liao Zhuo-ting (1984a). On the lower boundary of the Triassic in central and eastern Tethys. *Devs Geosci., Acad. Sinica*, 105–110.

Sheng Jin-zhang, Chen Chu-zhen, Wang Yi-gang, Rui Lin, Liao Zhuo-ting, Bando, Y., Ishii, K., Nakazawa, K. & Nakamura, J. (1984b). Permian and Triassic boundary in middle and eastern Tethys. *J. Fac. Sci. Hokkaido Univ.*, ser. IV, **21**(1): 133–81.

Srikantia, S. V. (1981). The lithostratigraphy, sedimentation and structure of Proterozoic-Phanerozoic formations of Spiti Basin in the Higher Himalaya of Himchal Pradesh, India. In *Contemporary Researches in Himalaya*, ed. A. K. Sinha, pp. 31–48. Dehra Dun: Bishen Singh Mahendra Pal Singh.

Sweet, W. C. (1970a). Permian and Triassic conodonts from a section at Guryul Ravine, Vihi district, Kashmir. *Pal. Contrib. Univ. Kansas*, **49**: 1–10.

Sweet, W. C. (1970b). Uppermost Permian and Lower Triassic conodonts of the Salt Range and Trans-Indus ranges, West Pakistan. In *Stratigraphic Boundary Problems: Permian and Triassic of West Pakistan*, ed. Kummel, B. & Teichert, C., pp. 207–75. (Dept. Geol., Univ. Kansas, Spec. Publ. 4). Lawrence: Univ. Press of Kansas.

Sweet, W. C. (1979). Graphic correlation of Permo-Triassic rocks in Kashmir, Pakistan, and Iran. *Geologica Palaeont.*, **13**(7): 231–48.

Sweet, W. C. (1988). A quantitative conodont biostratigraphy for the Lower Triassic. *Senckenbergiana lethaea*, **69**(3/4): 253–73.

Tokuoka, T. (1985). The Permian and Triassic boundary at Thini Khola, Thakkhola region, central Nepal. *Mem. Fac. Sci. Shimane Univ.*, **19**: 121–33.

Tozer, E. T. (1988). Definition of the Permian-Triassic (P-T) boundary – the question of the *Otoceras* beds. *Mem. Soc. Geol. Ital.*, **34**: 291–302.

Wang Yi-gang, Chen Chu-zhen & Rui Lin (1988). A potential stratotype of the P/T boundary. *Permophiles*, no. 13: 3–7.

Waterhouse, J. B. (1977). The Permian rocks and fauna of Dolpo, north-west Nepal. In: *Himalaya; Sciences de la Terre*, ed. Jest, C., pp. 479–96. Colloq. Internat. no. 268 (Ecologie et Geologie de l'Himalaya). Paris: C. N. R. S.

Waterhouse, J. B. (1978). Permian brachiopods and Mollusca from north-west Nepal. *Palaeontographica*, Abt. A, **160**: 1–175.

Waterhouse, J. B. (1979). Permian rocks of Kali Gandaki area (Thakkhola), north central Nepal. In: *Upper Palaeozoics of the Himalayas*, ed. Gupta, V. J., pp. 195–213. (Contributions to Himalayan Geology, 1). Delhi: Hindustan Publ. Co.

Yao Jianxin, & Li Zishun (1987). Permian-Triassic conodont faunas and the Permian-Triassic boundary at the Selong section in Nyalam County, Tibet, China. *Kexuo Tungbo*, **32**(22): 1556–60.

4 Permo-Triassic boundary on the Indian peninsula

R. S. TIWARI AND VIJAYA

Introduction

The Gondwana sequence of peninsular India is mainly terrestrial in nature and exhibits different patterns of sedimentation in each basin. This situation arose because deposition took place in linear, fault-bounded belts in which recurrent uplift or subsidence at varying rates created different tectonic regimes (Fig. 4.1). Consequently there are problems in interbasinal correlation. As a working system for correlation, however, various stratigraphic packages in a sequence are identified on the basis of lithic characters observed in discrete regions. These are then arrayed in a framework of greater spatial extent and this arrangement may be correlated precisely if biozones are available to use. However, biozones may or may not correspond with lithic units.

In recent years, a few widely spaced datum planes have been identified, along which biostratigraphic control matches well-delineated lithostratigraphic suites. The Permo-Triassic boundary has been one of the focal points for such an evaluation.

The conventional correlation of Gondwana formations is given in Fig. 4.2 (Datta & Mitra, 1982) and the Upper Permian–Lower Triassic interval is shown in greater detail in Fig. 4.3, which also includes information on general lithic characters. Some alterations based on recent data have been added.

The Upper Permian is characterized by diversification of the *Glossopteris* flora, which includes *Glossopteris*, *Vertebraria*, *Raniganjia*, *Schizoneura*, *Tryzygia*, *Asansolia*, and *Damudopteris*. The Lower Triassic flora records the advent of *Dicroidium* along with *Lepidopteris*, *Podozamites*, *Macrotaenopteris*, *Pseudoctensis*, and *Pterophyllum*. Palynologically, the Upper Permian assemblage is dominated by striate disaccate pollen (*Striatopodocarpites*, *Faunipollenites*, *Crescentipollenites*, *Striatites*, etc.) and qualifying taxa (*Densipollenites*, *Gondisporites*, *Thymospora*, *Guttulapollenites*, *Indospora*, etc.). On the other hand, the Lower Triassic assemblage is composed mainly of *Klausipollenites*, *Lunatisporites*, *Lundbladispora*, *Densoisporites*, *Callumispora*, *Verrucosisporites*, *Playfordiaspora*, *Alisporites*, and *Satsangisaccites*.

Generic percentage frequency and occurrence of species determine the relative positions of each palynoflora (Figs 4.4, 4.5). Some Lower Triassic forms (e.g. *Playfordiaspora*, *Lundbladispora*, *Lunatisporites*) first appear sporadically in the uppermost Permian, as precursors, whereas a few of the dominant Upper Permian forms (e.g. *Striatopodocarpites*, *Densipollenites*) continue, with diminishing frequency, into the Lower Triassic. The first consistent presence of specific forms has been given more weight than their first inconsistent occurrence.

There is no marine control for the Permo-Triassic boundary in peninsular India, hence both direct and inferential information from lithostratigraphy and biostratigraphy have been taken into consideration. Where available, the age of lithic suites has been determined from vertebrate fossils, plant mega- and microfossils, and estheriids. In cases in which narrowing of the Permo-Triassic transition interval is impossible due to lack of control, upper and lower boundaries of the transition interval have been determined and the systemic boundary has been 'inferred'.

In the Damodar Valley, Satpura, and Godavari Valley basins the level of the Permo-Triassic boundary is definite. In other basins, however, information that permits approximation of the boundary has been recognized but further resolution is awaited.

The basinal record

Damodar Valley basins

In the Damodar Valley basins, coal measures of the Upper Permian Raniganj Formation are overlain by the Triassic Panchet Formation, which consists of a thick cyclic sequence of khaki-greenish shales and sandstones followed by red, chocolate shales and arkosic sandstones (Fig. 4.3). The Permo-Triassic boundary is the lithic boundary between the two formations. In most of the basins the sequence is continuous, although there are some local gaps of restricted extent and small magnitude (Datta, Mitra & Bandyopadhaya, 1983). The Permian flora of the coal-bearing Raniganj Formation is dominated by the *Glossopteris* group. The Triassic flora of the lower Panchet is identified by the advent of *Dicroidium*. In a well-developed sequence, the latest Permian assemblage is dominated palynologically by *Striatopodocarpites*, with prominent *Densipollenites* and *Crescentipollenites*. The earliest Triassic assemblage, on the other hand, is marked by an impressive frequency of *Klausipollenites* and *Lunatisporites*, along with *Playfordiaspora*, *Lundbladispora*, *Densoisporites*, *Verrucosisporites*, *Callumispora*, and others. The transformation of palynoflora at the Permo-Triassic boundary is without a sharp or abrupt break (Fig. 4.5); however, the change is definitive in mode of alteration and quality of composition (Vijaya & Tiwari, 1987). The lithic boundary is shown in

37

Fig. 4.4 to be fairly close to the palynological boundary. The former is drawn above the sandstone unit of the top-most coal seam, whereas the latter is identified within the shale unit above the seam.

Occurrence of estheriids with restricted ranges in the Raniganj (*Monoleaia* Leaid-III, *Cyzicus–Lioestheria*) and in the Panchet (*Cyzicus–Euestheria, Estheriella*) supports assignment to Late Permian and Early Triassic, respectively (Ghosh *et al.*, 1987, 1988). Also, the presence of certain vertebrate fossils in the lower Panchet (*Lystrosaurus* and *Chasmatosaurus*) provides conclusive evidence of an Early Triassic age for this formation. Recent discoveries of bivalves in this basin, in the upper coal horizon of the Jharia and Raniganj coalfields, indicate an early 'Tatarian' age, but precise zonations have yet to be made (Chandra & Betekhtina, 1990). The two bivalve zones indicate a marine signature in the Upper Permian.

Godavari Valley basins

In the Godavari Valley basins, the Upper Permian is represented by the coal-free Barren Measures and part of the thick, overlying Kamthi Formation (Fig. 4.3). The Kamthi is divisible lithologically into three members: Lower Kamthi, with major coal deposits and carbonaceous facies; Middle Kamthi, dominated by red, green, and yellow shales and sandstones; and Upper Kamthi, with coarse, pebbly sandstones, siltstones, clasts, and red ferruginous shales and sandstones (Raja Rao, 1982). The lower lithic unit (or bed) of the Middle Member (Fig. 4.6) is Late Permian on the basis of the endothiodont *Cistece-phalus*, and captorhinomorphs, whereas the basal lithic unit of the Upper Kamthi is latest Permian or earliest Triassic because of the presence of dicynodont fossils (Kutty, Jain & Roychowdhury, 1988).

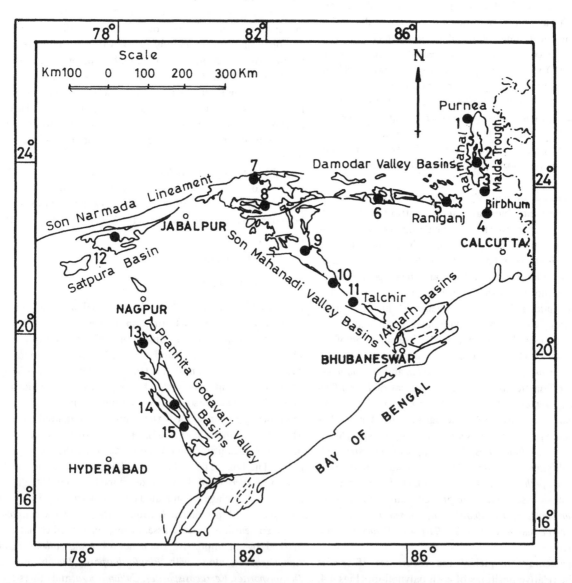

Fig. 4.1. Map showing location of different tectonic regimes in peninsular India and of sections that provide data base for Late Permian-Early Triassic biostratigraphy. 1, Purnea well, Galsi Basin; 2, Rajmahal Basin; 3, Diwanganj-Birbhum area; 4, eastern extension of Raniganj Basin; 5, West Raniganj Basin; 6, West Bokaro Coalfield; 7, Nidpur, Singrauli Basin; 8, Korar, Pali, Sohagpur area, South Rewa Basin; 9, Mand-Raigarh; 10, Handapa beds; 11, northwest part of Talchir Basin; 12, Bijori-Pachmarhi, Satpura Basin; 13, Wardha Basin; 14, 15, Godavari Basin.

Recent palynological studies of subsurface strata (Srivastava & Jha, 1990) provide information that will be useful in locating the Permo-Triassic boundary more precisely. That is, the lower member of the Kamthi contains an Upper Permian palynoflora, dominated by *Striatopodocarpites* and *Faunipollenites*. The lower lithic unit of the middle member yields successive, superposed assemblages rich in *Denisipollenites*, *Crescentipollenites*, and *Guttulapollenites*, which are of latest Permian age (analogous lithologically to the level of the endothiodont *Cistecephalus*). Higher in the Middle Kamthi, but separated from the previous assemblage by an unproductive 12 m thick interval, is an assemblage dominated by *Lunatisporites*, *Verrucosisporites*, and *Lundbladispora*, which is slightly younger than lowermost Triassic (equivalent to P–III in the Raniganj Basin; Tiwari & Singh, 1986). Thus it is confirmed that the Permo-Triassic boundary lies at the top of the basal lithic unit of the Middle Kamthi (Fig. 4.6). The palynologic boundary appears to precede the dicynodont level.

Rajmahal, Malda-Galsi, Purnea, and Birbhum basins

In the Rajmahal, Malda-Galsi, Purnea, and Birbhum basins, the palynoflora and plant fossils support the existence of

strata that might be equated with the upper Raniganj Formation of the Damodar Basin (Tiwari & Tripathi, 1984; Prasad, 1985; Tripathi, 1986, 1989). In some areas, the Triassic begins with a thick deposit of chocolate shales, green clay, and sandstones. In the Purnea Basin, Upper Permian and Lower Triassic palynofloras have been recovered and the change from one to the other has a Permo-Triassic aspect. However, the change from Permian to Triassic is sudden and this indicates a gap in the sequence (Venkatachala & Rawat, 1979).

The Dubrajpur Formation, which either overlies the Panchet Formation unconformably or overlaps older horizons, has yielded an Upper Triassic (Carnian-Norian) palynoflora (Tiwari, Kumar & Tripathi, 1984). Inferentially, the situation regarding the Permo-Triassic boundary is closely comparable in the Rajmahal, Purnea, and Birbhum basins to that in the Raniganj Basin – that is, the boundary coincides with the contact between the Raniganj and Panchet formations. However, precise placement of the boundary will require filling existing gaps in information currently available.

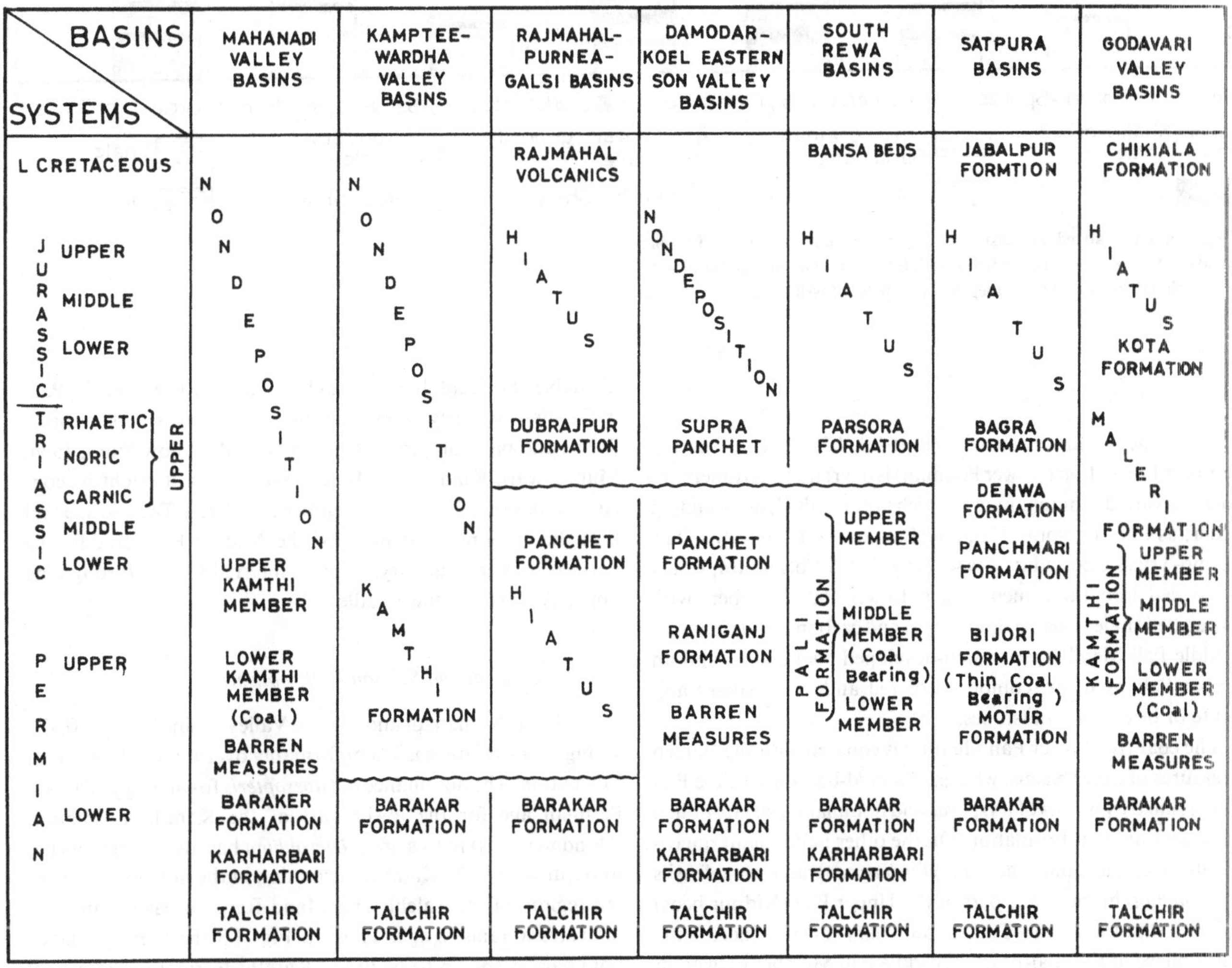

Fig. 4.2. **General correlation of Gondwana formations, depicting position of Permian and Triassic sequences. (After Datta & Mitra, 1982.)**

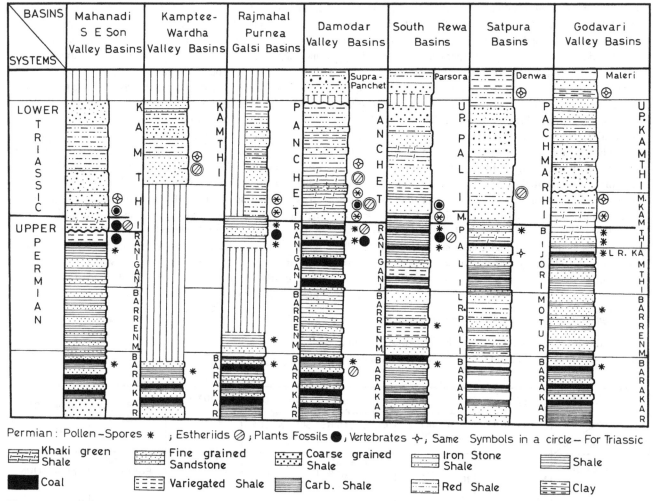

Permian : Pollen –Spores ✳ ; Estheriids ⊘ ; Plants Fossils ● ; Vertebrates ⊹ ; Same Symbols in a circle – For Triassic

| Khaki green Shale | Fine grained Sandstone | Coarse grained Shale | Iron Stone Shale | Shale |
| Coal | Variegated Shale | Carb. Shale | Red Shale | Clay |

Fig. 4.3. Generalized succession of Upper Permian and Lower Triassic strata in basins of peninsular India, with information on the distribution of depositional facies, plant megafossils, palynofossils, estheriids, and vertebrates.

South Rewa basins

In the South Rewa basins (Figs 4.2, 4.3), the coal-bearing Barakar Formation (Lower Permian) is overlain by sedimentary rocks of varied lithic character, which are classified, in ascending order, as Pali, Parsora, Tiki, and Bandogarh formations. The Permo-Triassic boundary lies within the Pali Formation, which is divided into three members: a Lower Pali Member, with variegated shales, coarse sandstones, and carbonaceous lenses; a Middle Pali Member, with well-developed coal seams; and an Upper Pali Member, comprised of red and green shales and white or brownish sandstones.

The coal-free Lower Pali Member is equated with the Barren Measures of other basins, whereas the coal-bearing Middle Pali is correlated by means of its spores and dominantly striate pollen with the Raniganj Formation. On the other hand, plant microfossils (taeniate pollen and *Playfordiaspora*) and megafossils (*Dicroidium*) in the lower part of the Upper Pali (Nidpur beds) indicate that the Permo-Triassic boundary must cut across the latest phase of the coal-bearing Middle Pali Member (Tiwari & Ram-Awatar, 1988, 1989; Ram-Awatar, 1988; Pal, pers. comm., 1988). No evidence for the age of the upper Upper Pali is

available at present, but the Tiki Formation, well above the Pali, is Carnian to Early Norian in age on the basis of vertebrate fossils and palynomorphs (Chatterjee & Roychowdhury, 1974; Maheshwari, Kumaran & Bose, 1978). Thus, it might be concluded that the Upper Pali Member is of Early Triassic age and that the carbonaceous units of the Middle Pali straddle the Permo-Triassic boundary, which is inferred to pass through the uppermost part of the Middle Pali.

Mahanadi and SE Son Valley basins

In the Mahanadi and SE Son Valley basins (Figs. 4.3), the Raniganj Formation is unconformably overlain by the Kamthi Formation. The dominance of *Glossopteris* fossils suggest a Late Permian age for the lower part of the Kamthi Formation (Handapa beds) in this area. *Dicroidium* has also been reported to occur within the Kamthi Formation in the northwest part of the region, in the Talchir and Mand-Raigarh basins, and this indicates a Triassic age. Thus, in Fig. 4.3 the Permo-Triassic boundary is shown to be in the Kamthi Formation, at a level slightly above the beds with profuse *Glossopteris* (equivalent to the Handapa beds).

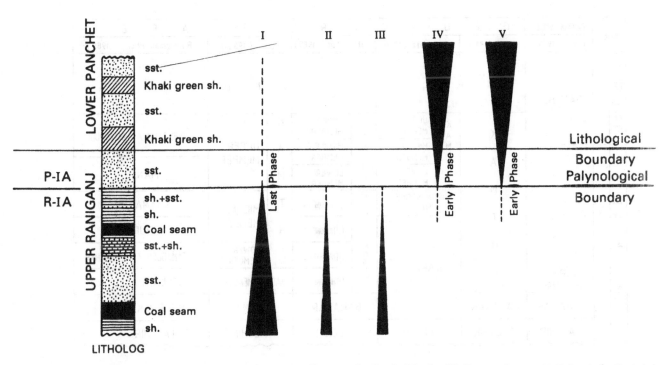

Fig. 4.4. Range and relative abundance of representative spore-pollen species in a standard section including the Raniganj–Panchet boundary in the Raniganj Coalfield, Damodar Basin. I = *Densipollenites invisus, D. indicus, D. densus*; II = *Densipollenites magnicorpus*; III = *Gondisporites raniganjensis*; IV = *Lundbladispora microconata, L. brevicula, Densoisporites contactus, Playfordiaspora cancellosa*; V = *Lunatisporites ovatus, L. diffusus, L. pellucidus*. (After Vijaya & Tiwari, 1987.)

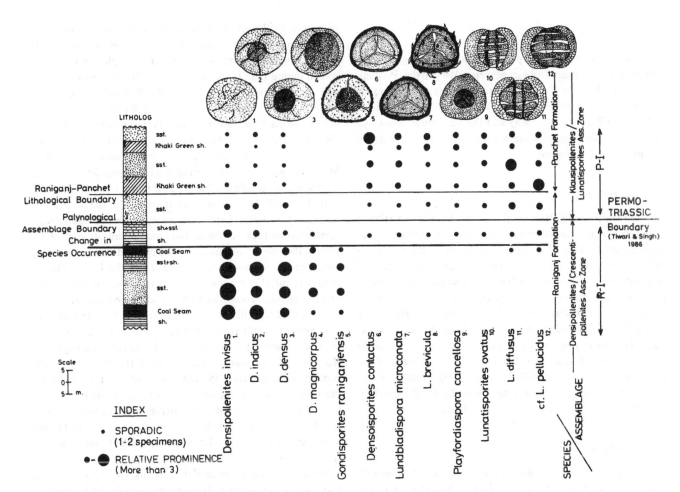

Fig. 4.5. Transformation of palynoflora at Raniganj–Panchet boundary in Damodar Basin. (After Vijaya & Tiwari, 1987.)

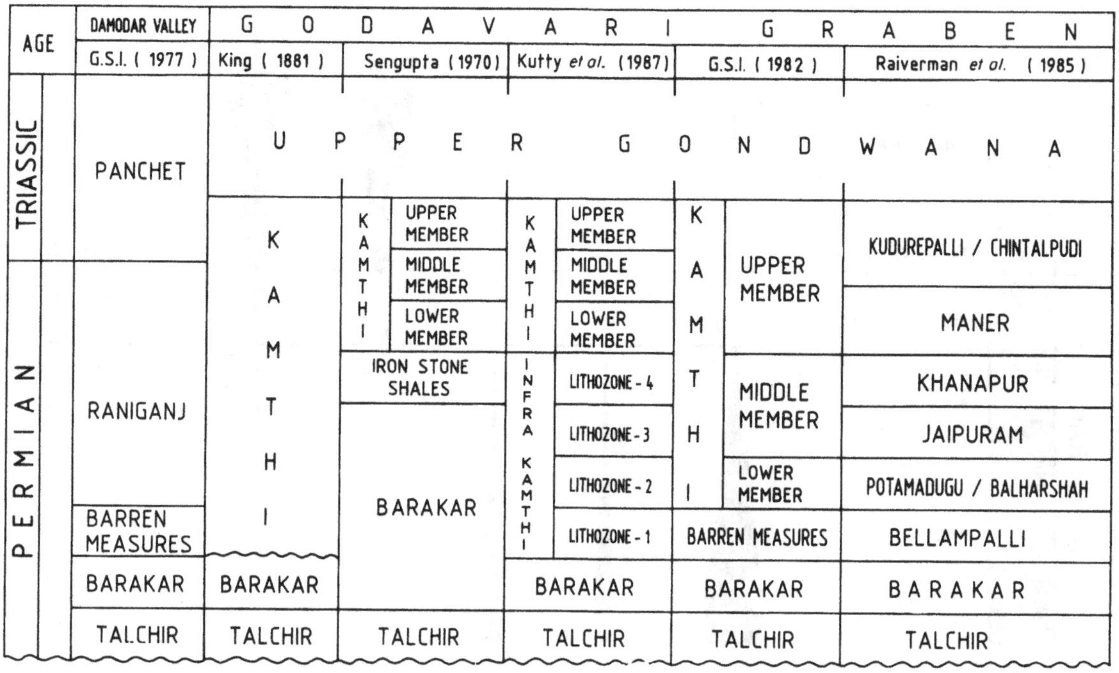

AGE	DAMODAR VALLEY G.S.I. (1977)	G O D A V A R I G R A B E N				
		King (1881)	Sengupta (1970)	Kutty *et al.* (1987)	G.S.I. (1982)	Raiverman *et al.* (1985)
TRIASSIC — PANCHET	PANCHET	U P P E R G O N D W A N A				
PERMIAN — RANIGANJ	RANIGANJ	K A M T H I	KAMTHI — UPPER MEMBER	KAMTHI — UPPER MEMBER	KA — UPPER MEMBER	KUDUREPALLI / CHINTALPUDI
			KAMTHI — MIDDLE MEMBER	KAMTHI — MIDDLE MEMBER	A M T — UPPER MEMBER	MANER
			KAMTHI — LOWER MEMBER	KAMTHI — LOWER MEMBER		
			IRON STONE SHALES	INFRA KAMTHI — LITHOZONE-4	H I — MIDDLE MEMBER	KHANAPUR
			BARAKAR	LITHOZONE-3		JAIPURAM
	BARREN MEASURES			LITHOZONE-2	LOWER MEMBER	POTAMADUGU / BALHARSHAH
				LITHOZONE-1	BARREN MEASURES	BELLAMPALLI
	BARAKAR	BARAKAR		BARAKAR	BARAKAR	B A R A K A R
	TALCHIR	TALCHIR	TALCHIR	TALCHIR	TALCHIR	TALCHIR

Fig. 4.6. Lithostratigraphic correlation schemes for Permian and Lower Triassic rocks in the Godavari graben. Palynological studies support the proposal of Sengupta (1970), in which the Permo-Triassic boundary is located in the Middle Kamthi Member (After Srivastava & Jha, 1988).

Kamptee-Wardha Valley basins

In basins in the Kamptee-Wardha Valley region (Fig. 4.3), a nondepositional phase occurred late in the Permian Period. The Kamthi Formation, which overlaps the Barakar coal measures, contains in the Mangli bed, Early Triassic estheriids (Ghosh *et al.*, 1988) and the vertebrate *Brachyops laticeps*. Triassic deposits are recorded first in the Kamthi Formation in these basins.

Satpura Basin

In the Satpura Basin (Fig. 4.3), the Bijori Formation is Late Permian and characterized by the fossil amphibian, *Gondwanosaurus bijoriensis*. The uppermost part of the Bijori Formation has yielded a latest Permian palynoassemblage containing *Striatopodocarpites*, *Densipollenites*, *Guttulapollenites*, and a few elements of Early Triassic affinity, such as *Klausipollenites*, *Lunatisporites*, *Lundbladispora*, which suggest proximity of the Permo-Triassic boundary (Bharadwaj, Tiwari & Anand-Prakash, 1978). Overlying the Bijori Formation is the Pachmarhi Formation, which is dated as Early Triassic on the basis of the conchostracan *Estheriella* (Ghosh *et al.*, 1988). Thus the systemic boundary may coincide with the well-defined boundary between the Bijori and Pachmarhi formations.

Synthesis

Recent studies of palynomorphs and estheriids enable definitive placement of the Permo-Triassic boundary in continental deposits of peninsular India. In the Damodar graben, the

paleontological level of the boundary nearly coincides with the contact between the Raniganj and Panchet formations. In general, the plant megafossils and estheriids play an important role in dating Late Permian and Early Triassic sediments, but palynology narrows estimation of the Permo-Triassic boundary. In the Satpura Basin, the Bijori–Pachmarhi contact is inferred to mark the transition from Permian to Triassic because the top of the Bijori yields a palynoflora of latest Permian age that also contains elements that foreshadow the Triassic. Moreover, a Permian age for the Bijori is controlled by animal fossils, and a Triassic age for the Pachmarhi by estheriids.

In the South Rewa basins, the upper part of the Middle Pali Formation yields a Late Permian spore-pollen assemblage; carbonaceous rocks continue into the uppermost part of the Middle Pali, on top of which taeniate pollen and the *Dicroidium* flora begin a consistent occurrence. Thus, the Permo-Triassic boundary cuts through the uppermost reaches of the Middle Pali and the carbonaceous lithic unit straddles the boundary.

In the Godavari Basin, the Permo-Triassic boundary appears not to coincide with any distinct lithic unit, but to pass through the middle part of the Middle Kamthi. The coal-bearing Lower Kamthi and the lower, green facies of the Middle Kamthi, contain a characteristic Permian palynoflora. The top of the greenish shale unit in the upper part of the Middle Kamthi has yielded an assemblage of palynofossils rich in *Lundbladispora* and *Lunatisporites*, which is reminiscent of the Lower Triassic assemblage in the Panchet Formation of the Damodar Basin (P–III of Tiwari & Singh, 1986; Table 4.1).

In the Kamptee-Wardha Valley basins, the Upper Permian and part of the Lower Triassic are missing as a result of nondeposition, and in the Mahanadi–SE Son Valley basins, the

Table 4.1. *Details of palynological assemblages in the Permo-Triassic transitional interval of the Damodar Valley basins*

System	Stage	Formation	Assemblage	Sub-Assemblage		Additional Quantitatively Important Miospores	Other Associated Miospores may be present (Quantitative)
LOWER TRIASSIC	T₁	PANCHET FM	P-IV LUNDBLADISPORA DENSOISPORITES	B	Lundbladispora Lunatisporites	Densoisporites, Verrucosisporites, Callumispora, Alisporites, Osmundacidites, Crescentipollenites.	Parasaccites, Callumispora, Goubinispora, Triadispora, Ringosporites, Simeonospora, Indotriradites, Striatopodocarpites, Densostriapollis.
				A	Lundbladispora Densoisporites	Lunatisporites, Klausipollenites, Playfordiaspora, Crescentipollenites, Striatopodocarpites, Faunipollenites.	Cyathidites, Callumispora, Indotriradites, Chordasporites, Striatopodocarpites.
			P-III LUNATISPORITES VERRUCOSISPORITES	B	Lunatisporites Lundbladispora	Callumispora, Osmundacidites, Densoisporites, Playfordiaspora, Faunipollenites, Striatopodocarpites.	Indotriradites, Latosporites, Klausipollenites, Alisporites, Striatites, Goubinispora, Crescentipollenites, Pyramidosporites.
				A	Lunatisporites Verrucosisporites	Alisporites, Lundbladispora, Densoisporites, Falcisporites, Playfordiaspora, Guttatisporites.	Striatopodocarpites, Crescentipollenites, Brevitriletes, Chordasporites, Inaperturopollenites.
			P-II VERRUCOSISPORITES CALLUMISPORA	B	Verrucosisporites Lundbladispora	Densoisporites, Alisporites, Nidipollenites, Guttatisporites, Simeonospora, Callumispora.	Striatopodocarpites, Crescentipollenites, Apiculate Triletes, Sahnites, Lunatisporites.
				A	Verrucosisporites Callumispora	Lundbladispora, Guttatisporites, Densoisporites, Striatopodocarpites.	Alisporites, Lunatisporites, Playfordiaspora, Inaperturopollenites, Densipollenites, Sahnites.
			P-I STRIATOPODOCARPITES KLAUSIPOLLENITES	B	Striatopodocarpites cf. Lunatisporites	Klausipollenites, Alisporites, Faunipollenites, Chordasporites.	Inaperturopollenites, Weylandites, Densipollenites, Crescentipollenites, Lundbladispora, Densoisporites, Playfordiaspora, Goubinispora, Simeonospora.
				A	Striatopodocarpites Klausipollenites	Nidipollenites, Falcisporites, Alisporites, Densipollenites, Weylandites, Striatites.	Guttatisporites, Satsangisaccites, Densosporites, Lundbladispora, Faunipollenites, Apiculate Triletes, Playfordiaspora, Verticipollenites.
UPPER PERMIAN	P₂	RANIGANJ FM	R-I STRIATOPODOCARPITES DENSIPOLLENITES	B	Striatopodocarpites Crescentipollenites	Verticipollenites, Striatites, Lahirites, Densipollenites, Faunipollenites, Scheuringipollenites, Klausipollenites.	cf. Lunatisporites, Gondisporites, Inaperturopollenites, Lophotriletes, Verrucosisporites, Striatosporites, Lundbladispora, Falcisporites.
				A	Striatopodocarpites Densipollenites	Verrucosisporites, Lahirites, Faunipollenites, Crescentipollenites, Scheuringipollenites.	Gondisporites, Ephidripites, Striatosporites, Lundbladispora, Klausipollenites, Playfordiaspora, Chordasporites, Lunatisporites, Corisaccites, Indospora.
			R-II STRIATOPODOCARPITES FAUNIPOLLENITES	B	Striatopodocarpites Gondisporites	Verticipollenites, Lahirites, Faunipollenites.	Apiculate Triletes, Gnetaceaepollenites, Densipollenites, Crescentipollenites, Verrucosisporites, Latosporites, Guttulapollenites.
				A	Striatopodocarpites Faunipollenites	Verticipollenites, Lahirites, Hindipollenites, Striaites, Crescentipollenites, Faunipollenites, Scheuringipollenites.	Apiculate Triletes, Thymospora, Gondisporites, Densipollenites, Lunatisporites, Callumispora, Striatosporites, Gnetaceaepollenites, Welwitschiapites, Verrucosisporites, Guttulapollenites.

Source: (After Tiwari & Singh, 1986.)

picture has still to emerge. As local nomenclature goes, the Raniganj Formation is unconformably overlain by the Kamthi Formation, which has not yet been subdivided. The lower part of the Kamthi sequence contains a rich *Glossopteris* flora, which indicates a Permian age. *Dicroidium* appears somewhere above this bed, hence the Permo-Triassic boundary is in the lower portion of the Kamthi – not, however, at the Raniganj/Kamthi contact, which is an unconformity.

In the Rajmahal, Purnea, Galsi, and Birbhum basins, latest Permian palynomorphs and specimens of *Glossopteris* occur in a shale-sandstone unit in the Dewanganj subsurface sequence and in the Brahmini River coalfield. The palynology of Raniganj and Panchet deposits in a Purnea well also suggest a position for the Permo-Triassic boundary comparable to that in the Raniganj Basin.

The palynological pattern outlined in the preceding discussion is not precisely commensurate with that in the well-documented marine sequences that span the Permo-Triassic boundary in the Salt Range (Balme, 1970) and in Australia (Dolby & Balme, 1976) because the marine sequences are broken by intervals of nondeposition. Nevertheless, major episodes are remarkably similar at control points (Helby, Morgan & Partridge, 1987). For example, proliferation of striate disaccates is well established in the Late Permian and occurrence of taeniate (ribbed) disaccates and cavates is a feature of the Early Triassic. In the Salt Range, the middle unit of the Kathwai Member (Mianwali Formation) starts slightly above the level of the *Otoceras* beds (E2) of Kashmir, hence does not represent the lowermost Triassic (Kapoor, this volume). Its palynoflora is dominated by cavate spores and taeniate pollen, a floral composition that is also found well above the Permo-Triassic boundary in peninsular India (P–III of Table 4.1, Tiwari & Singh, 1986). However, because the freshwater depositional sequence of peninsular India is continuous in most areas, two additional assemblages (P–I and P–II of Table 4.1) are identified in rocks below equivalents of the middle unit of the Salt Range Kathwai. Assemblages P–I and P–II are thus the earliest Early Triassic palynozones of the peninsula.

The *Playfordiaspora crenulata* Zone (= *P. microcorpus* Zone) of Australia (Foster, 1982) has some similarity to the latest Permian zones of peninsular India. That is, *Playfordiaspora* and *Densoisporites*, which characterize the former, are sporadically represented in the R–I Zone of the latter. Forms such as *Triquitrites* and *Triplexisporites*, which occur in the Permo-Triassic successions in the Salt Range and Australia, are not known at this level in peninsular India.

For long-distance correlations dominance, subdominance, or rarity of taxa cannot be taken into account because of variations in ecologic and environmental conditions in widely separated areas. Nevertheless, the general compositions of floras are comparable, as are their evolutionary trends. The peninsular record of floral assemblages is an unbroken continuum, which exhibits a distinctive change at the Permo-Triassic boundary (Fig. 4.7).

References

Balme, B. E. (1970). Palynology of Permian and Triassic strata in the Salt Range and Surghar Range, West Pakistan. In: *Stratigraphic*

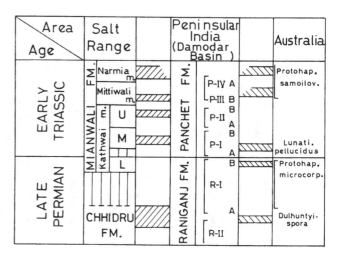

Fig. 4.7. **Broad comparison of Late Permian–Early Triassic palynofloral zones of Damodar Valley basins (Tiwari & Singh, 1986) with those in the Salt Range (Balme, 1970) and Australia (Helby *et al.*, 1987). Importance has been given to compositional trends rather than exclusively to abundance or sporadic first appearance of an element. Obliquely lined intervals indicate levels of comparability in spore-pollen zones. P–I to P–IV are assemblages in the Panchet Formation in Damodar Valley basins; R–I and R–II identify assemblages in the upper Raniganj Formation.**

Boundary Problems: Permian and Triassic of West Pakistan, ed. Kummel, B. & Teichert, C., pp. 305–453. (Dept. Geol. Spec. Pub. 4). Lawrence: Univ. Kansas Press.

Bharadwaj, D. C., Tiwari, R. S. & Anand-Prakash (1978). Palynology of Bijori Formation (Upper Permian) in Satpura Gondwana Basin, India. *Palaeobotanist*, **25**: 70–8.

Chandra, S. & Betekhtina, O. A. (1990). Bivalves in the Indian Gondwana coal measures. *Indian J. Geol.*, **62**(1): 18–26.

Chatterjee, S. & Roychowdhury, T. (1974). Triassic Gondwana vertebrates from India. *Indian J. Earth Sci.*, **1**(1): 96-112.

Datta, N. R. & Mitra, N. D. (1982). Gondwana geology of Indian plate – its history of fragmentation and dispersion. *Int. Centennial Symp. Geol. Surv. Japan, Tsukuba*, preprint, 1–31.

Datta, N. R., Mitra, N. D. & Bandyopadhaya, S. K. (1983). Recent trends in the study of Gondwana basins of peninsular and extra-peninsular India. *Petroliferous Basins of India. Petrol. Asia J.*, pp. 159–69.

Dolby, J. & Balme, B. E. (1976). Triassic palynology of the Carnarvon Basin, Western Australia. *Rev. Palaeobot. Palynol.*, **22**: 105–68.

Foster, C. B. (1982). Spore pollen assemblages of the Bowen Basin, Queensland (Australia): their relationship to the Permian/Triassic boundary. *Rev. Palaeobot. Palynol.*, **36**: 165–83.

Ghosh, S. C., Bhattacharji, T. K., Dutta, A., Sen, C. R. & Dutta, N. R. (1987). A note on the biostratigraphy of Panchet Formation around Andal area, eastern part of Raniganj Coalfield, West Bengal. *Rec. Geol. Surv. India*, Spl. Pub., **11**(1): 233–41.

Ghosh, S. C., Dutta, A., Nandi, A. & Mukhopadhyay, S. (1988). Estheriid zonation in the Gondwanas. *Palaeobotanist*, **36**: 143–53.

Helby, R., Morgan, R. & Partridge, A. D. (1987). A palynological zonation of the Australian Mesozoic. In: *Studies in Australian Mesozoic Palynology*, ed. Jell, P. A., pp. 1–85. Sydney: Assoc. Australian Palynologists.

Kutty, T. S., Jain, S. L. & Roychowdhury, T. (1988). Gondwana sequence of the northern Pranhita-Godavari Valley. *Palaeobotanist*, **36**: 214–29.

Maheshwari, H. K., Kumaran, K. P. N. & Bose, M. N. (1978). The age of the Tiki Formation with remarks on the miofloral succession in the Triassic Gondwana of India. *Palaeobotanist*, **25**: 254–65.

Prasad, B. (1985). A new lithostratigraphic unit in the Lower Gondwana succession of Pachwara Coalfield, Rajmahal Hills, India. *Geophytology*, **25**(2): 110–12.

Raja Rao, C. S. (1982). Coal resources of Tamil Nadu, Andhra Pradesh, Orissa and Maharastra. *Bull. Geol. Surv. India*, Ser A (45) *Coalfields of India*, **2**: 9–40.

Ram-Awatar (1988). Palynological dating of Supra-Barakar formations in Son Valley graben. *Palaeobotanist*, **36**: 133–7.

Sengupta, S. (1970). Gondwana sedimentation around Bheemaram, Pranhita, Godavari Valley, India. *J. Sed. Petrol.*, **40**: 140–70.

Srivastava, S. C. & Jha, N. (1988). Palynology of Kamthi Formation in Godavari graben. *Palaeobotanist*, **36**: 123–32.

Srivastava, S. C. & Jha, N. (1990). Permo-Triassic palynofloral transition in Godavari graben, Andhra Pradesh. *Palaeobotanist*, **38**: 92–7.

Tiwari, R. S., Kumar, P. & Tripathi, A. (1984). Palynodating of Dubrajpur and intertrappean beds in subsurface strata of northwestern Rajmahal Basin. *V. Indian Geophytol. Conf., Lucknow (1983)*, Spl. Pub., 207–25.

Tiwari, R. S. & Ram-Awatar (1988). The Nidpur plant beds, their palynological contents and age connotation (Abstract). *Proc. Symp. Vistas in Indian Palaeobotany, Sahni Inst., Lucknow, 1988*, p. 78.

Tiwari, R. S. & Ram-Awatar (1989). Sporae-dispersae and correlation of Gondwana sediments in Johilla Coalfield, Son Valley graben, M. P., India. *Palaeobotanist*, **37**(1): 94–114.

Tiwari, R. S. & Singh, Vijaya (1986). Palynological evidences for Permo-Triassic boundary in Raniganj Coalfield, Damodar Basin, India. *Bull. Geol. Min. Metall. Soc. India*, **54**: 256–64.

Tiwari, R. S. & Tripathi, A. (1984). A report of Raniganj mioflora from sediments of Durajpur Formation in Brahmini Coalfield, Rajmahal Basin. *Geophytology*, **14**: 244–5.

Tripathi, A. (1986). Upper Permian palynofossils from the Rajmahal Basin, Bihar. *Bull. Geol. Min. Metall. Soc. India*, **54**: 265–71.

Tripathi, A. (1989). Palynological evidence for the presence of Upper Permian sediments in the northern part of Rajmahal Basin, Bihar. *J. Geol. Soc. India*, **34**: 198–207.

Venkatachala, B. S. & Rawat, M. S. (1979). Early Triassic palynoflora from the subsurface of Purnea, Bihar, India. *J. Palynol.*, **14**(1): 59–70.

Vijaya, & Tiwari, R. S. (1987). Role of spore-pollen species in demarcating the Permo-Triassic boundary in Raniganj Coalfield, West Bengal. *Palaeobotanist*, **35**: 242–8.

5 The Permo-Triassic boundary in the southern and eastern USSR and its international correlation

YURI D. ZAKHAROV

Introduction

Transcaucasia and Primorye are regions in which problems of Permian and Triassic biostratigraphy were studied long ago. The first description of Late Permian invertebrates from Transcaucasia was made by Abich (1878) who considered them to be Early Carboniferous. After finding some fossils in his collection that were later described by Stoyanow (1910) as *Paratirolites*, Mojsisovics (1879) erroneously correlated the appropriate Transcaucasian sediments with the Lower Triassic *Tirolites* beds in the Alps. After the discovery of the *Otoceras* beds in the Himalaya (Griesbach, 1880) new possibilities appeared for correlation of the Permo-Triassic boundary beds in Transcaucasia.

Griesbach (1880) considered the *Otoceras* beds to be earliest Triassic. Later he began to think that they might be intermediate in age between Permian and Triassic. The ammonoids described by Abich and later named *Araxoceras*, *Prototoceras*, and *Vescotoceras* by Spath (1930) and Ruzhencev (1959, 1962), were considered by Griesbach to be close to, or identical with *Otoceras*. However, on the basis of suture-line data, Mojsisovics (1892) determined '*Otoceras*' from Transcaucasia to be more primitive than *Otoceras* from the Himalayas. In Mojsisovics' opinion, the Transcaucasian forms seemed to be Permian and the Himalayan forms Triassic. Later this idea came to be generally accepted, but it was also discovered that the Transcaucasian rocks (Dzhulfian Stage) with otoceratids lie below Dorashamian sediments (Rostovcev & Azaryan, 1974), which form the uppermost part of the Permian. In southeast China the Changxing Formation occupies this position (Chao, 1965).

Permian ammonoids are not diverse in Primorye. The first Permian ammonoid (*Daubichites*) was discovered in South Primorye in 1961 (Popov, 1963). Then, some Late Permian ammonoids were described by Ruzhencev (1976), Pavlov, and Zakharov (Zakharov & Pavlov, 1986a,b).

At the outset of investigations of the Lower Triassic, Diener (1895), an expert on Himalayan geology, attached great importance to a study of the Early Triassic fauna of the Ussurian region (Primorye), which, in his opinion, closely resembles the fauna of the Himalayas. Significant contributions to studies of ammonoids and biostratigraphy of the Lower Triassic in Primorye were made by Kiparisova (1961) and Kiparisova & Popov (1956, 1964). Many new Early Triassic ammonoids have been described from Primorye in the last two decades (Zakharov, 1968;

Burij & Zharnikova, 1970, 1980; Zharnikova, 1985) and a special scale based on conodonts (Burij, 1979) has recently been made more exact (Zakharov & Rybalka, 1991). Recently, a Lower Triassic ammonoid scale for South Primorye has been detailed and sections in this region have been proposed as stratotypes for two uppermost Lower Triassic stages, the Ayaxian and Russian (Zakharov, 1973, 1978).

Data are contradictory on the age of Paleozoic rocks beneath the Lower Triassic of the North Caucasus (Kulikov & Tkachuk, 1979; Kotljar, Zakharov, Koczyrkevicz et al., 1983), the South Pamirs (Miklukho-Maklai, 1963; Kotljar, Zakharov, Koczyrkevicz et al., 1983) and the Mangyshlak Peninsula (Astakhova, 1956; Shevyrev, 1968; Gavrilova, 1982).

Permian and Lower Triassic biostratigraphy

Transcaucasia

The upper part of the Permian in Transcaucasia is represented by limestones with chert nodules of the Arpinian Formation; shales and limestones with chert lenses and nodules of the Khatchikian Formation; and limestones and calcareous clayey sediments of the Akhurinian Formation and the lower part of the Karabaglyarian Formation. Twelve zones of the Midian, Dzhulfian and Dorashamian stages are recognized within these sequences: (1) *Chusenella abichi*, (2) *Baisalina pulchra*, (3) *Hemigordiopsis orientalis*, (4) *Pseudodunbarula arpaensis–Araxilevis intermedius* (with beds of *Vescotoceras* sp. in the uppermost part) (Midian Stage); (5) *Araxoceras latissimum*, (6) *Vedioceras ventrosulcatum* (Dzhulfian Stage); (7) *Phisonites triangulus*, (8) *Iranites transcaucasius*, (9) *Dzhulfites spinosus*, (10) *Shevyrevites shevyrevi*, (11) *Paratirolites kittli*, (12) *Pleuronodoceras occidentale* (Dorashamian Stage) (Kotljar, Zakharov, Koczyrkevicz et al., 1983; Kotljar, Zakharov, Vuks et al., 1991).

Early Midian cephalopods are rare in Transcaucasia. Only a few nautiloids (*Pleuronautilus dzhagadzurensis* Zakharov) have been recognized within the *Chusenella abichi* Zone (upper part of Arpinian Formation). In the preceding two zones, cephalopods are unknown in this region, but in central Iran this interval has yielded ammonoids referred to *Xenodiscus carbonarius* Waagen and *Cyclolobus* sp. (Bando, 1979). The suture of the *Cyclolobus* is not known, however, so generic identity of the specimen must be confirmed.

Cephalopods are also rare in the *Pseudodunbarula arpaensis–*

46

Araxilevis intermedius Zone. They have been collected only from the uppermost part of the zone. I have discovered a single example of *Vescotoceras*, close to *V. parallelum* Ruzhencev (its suture line is unknown) in the Dorsham I–2 section, 9–10 m below the top of the zone (Kotljar, Zakharov, Koczyrkevicz et al., 1983). Within the upper part of the zone rare specimens of *Pseudogastrioceras* (Dorasham II–1) and *Permonautilus* (Dorasham II–2, in block) have also been found. Later, I discovered two additional specimens of *Vescotoceras* sp. 6–7 m below the top of the zone in the Karabagljar-1 section. In rocks of the same age in Iran (*Araxilevis* beds) a single araxoceratid ammonoid, *Araxoceras* sp. aff. *A. rotoides* Ruzhencev, has been discovered (in a block) (Bando, 1979; Iranian–Japanese Research Group, 1981).

In Transcaucasia, the Zone of *Araxoceras latissimum* contrasts sharply with the one beneath in quantity and diversity of fossil cephalopods. Of the 28 ammonoid and 39 nautiloid species known from this zone, none has been reported from older sediments. In addition, this level records the first appearance of several nautiloid genera and such ceratitoid genera as *Prototoceras*, *Pseudotoceras*, *Rotaraxoceras*, and *Urartoceras*.

Bando (1979) published some data on the discovery of *Eoaraxoceras ruzencevi* Spinosa, Furnish, & Glenister, a representative of the ancestral group of araxoceratid ammonoids, in the lower part of the *Araxoceras* Zone in Iran. However, based on morphology of the shell wall and septa, Bando's specimen may be considered to represent a new genus close to *Prototoceras*. The real *Eoaraxoceras* is associated with *Timorites* in Mexico.

In Transcaucasia representatives of the Dzhulfitidae are the dominant Dorashamian ammonoids. In the lower part of the argillaceous rocks of the Karabagljarian Formation (*Pleuronodoceras occidentale* Zone), which are transitional Permo-Triassic beds, we have recently discovered *P. occidentale* Zakharov and some other ammonoids, typical Permian nautiloids (*Lopingoceras* sp.), and Permian-type brachiopods identified preliminarily by G. V. Kotljar as *Haydenella* sp. (Zakharov, 1985; Zakharov & Rybalka, 1991).

In descending order, the sequence of Permo-Triassic boundary beds in the hypostratotype section Dorasham II–3 is:

Lower Triassic, Induan
Lower Karabagljarian Formation (*Claraia* beds)

(34) Greyish-yellow, medium-bedded limestone . . . . . **9.00 m**
Contains *Claraia clarai* Emmr., *Isarcicella? parva* (Kozur and Pjatakova), *Hindeodus turgidus* (Kozur, Mostler & Rahimi-Yazd), *Isarcicella isarcica* (Huckriede), *Hindeodus minutus* (Ellison) (Kozur et al., 1978). In the upper part of the *Claraia* beds (limestone, argillaceous limestone, and marl), *C. stachei* Bittner, *C. aurita* Hauer (Kotljar, Chedija, Kropatcheva et al., 1983) and *Koninckites* sp. (Kozur et al., 1978) have been recognized.

(33) Grey argillite intercalated with medium-bedded, light grey and pink argillaceous limestone . . . . **0.60 m**

(32) Light grey, massive, argillaceous limestone . . . . . **0.17 m**
Some conodonts (*Isarcicella? parva* (Kozur & Pjatakova, *Hindeodus turgidus* (Kozur, Mostler & Rahimi–Yazd), *Isarcicella isarcica* (Huckriede) have been found, apparently within this layer (Kozur et al., 1978).

(31) Greyish-green, thin-bedded marl, intercalated with greyish-green argillite . . . . **0.30 m**

(30) Greyish-green argillite with lenses of marl . . . . . . **0.07 m**
Some specimens of *Claraia intermedia* Bittner have been collected from the argillite.

(29) Greyish-green argillaceous marl, intercalated with thin-bedded argillite . . . . **0.23 m**
Contains rare *Claraia intermedia* Bittner.

(28) Greyish-green argillite . . . . **0.25 m**
Yields numerous specimens of *Claraia intermedia* Bittner and *Ophiceras* (*Lytophiceras*) sp.

(27) Greyish-green marl, intercalated with thin-bedded argillite . . . . **0.30 m**

(26) Reddish-brown marl, intercalated with limestone
. . . . **0.08 m**

(25) Reddish-brown argillite . . . . **0.04 m**
Yields *Ophiceras* (*Lytophiceras*) sp.

(24) Reddish-brown lumpy limestone . . . . **0.03 m**

(23) Greyish-green argillite . . . . **0.04 m**

(22) Greyish-green marl. . . . . **0.04 m**
Isarcicella? parva (Kozur & Pjatakova) was apparently found in this layer (Kozur et al.,1978). According to G. V. Kotljar (pers. comm.), some representatives of this species were also recovered from reddish brown argillite probably below this layer.

(21) Reddish-brown argillite . . . . **0.20 m**
Contains rare *Claraia* sp., abundant ostracodes.

(20) Light green argillite . . . . **0.03 m**

(19) Grey stromatolitic limestone . . . . **0.10 m**

(18) Dark grey and greyish green argillite . . . . **0.02 m**

(17) Light grey limestone, with ochre in cracks . . . . . . **0.10 m**

(16) Thin-bedded alternation of yellowish green argillite and light grey marl, with ochre in cracks; containing an abundance of ostracodes . . . . **0.10 m**

(15) Light brown nodular sandy shale, with abundant ostracodes . . . . **0.04 m**

(14) Light green argillite . . . . **0.08 m**
Yields *Claraia intermedia* Bittner at the base.

Upper Permian, Dorashamian
Pleuronodoceras occidentale Zone

(13) Reddish brown argillite; abundant ostracodes . . . **0.20 m**
Neogondolella ex gr. *N. orientalis* (Barskov & Koroleva) was found in reddish-brown argillite at this or a somewhat lower level (Kozur et al., 1978).

(12) Greyish-green argillite . . . . **0.04 m**

(11) Reddish-brown argillite, with lens (2–3 cm) of greyish-green argillite in middle part of bed; ostracodes abundant
. . . . **0.20 m**

(10) Brown and dark grey marl . . . . **0.12 m**

(9) Reddish-brown argillite . . . . **0.15 m**

(8) Light green argillite . . . . **0.02 m**

(7) Reddish-brown, thin-bedded argillite, with 2 cm lens of light green argillite . . . . **0.40 m**

(6) Reddish-brown argillite with small bellerophontid mollusks. . . . . **0.40 m**

(5) Thin-bedded alternations of reddish brown argillite and argillaceous limestone . . . . **0.20 m**
Argillite contains *Xenodiscus* sp.

(4)   Greyish-brown marl with small gastropods, intercalated
       with reddish-brown, thin-bedded argillite . . . . . . **0.15 m**
(3)   Greyish-brown marl . . . . **0.05 m**
(2)   Dark grey and brown argillite with lens (4 cm) of light grey
       marl . . . . **0.09 m**

Argillite yields *Pleuronodoceras occidentale* Zakharov.

Paratirolites kittli Zone (upper part)

(1)   Reddish-brown nodular limestone intercalated with
       yellowish-brown, thin-bedded argillite. Characterized by
       Hemigordius sp., *Streblospira minima* Vuks, *Nodosaria* ex
       gr. *hoi* (Trifonova) amongst nonfusulinid foraminifers;
       Michelina vaga Tchudinova, *M. parva* Tchudinova,
       Pentaphyllum dzhulfense (Iljina), *Ufimia differentiata*
       (Iljina), amongst corals; *Haydenella minuta* Sarytcheva,
       Araxathyris araxensis minor Grunt amongst
       brachiopods; *Paratirolites kittli* Stoyanow, *P.
       trapezoidalis* Shevyrev, *P. dieneri* Stoyanow, *Abichites
       mojsisovicsi* (Stoyanow), *A. stoyanowi* (Kiparisova)
       amongst ammonoids; *Neogondolella orientalis* (Barskov
       & Koroleva), *N. carinata* (Clark), *N. carinata subcarinata*
       Sweet, *Hindeodus minutus* (Ellison) amongst conodonts.
       Paratirolites sp. was collected 0.5 m below the top of the
       zone, and the conodont form intermediate between
       Hindeodus minutus and *I.? parva* was also discovered in
       this part of the section (Kozur *et al.*, 1978; Kotljar,
       Zakharov, Koczyrkevicz *et al.*, 1983).

North Caucasus

In the opinion of some specialists (Miklukho-Maklai,
1954; Kotljar, Zakharov, Koczyrkeivicz *et al.*, 1983), foraminiferal associations in the Nikitinian Formation of the north
Caucasus are close to those of the Changxingian of South China.
At the same time, the upper part of the Urushtenian Formation
(clay facies), which overlies the Nikitinian, contains an abundant aggregation of cephalopods identified as Midian (Zakharov, 1985) (Pseudothorhoceratidae, *Pseudotemnocheilus*? sp.,
Tainoceras sp. indet., *Neogeoceras* sp., *Neocrimites* (*Neocrimites*) sp., *Waagenina caucasica* Zakharov, *Tauroceras* n. sp.,
Xenodiscus koczyrkeviczi Zakharov). New data confirm the
suggestion of a relatively ancient age for this part of the Permian
section made by Kulikov & Tkachuk (1979) and testify to the
fact that basal Triassic deposits in the north Caucasus (Malaya
Laba river) lie with significant unconformity on pre-Dzhulfian
strata. The suggestion of Kulikov & Tkachuk (1979) that
bivalves assigned to *Claraia* appear in sediments corresponding
in age to the Guadalupian Series in North America is also
confirmed, as is the necessity to re-evaluate the significance of
some groups of invertebrates recognized in the Changxing. The
first steps in this direction were taken by Zhao, Liang & Zheng
(1978) and Chedija & Davydov (1980), who showed that ancient
representatives of *Palaeofusulina* and *Colaniella* ex gr. *C. parva*
(Colani) may be found in pre-Changxingian sediments. The
former was discovered in association with *Codonofusiella* in
Wujiaping strata in China, beneath the *Konglingites* beds of the
Dalong Formation; the latter in the Kastanatdzhilginian Formation (Midian) of the Pamirs. In Zheng's opinion (Zhao, Liang

& Zheng, 1978), the lower boundary of the Changxing Formation corresponds to the appearance of numerous and highly
developed *Palaeofusulina*, which coordinates with data on the
distribution of ammonoids in southeast China.

Pamirs

The Permian is represented for certain in the southeastern
Pamirs by Asselian (upper Bazardarinian beds), Sakmarian,
Bolorian (Sulistykian beds), Kubergandinian, Murgabian
(*Neoschwagerina simplex* Zone, *N. craticulifera* Zone, *N. mar-
garitae* Zone) (Kotljar, Vuks, Kropatcheva *et al.*, 1991), and
Midian strata (upper part of Ganskian Formation, Takhtabulakian Formation). Some years ago, a goniatite (*Subeothinites
pamiriensis* Zakharov) of the family Eothinitidae was discovered
in the middle part of the Takhtabulakian Formation in association with brachiopods (Kotljar, Zakharov, Koczyrkevicz *et al.*,
1983). No representatives of this family have ever been found in
the Dzhulfian or Dorashamian of Transcaucasia, Iran, or
China. Therefore, the age of these sediments seems to be pre-
Dzhulfian. No fossils have been found in the upper part of the
Takhtabulakian Formation, except in the weathering crust
(siallitic beds, 5–6 m), which contain some miospores (*Rhizo-
mospospora delicata* Jard, *Puncatisporites* sp. and others) (Kotljar, Zakharov, Koczyrkevicz *et al.*, 1983). The lowermost part of
the Triassic (Karatashian Formation) is represented by *Claraia*
beds (Djufur, Dronov & Kushlin, 1958; Kushlin, 1973; Dagys,
Arkhipov & Bytchkov, 1979). No Induan ammonoids or conodonts are known from this region.

Mangyshlak peninsula

In Mangyshlak, the unfossiliferous Upper Permian Otpa-
nian Formation is overlain by Dolnapinian sediments, of
Induan and lower Ayaxian age, which contain the bivalves
Claraia sp. aff. *C. stachei* (Bittner), *Ornithopecten temirbabensis*
Kiparisova, *Eumorphotis* ex gr. *E. multiformis* (Bittner), *Eumor-
photis* sp., *Anodontophora* sp. aff. *A. canalensis* (Cat.), and
Mitilus tuarkyrensis Kiparisova; the conchostracans *Pseudes-
theria tumariana* Novojilov, *P. pliciferina* Novojilov, and *P.
timanensis* Molina; ostracodes of the genera *Darwinula* and
Triassinella; and the miospores *Platysaccus* and *Angustisulcites*
(Gavrilova, 1982). The middle and upper parts of the Lower
Triassic are represented in this region by the Tartalian *Tirolites
cassianus–Doricranites* and *Columbites–Tirolites* beds (Asta-
chova, 1956; Shevyrev, 1968) and by Karaczhatykian strata
(*Arnautoceltites* beds).

South Primorye

Permian ammonoids of Primorye have been studied quite
recently. They occur in volcanogenic Murgabian rocks (Vladivostokian Formation) (Popov, 1963), lower Midian limestones
(Chandalazian Formation) (Zakharov & Pavlov, 1986b), and
upper Midian, Dzhulfian and Dorashamian clays (Ludjanzian)
(Ruzhencev, 1976; Zakharov & Pavlov, 1986a, b). The study of
the Lower Triassic ammonoids has taken place over a long
period of time.

The following ammonoid zones and beds are recognized within the Middle and Upper Permian and the Lower Triassic in South Primorye (Zakharov, 1968, 1978; Zakharov & Pavlov, 1986a,b; Zakharov & Rybalka, 1991; Zakharov et al., 1991):

Lower Triassic
Russian Stage (upper Olenekian)
(16) *Subcolumbites multiformis* Zone
(15) *Neocolumbites insignis* Zone

Ayaxian Stage (lower Olenekian)
(14) *Tirolites–Amphistephanites* Zone
    (b) *Tirolites ussuriensis* beds
    (a) *Bajarunia dagysi* beds
(13) *Anasibirites nevolini* Zone
(12) *Hedenstroemia bosphorensis* Zone

Induan Stage
(11) *Gyronites subdharmus* Zone
(10) *Glyptophiceras ussuriense* beds

Upper Permian
Dorashamian Stage (upper Ludjanzian Formation)
(9) *Liuchengoceras*–Pleuronodoceratidae beds (corresponding in their lower part to the *Colaniella parva* beds)
(8) *Iranites*? sp. beds
Dzhulfian Stage (upper Ludjanzian Formation)
(7) *Eusanyangites bandoi* beds
(6) Beds with no ammonoids

Middle Permian
Midian Stage
(5) *Cyclolobus kiselevae* beds (middle and lower Ludjanzian Formation)
(4) *Xenodiscus subcarbonarius* beds (corresponding to upper part of the *Nodosaria dzhulfensis* Zone; upper Chandalazian Formation)
(3) *Stacheoceras orientale* beds (corresponding to the middle part of the *Nodosaria dzhulfensis* Zone; upper Chandalazian Formation)
(2) Beds with no ammonoids (corresponding to the lower part of *Nodosaria dzhulfensis* Zone and Zone of *Metadoliolina lepida–Lepidolina kumaensis*; middle Chandalazian Formation)
(1) *Neocrimites* (*Neocrimites*) *kropatchevae* beds (corresponding to *Parafusulina stricta* Zone; lower Chandalazian Formation, with exception of basal beds).

The oldest Permian ammonoid known in Primorye is *Daubichites orientalis* Popov, type of *Daubichites* (Popov, 1963). The holotype of this species was collected from alluvium of the Lagernyj stream (Ussuri River basin) in the area of distribution of exclusively lower Vladivostokian sediments (Murgabian Stage). *Roadoceras subroadense* (Zakharov & Pavlov) occurs in the upper part of the Vladivostokian series, about 250 m above the bed containing Murgabian *Stenopora clara* Kiseleva (Zakharov & Pavlov, 1986b).

Of the Midian goniatites of Primorye, the most interesting are *Timorites markevichi* Zakharov, from the lower Chandalazian (Kaluzin Cape beds), and *Cyclolobus kiselevae* Zakharov from the Ludjanzian, which is associated with *Timorites* sp. (V. E. Ruzhencev's identification) in the upper part of the *Cyclolobus kiselevae* beds. Contrary to the present author's opinion, some workers (Kotljar, Vuks, Kropatcheva et al., 1991) consider the *Cyclolobus kiselevae* beds to be of Dzhulfian age.

No early Dzhulfian ammonoids have been discovered in Primorye. About 40 m of unfossiliferous rocks between the *Cyclolobus kiselevae* beds and the upper Dzhulfian apparently represent the lower Dzhulfian. The late Dzhulfian ammonoid complex of Primorye includes representatives of the Konglingitinae (*Eusanyangites bandoi* Zakharov & Pavlov) and the Medlicottiidae (*Eumedlicottia* ex gr. *E. primas* Waagen, *Propinacoceras hidium* Ruzhencev, *Neogeoceras thaumastum* Ruzhencev).

No certain early Dorashamian ammonoids are known from Primorye, although *Iranites*? sp. may be of this age. Late Dorashamian ammonoids include recently discovered representatives of *Liuchengoceras melnikovi* Zakharov, and a probable new genus of the Pleuronodoceratidae (Kotljar, Zakharov, Vuks et al., 1991).

Lower Triassic strata rest on rocks of different ages in Primorye, including late Paleozoic granitoids and Middle Permian volcanogenic rocks; but they lie just above the upper Dorashamian only in the Artemovka River basin.

Lower Induan *Glyptophiceras ussuriense* beds have recently been recognized on the west coast of Ussuri Bay (Zakharov & Rybalka, 1991). In the same section a new species of the subgenus *Lytophiceras* was found together with *Proptychites hiemalis* Diener in the lower part of the *Gyronites subdharmus* Zone. As a result of new discoveries of ammonoids in South Primorye, the lower Induan, consisting mainly of coarse clastic rocks (*Glyptophiceras ussuriense* beds, lower part of *Gyronites subdharmus* Zone) may be correlated with the upper part of the *Otoceras woodwardi* Zone and with the *Ophiceras tibeticum* Zone.

The next three Lower Triassic zones make up the Ayaxian. *Tirolites* beds (*Tirolites–Amphistephanites* Zone), recently discovered in Primorye, are of considerable interest. Originally they were included in the *Tirolites cassianus* Zone (Burij, Zharnikova & Burij, 1976); Burij & Zharnikova, 1980). However, it should be noted that representatives of *Tirolites* have relatively long ranges. In the Mangyshlak Peninsula they are found in both the *Columbites* beds and underlying sediments (Shevyrev, 1968). In the Alps, beds with *Tirolites* make up the upper part of the Lower Triassic and two zones are recognized at this level: *Tirolites cassianus* and *T. carniolicus*. These have been correlated by Krystyn (1974) with the *Tirolites–Columbites* and *Subcolumbites* zones, respectively, of western North America. In Primorye, I first recognized a representative of *Tirolites* in the upper part of the *Neocolumbites insignis* Zone (Zakharov, 1968). Subsequently, a large *Tirolites* fauna has been collected in Primorye from strata between beds yielding *Anasibirites* and *Neocolumbites*. I have restudied ammonoids identified as *Tirolites cassianus* by Burij & Zharnikova (1980) and have described them as a new species. Of great importance is the discovery of the recently described genus *Bajarunia* in the lower part of the *Tirolites–Amphistephanites* Zone in Primorye. This find permits correlation of the lower part of this zone with the lower part of the beds with *Dieneroceras demokidovi* (Dagys, 1983) in the

Boreal realm. No ammonoids are known at this level in the Alps or Salt Range.

Prohungarites popowi Kiparisova and *Prohungarites.*? sp. have been found at the very base of the Anisian in Primorye, where they are associated with such species as *Ussuriphyllites amurensis* (Kiparisova), *Leiophyllites praematurus* Kiparisova, *Megaphyllites atlasoviensis* Zakharov, *Lenotropites*? *solimani* (Toula), *Arctohungarites primoriensis* Zakharov, and an indeterminate representative of the Parapopanoceratidae (Kiparisova, 1961; Zakharov, 1968; Zakharov & Rybalka, 1991). Rocks yielding these ammonoids lie just above the Lower Triassic *Subcolumbites multiformis* Zone.

Observation shows some features of the faunal succession of adjacent Lower and Middle Triassic zones in Primorye. That is, *Leiophyllites praematurus* has been found in both zones; Anisian *Megaphyllites atlasoviensis* closely resembles Early Triassic *M. immaturus*; *Ussuriphyllites*, obviously descended from *Eophyllites* probably appeared at the very beginning of the Middle Triassic; and *Prohungarites* is common in the Lower–Middle Triassic boundary beds. Lower Triassic strata must be regarded as rather complete in South Primorye.

As mentioned previously, Permo-Triassic boundary beds of Primorye are most nearly complete in the Artemovka River basin, where they are represented by the following:

Lower Induan
(21) Dark grey shale and grey sandstone........**500–600 m**
(20) Greyish-green, thin-bedded shale containing rare marl concretions.............................**31.0 m**
    In the basal part of the unit, I. V. Burij and N. K. Zharnikova collected *Claraia* sp., *Posidonia ussurica* Kiparisova, and two small ammonoids ('*Xenodiscus*' sp. and '*Dieneroceras*'). The latter specimen retains a suture line and I have identified it as *Ophiceras* (*Lytophiceras*) sp.
(19) Black, thin-bedded shale containing small nodules of clay and marl, intercalated with grey sandstone with plant remains................................**8.8 m**
    Previously, thickness of this unit has erroneously been given as 17.0 m

Upper Dorashamian
(18) Greyish-green sandstone ....**1.1–1.4 m**
    Contains an assemblage of bryozoans referred by A. V. Kiseleva (pers. comm.) to *Fistulipora*, *Eridopora*, and *Stenodiscus*, and abundant siltstone debris in the lower part. I. V. Burij and N. K. Zharnikova (in Zakharov & Pavlov, 1986a) found a limestone boulder (?) with Permian invertebrates in the basal part of this unit.
(17) Dark grey sandy shale ....**60.0 m**
    With bivalves (determined by N. K. Zharnikova as *Aviculopecten* sp.), gastropods, nautiloids, and remains of carbonized wood. A nodule with *Liuchengoceras melnikovi* Zakharov was found by N. G. Melnikov about 20 m below the base of the Lower Triassic. Two other ceratitoid ammonoids, apparently belonging to a new genus of the Pleuronodoceratidae, were collected by N. K. Zharnikova at the base of this unit.

Lower Dorashamian (?)
(16) Dark grey sandy shale intercalated with grey sandstone

and dark grey mudstone, sandy mudstone and siltstone yielding *Neogeoceras* sp ....**38.0 m**
(15) Dark grey sandy shale ....**18.0 m**
    Contains Cyclolobidae gen. et sp. indet. (*Changsingoceras*?) (Zakharov & Pavlov, 1986a).
(14) Grey sandstone........................**11.0 m**
(13) Dark grey mudstone and siltstone ....**6.0 m**
(12) Dark grey mudstone, siltstone and sandy shale ..**19.5 m**
    Yields *Neogeoceras thaumastum* Ruzhencev, *Eumedlicottia* sp., *Iranites*? sp. (Zakharov & Pavlov, 1986a).

Upper Dzhulfian
(11) Dark grey mudstone and siltstone with numerous nodules of marl ....**1.1 m**
    Contains plant remains, bivalves, gastropods, and cephalopods (Pseudorthoceratidae gen. et sp. indet., *Eumedlicottia* ex gr. *primas* Waagen, *Propinacoceras hidium* Ruzhencev, *Neogeoceras thaumastum* Ruzhencev, *Eusanyangites bandoi* Zakharov & Pavlov (Ruzhencev, 1976; Zakharov & Pavlov, 1986a).
(10) Dark grey sandy mudstone with rare marl nodules **9.5 m**
    Contains *Eumedlicottia* ex gr. *E. primas* Waagen, *Neogeoceras thaumastum* Ruzhencev (Zakharov & Pavlov, 1986a).

Lower Dzhulfian (?)
(9) Dark grey siltstone, mudstone and sandy shale
    ....**380–400 m**

Upper Midian
(8) Greyish-green sandstone, dark grey siltstone and mudstone with rare nodules of marl.......**160–200 m**
    Contains Pseudorthoceratidae gen. indet., and *Cyclolobus*? sp.
(7) Greyish-green tuffaceous sandstone ....**9.5 m**
    Yields bryozoans identified by Kiseleva (1982) as *Fistulipora*, *Eridopora*, *Stenodiscus*, *Dyscritella*, *Pseudobatostomella*, *Rhabdomeson*, *Streblascopora*, *Claustotrypa*, *Fenestella*, *Penniretepora*, *Polypora*, *Acanthocladia*, and *Girtyoporina*, and rare lenses of siltstone.
(6) Grey sandstone and conglomerate, with bryozoans, crinoids, and brachiopods ....**60–70 m**
    The part of the Permian Ludjansian Formation studied in this section is 232–235 m thick.

Physical and Chemical Characteristics of the Permo-Triassic boundary beds

Some paleomagnetic results

Studies of the terrestrial Permo-Triassic sediments of the Urals and the Russian Platform show that the lower boundary of the Illawara hyperzone is at the base of the Upper Tatarian. However, location of this boundary has not been determined exactly in marine sediments in Transcaucasia, the Pamirs, North America, or southeast China. We know only that, in Transcaucasia the upper part of the Arpian Formation (upper part of *Chusenella abichi* Zone) corresponds at least to the zone of normal polarity (Kotljar *et al.*, 1984). In the Pamirs, the boundary lies no higher than the base of the Kutalian subformation (Davydov *et al.*, 1982), which is considered to be the lower member of the Midian. In North America, the base of the Illawara hyperzone may be located within the Capitanian Gerster Formation.

In Transcaucasia, the Permian part of the Illawara hyperzone contains no fewer than four zones of reversed polarity (Kotljar *et al.*, 1984), two of which correspond to the Dzhulfian and Dorashamian. More detailed data have been obtained recently on the Upper Permian and Permo-Triassic boundary beds (Zakharov *et al.*, 1991). It has been found that the Upper Permian reversed intervals correspond to all but the uppermost parts of the *Vedioceras ventrosulcatum* and *Paratirolites kittli* zones. However, the Permo-Triassic boundary beds in both Transcaucasia (upper part of *Paratirolites kittli* Zone, *Pleuronodoceras occidentale* Zone, and lower part of *Claraia* beds) and southeast China (Zhang, 1987) represent a single zone of normal polarity. It is interesting that in southeast China, a zone of reversed polarity has been recognized in the *Konglingites* beds of the Dalong Formation (Zhang, 1987), which correlates with the *Vedioceras ventrosulcatum* Zone of Transcaucasia. The upper Dalong Formation and the lower Feixianguan Formation are included by Zhang (1987) in a zone of mixed magnetic polarity. However, Heller *et al.* (1988) and Steiner *et al.* (1989) show that polarity of the upper Dalong is mostly reversed and that of the lower Feixianguan dominantly normal.

Carbon-isotopic composition and iridium content of Permo-Triassic strata

Unusually high $\delta^{13}C$ values have been recognized in well-preserved brachiopod shells from the Upper Midian ($+1.0^o/_{oo}$: Akhura, *Hemigordius irregulariformis–Orthotetina arakeljani* Zone; $+2.1^o/_{oo}$: Vedi, *Hemigordius irregulariformis–Orthotetina arakeljani* Zone; $+1.3^o/_{oo}$: Ogbin, *Pseudodunbarula arpaensis–Araxilevis intermedius* Zone) and Dzhulfian ($+2.5^o/_{oo}$: Dorasham II–2, *Araxoceras latissimum* Zone). New data on the carbon-isotope composition of Permian and Triassic rocks in different territories obtained recently by Chen *et al.* (1984), Sun *et al.* (1984), Baud, Holser & Magarits (1986), and Yang *et al.* (1987) confirm Holser's (1984) idea about an abrupt decrease in $\delta^{13}C$ values in carbonate rocks at the Permo-Triassic boundary.

The iridium content of strata in the Permo-Triassic boundary beds (upper part of *Paratirolites kittli* Zone, *Pleuronodoceras occidentale* Zone, and lower part of *Claraia* beds in Transcaucasia ranges from 0.003 to 0.004 ppb (Alekseev *et al.*, 1983). The maximum concentration of iridium recognized is not very high and is in clay and marl sediments 10 cm below the base of a bed of stromatolitic limestone. At the same level and in the same section (Sovetoshen), I collected two Permian brachiopods, which have been determined as *Haydenella* sp. by G. V. Kotljar. Thus this part of the section appears to be the upper part of the *Pleuronodoceras occidentale* Zone. An iridium anomaly has been recognized at the Permo-Triassic boundary in southeast China (Sun *et al.*, 1984) and in Italy (Brandner *et al.*, 1986), which suggests a global distribution. However, Clark *et al.* (1986) and M. A. Nasarov (pers. comm.) now have doubts about this.

The problem of stadial division and correlation of the Permian and Lower Triassic

The Permian System was established on the Russian Platform and in the western Urals, where it consists of alternating marine, lagoonal, and terrestrial sediments. Paleogeographic conditions at the end of the Permian, which were caused by frequent changes of climate and by numerous fluctuations in sea level, and long isolation of some marine basins, are reasons for difficulties in working out parallel stratigraphic scales in different regions of the world: Russian Platform (7 stages), Tethys (9 regional stages), and North America (5 regional stages). A need for establishing regional stages for the Lower Triassic is being discussed (Zakharov, 1991).

One of the more difficult questions of Permian biostratigraphy is correlation of marine, lagoonal, and terrestrial sediments. Also, there are contradictory opinions about the correlation of continental Tatarian strata in the Boreal realm with their marine equivalents in the Tethyan realm. In this connection, new data on paleomagnetic hyperzones are of great interest, for these permit correlation of the Upper Tatarian with at least Dzhulfian–Dorashamian and Midian rocks.

The *Cyclolobus* beds were long considered to crown the Permian section (Ruzhencev, Sarytcheva & Shevyrev, 1965). More recently Ruzhencev (1976) came to the conclusion that '*Cyclolobus*' *teicherti* Furnish & Glenister, described from the Chhidru Formation in the Salt Range, is really a representative of *Timorites*. At the same time, however, he wrote that it was difficult to account for its high stratigraphic position.

A recent study in northeast India (Gungri Formation) shows that specimens of *Cyclolobus* have indeed been discovered 0.45 m below the base of the *Otoceras* beds (Waterhouse & Gupta, 1985). At the same time, these authors recognized that the contact between the Gungri Formation and the overlying *Otoceras* beds is a laterite 10–12 cm thick. This limonitic pebbly layer suggests a hiatus and thus a probable break in sedimentation and an interval of subaerial weathering.

Cyclolobus kiselevae Zakharov (1983) from South Primorye is very close to *C. oldhami* Diener from the Himalayas and this permits correlation of the *C. kiselevae* beds with the upper Chhidruan. But the *Cyclolobus* beds of Primorye do not cap the Permian section. Just above them, there are Dzhulfian and Dorashamian beds, corresponding apparently to the hiatus between the Gungri Formation and *Otoceras* beds in the Himalayas. The *C. kiselevae* beds in Primorye may thus be of late Midian age.

Based on new data on the distribution in the lower, middle, and upper parts of the Midian in the eastern Tethyan region (Ruzhencev, 1976; Liang, 1983; Zakharov, 1983; Ehiro, Shimoyama & Murata, 1986; Kotljar, Zakharov, Vuks *et al.*, 1991; Zakharov, *et al.*, 1991) of representatives of *Timorites*, a typical Capitanian element, one may conclude that the Capitanian (*Timorites* Zone) corresponds to at least a considerable part of the Midian Stage.

There is no generally accepted view as to the extent of the Dzhulfian, the stratotype of which is in Transcaucasia. Originally, it was established for the *Prototoceras* beds (Schenck *et al.*, 1941). Later, Ruzhencev (in Ruzhencev *et al.*, 1965) included

Table 5.1. *A proposed standard for the Lower Triassic*

Stage	Substage	Stratotypes			Hypostrato-type	Regional Stage		Stratotypes		
		Salt Range (Waagen, 1895)	South Primorye (Zakharov, 1968, 1978; Zakharov, Rybalka, 1991)		Kashmir (Diener, 1895)			Arctic Canada (Tozer, 1961, 1967)	Siberia (Dagys, Kazakov, 1984)	
Russian	—		*Subcolumbites multiformis*				Spathian		?		
			Neocolumbites insignis						*Olenikites spiniplicatus*	*Keyserlingites subrobustus*
						Olenekian				*Parasibirites grambergi*
Ayaxian	Schmidt-ian		*Tirolites-Amphi-stephanites*	*Tirolites ussuriensis* beds			?		*Dieneroceras demokidovi*	*Nordophiceras euomphalus*
				Bajarunia dagysi beds						*Bajarunia contrarium*
	Tobizin-ian		*Anasibirites nevolini*				Smithian		*Wasatchites tardus*	
			Hedenstroemia bosphorensis						*Hedenstroemia hedenstroemi*	
Induan (Brahman-ian)	Gandar-ian	*Prionolobus rotundatus*				Dienerian		*Vavilovites sverdrupi*		
		Gyronites frequens						*Proptychites candidus*		
	Ganget-ian				*Ophiceras tibeticum*	Griesbachian		*Proptychites striatus*		
								Ophiceras commune		
					Otoceras woodwardi			*Otoceras boreale*		
								Otoceras concavum		

Only stratotype and some hypostratotype zones are shown.

both the *Araxilevis* and *Conodofusiella–Reichelina* beds in the Dzhulfian. Many specialists now use Dzhulfian for the *Araxilevis* beds, the *Araxoceras* Zone, and the *Vedioceras* Zone (Leven, 1980; Iranian–Japanese Research Group, 1981). However, a few years ago (Kotljar, Chedija, Kropatcheva *et al.*, 1983; Kotljar, Zakharov, Koczyrkevicz *et al.*, 1983) strata characterized by the *Araxilevis* and *Codonofusiella* fauna were shown to be a single biostratigraphic member in the stratotype of the *Codonofusiella* and *Reichelina* beds, and the lower boundary of the *Araxilevis* beds was shown to be diachronous in Transcaucasia. Thus the *Araxilevis* beds were excluded from the Dzhulfian and were grouped with the *Codonofusiella* and *Reichelina* beds as an upper member of the Midian. At present, however, Kotljar, Vuks, Kropatcheva & Chedia (in Kotljar, Zakharov, Vuks *et al.*, 1991) are inclined to place this member (the *Pseudodunbarula arapensis–Araxilevis intermedius* Zone) in the Dhulfian, taking into account the well-known data on foraminifers and brachiopods of this region. But, in my opinion, the lower boundary of the Dzhulfian seems to be connected first of all with a sharp increase in taxonomic diversity of ceratitoid ammonoids (Araxocerati-dae), which is noticeable first at the base of the *Araxoceras latissimum* Zone. This agrees with the interpretation of Schenck *et al.* (1941), even though the first rare araxoceratid ammonoids

probably appeared in association with *Timorites* in Late Midian time.

Discoveries in Transcaucasia of *Pleuronodoceras occidentale* Zakharov (1985) in Upper Dorashamian clays, and of *Liuchengoceras melnikovi* Zakharov (Kotljar, Zakharov, Vuks *et al.*, 1991) in the Upper Ludjanzian of Primorye, permit correlation of these strata with the upper Changxingian of southeast China.

It is important that stratotypes for Lower Triassic stages be chosen in the Tethyan, rather than the Boreal realm, for historically stratotypes of other Triassic stages have been situated within the Tethyan realm.

At least three major stages in evolution of the Early Triassic fauna should be taken into account in discussions of the standards for Lower Triassic stages. Within the Tethyan realm, the initial stage has been most thoroughly studied in the Himalayas, but the rich cephalopod faunas of the succeeding two stages – Early Olenekian (or Ayaxian) and Late Olenekian (or Russian) – have been recognized in Primorye. The first stage is characterized by the appearance and development of the Ophiceratidae and by the appearance of the Meekoceratidae, Paranoritidae, and Proptychitidae. The second stage is distinguished by the appearance first of the Hedenstroemiidae, Ussuriidae, Prosphingitinae, Owenitidae, Dieneroceratidae,

Flemingitidae, and Xenoceltitidae, and then by the Kashmiritidae, Sibiritidae, Palaeophyllitidae, Tirolitidae, Doricranitidae, and Dinaritidae. The third stage is characterized by the appearance of the Megaphyllitidae, and by the appearance and development of the Columbitidae and Keyserlingitidae. All representatives of these major Early Triassic ammonoid stocks, with the exception of the Otoceratidae and Doricranitidae, are known in Primorye. Thus, in this respect, only sections in western North America may compete with those in South Primorye.

In the light of the data just summarized, it is proposed (Table 5.1) that the standard Lower Triassic stadial sequence be based on sequences from similar provinces in the Tethyan realm and consist of an Induan (or Brahmanian) Stage founded on stratotypes in the central Himalayas and the Salt Range, succeeded by Ayaxian and Russian stages, with stratotypes in the Ussurian province. If sections in Primorye are inadmissible as stratotypes for the Ayaxian and Russian stages because of their relatively difficult accessibility, those sections might be used as hypostratotypes for upper stages of the Lower Triassic, which are rather poorly characterized paleontologically in the Himalayan province.

It is also proposed that Lower Triassic stages based by Tozer (1961, 1967) on stratotypes in the Boreal realm be regarded as regional stages and substages. The boreal Griesbachian and Dienerian correspond to the Gangetian and Gandarian substages of the Induan (or Brahmanian), respectively. The Smithian and Spathian stages of Tozer seem to represent an incomplete sequence in Arctic Canada and are thus regarded as regional substages of the Olenekian because correlatives of the *Dieneroceras demokidovi* Zone (*Bajarunia euomphalus* beds and *Nordophiceras contrarium* beds) of Siberia and the *Tirolites–Amphistephanites* Zone (*Bajarunia dagysi* beds, *Triolites ussuriensis* beds) of South Primorye are unknown in Arctic Canada.

On the Permo-Triassic boundary

There are many points of view as to where the Permo-Triassic boundary should be placed. A traditional view, mentioned above and supported by most recent authors, places the boundary at the base of the *Otoceras* beds (Griesbach, 1880; Mojsisovics, 1892). However, a growing number of workers (e.g., Kozur & Mostler, 1976; Kozur, 1986; Budurov *et al.*, 1986; Fuglewicz, 1987) suggest that the boundary be placed somewhat higher, at the base of the *Isarcicella isarcica* Zone, which overlies the *Otoceras* beds. Newell (1986) suggests that the boundary be situated even higher, at the top of the Griesbachian Stage.

In my opinion, the following arguments favor the traditional point of view:

1 In many parts of the world, the *Otoceras* beds and their equivalents lie unconformably on Paleozoic rocks that may be connected with a major eustatic fluctuation of sea level and, in particular, with a regression.

2 Dzhulfian and Dorashamian cephalopod faunas have not been discovered in the Boreal realm, which was probably isolated in the Late Permian so that cephalopod faunas are absent there. It appears more reasonable to connect both the major invasion of the *Otoceras* fauna into the Boreal realm and the global transgression of this time with the beginning of the Triassic, than with the very end of the Permian.

3 In the Boreal realm, *Otoceras* is in some places associated with *Ophiceras* (*Lytophiceras*) sp. (Zakharov, 1971) and *Glyptophiceras* (*Tompophiceras*) *nielsoni* Spath (Kortchinskaya & Vavilov, 1991), which are common elements in overlying Induan strata.

4 Good evidence of global events marking the Paleozoic–Mesozoic boundary seems to be provided by an iridium anomaly and an abrupt decrease of $\delta^{13}C$ somewhat below the *Isarcicella isarcica* Zone in the central Tethys. This level may correspond to the base of the *Otoceras* beds in the Boreal realm and Himalayan province.

References

Abich, H. W. (1878). *Geologische Forschungen in den kaukasischen Ländern. Th. 1, Eine Bergkalkfauna aus der Araxesenge bei Djoulfa in Armenien.* Wien: Alfred Holder.

Alekseev, A. S., Barsukova, L. D., Kolesov, G. M. *et al.* (1983). The Permian–Triassic boundary event: Geochemical investigation of the Transcaucasia section. *Lunar and Planetary Sci., XIV Conference*, pp. 7–8.

Astakhova, T. V. (1956). On the stratigraphical position of the *Doricranites* beds. *Dokl. Akad. Nauk USSR*, **111**(5): 1065–7. (In Russian.)

Bando, Y. (1979). Upper Permian and Lower Triassic ammonoids from Abadeh, central Iran. *Mem. Fac. Educ. Kagawa Univ.*, **29**(2): 103–38.

Baud, A., Holser, W. T. & Magarits, M. (1986). Carbon-isotope profiles in the Permian–Triassic of the Tethys from the Alps to the Himalayas [Abstract]. In *Field Conference on Permian and Permian-Triassic Boundary in the South-Alpine Segment of the Western Tethys, and Additional Regional Reports*, Abstracts vol., p. 14. Pavia: Soc. Geol. Ital. and IGCP Project 203.

Brandner, R., Donofrio, D. A., Krainer, K., Mostler, H., Nazarov, M. A., Resch, W., Stingl, V. & Weissert, H. (1986). Events at the Permian–Triassic boundary in the Southern and Northern Alps [Abstract]. In *Field Conference on Permian and Permian-Triassic Boundary in the South-Alpine Segment of the Western Tethys, and Additional Regional Reports*, Abstracts vol., p. 15. Pavia: Soc. Geol. Ital. and IGCP Project 203.

Budurov, K. J., Gupta, V. J., Kachroo, R. K. & Sudar, M. N. (1986). Problems of the Lower Triassic conodont stratigraphy and the Permian/Triassic boundary [Abstract]. In *Field Conference on Permian and Permian–Triassic Boundary in the South-Alpine Segment of the Western Tethys, and Additional Regional Reports*, Abstracts vol., p. 19. Pavia: Soc. Geol. Ital. and IGCP Project 203.

Burij, G. I. (1979). *Lower Triassic conodonts of South Primorye.* Moskva: 'Nauka'. (In Russian.)

Burij, I. V. & Zharnikova, N. K. (1970). New Lower Triassic ceratites of South Primorye. In *New Species of the Ancient Plants and Invertebrates of the USSR*, pp. 150–60. Moskva: 'Nauka'. (In Russian.)

Burij, I. V. & Zharnikova, N. K. (1980). Ammonoids from the *Tirolites* Zone of South Primorye. *Paleont. Zhurn.*, 1980(3): 61–9. (In Russian.)

Burij I. V., Zharnikova, N. K. & Burij, G. I. (1976). On the stage division of the Lower Triassic in South Primorye. *Geologija i geophizica*, **1976**(7): 150–6. (In Russian.)

Chedija, I. O. & Davydov, V. I. (1980). On stratigraphical distribution of *Colaniella* (Foraminifera). *Dokl. Akad. Nauk USSR*, **252**(4): 948–51. (In Russian.)

Chen Jing-Shi, Shao Maoreng, Huo Weigue & Yao Yuyuan (1984). Carbon isotope of carbonate strata at the Permian–Triassic boundary in Changxing, Zhejiang. *Sci. geol. sin.*, **1**:88–93.

Chao King-koo (1965). The Permian ammonoid-bearing formations of South China. *Sci. sinica*, **14**(12): 1813–26.

Clark, D. L., Wang Cheng-Yuan, Orth, C. J. & Gilmore, J. S. (1986). Conodont survival and low iridium abundances across the Permian–Triassic boundary in South China. *Science*, **233**: 984–6.

Dagys, A. S. (1983). Morphology, classification and evolution of the genus *Nordophiceras* (ammonoids). *Trudy IGG SO Akad. Nauk USSR*, **538**: 37–51. (In Russian.)

Dagys, A. S., Arkhipov, Yu. V. & Bytchkov, Yu. M. (1979). Stratigraphy of the Triassic System of northeastern Asia. *Trudy IGG SO Akad. Nauk USSR*, **447**: 1–244. (In Russian.)

Davydov, V. I., Komissarova, R. A., Khramov, A. N. & Chedija, I. O. (1982). About the paleomagnetic characteristic of the Upper Permian of South East Pamirs. *Dokl. Akad. Nauk USSR*, **267**(5): 1177–81. (In Russian.)

Diener, C. (1895). Triadische cephalopodenfaunen der ostsibirischen Kustenprovinz. *Mem. Com. Geol.*, **14**(3): 1–59.

Djufur, M. S., Dronov, V. I. & Kushlin, B. K. (1958). On Triassic stratigraphy of southeastern Pamirs. *Dokl. Akad. Nauk USSR*, **123**(3): 523–6.

Ehiro, M., Shimoyama, S. & Murata, M. (1986). Some Permian Cyclolobaceae from the southern Kitakami Massif, northern Japan. *Trans. Proc. Palaeont. Soc. Japan*, **142**: 400–8.

Fuglewicz, R. (1987). Permo-Triassic boundary in Poland. In *Final Conference on Permo-Triassic Events of East Tethys Region and Their Intercontinental Correlations*, Abstracts vol., p. 6. Beijing: IGCP Project 203.

Gavrilova, V. A. (1982). Lower Triassic stratigraphy and ammonoids of Mangyshlak. *Avtoreferat diss. na soisk, utch. step. kand. geol.-min. nauk.*, Leningrad. (In Russian.)

Griesbach, C. L. (1880). Palaeontological notes on the Lower Trias of the Himalayas. *Rec. Geol. Surv. India*, **13**(2): 94–113.

Heller, F., Lowrie, W., Li Huanmei & Wang Junda (1988). Magnetostratigraphy of the Permo-Triassic boundary section at Shangsi (Guangyuan, Sichuan Province, China). *Earth and Planetary Sci. Letters*, **88**(1988): 348–56.

Holser, W. T. (1984). Gradual and abrupt shifts in ocean chemistry during Phanerozoic time. In *Pattern of Change in Earth Evolution*, ed. H. D. Holland & A. F. Trendall, pp. 123–43. Berlin: Springer-Verlag.

Iranian–Japanese Research Group (1981). The Permian and Lower Triassic System in Abadeh Region, Central Iran. *Mem. Fac. Sci. Kyoto Univ., ser. Geol. Mineral.*, **47**(2): 61–133.

Kiparisova, L. D. (1961). *Paleontological Basis of Triassic Stratigraphy of Primorye region. Part I. Cephalopods*, Trudy VSEGEI, n. ser., 48. Leningrad: Gosgeoltekhizdat. (In Russian.)

Kiparisova, L. D. & Popov, Yu. N. (1956). Stage division of the Lower Triassic. *Dokl. Akad. Nauk USSR*, **104**(4): 842–5. (In Russian.)

Kiparisova, L. D. & Popov, Yu. N. (1964). Project of the stage division of the Lower Triassic. In *22 ses. Mezhdunar. geol. kongr., probl. 16a*, Dokl. sov. geol., pp. 91–9. Moskva: 'Nedra'. (In Russian.)

Kiseleva, A. V. (1982). *Late Permian Bryozoans of South Primorye.* Moskva: 'Nauka'. (In Russian.)

Kortchinskaya, M. V. & Vavilov, M. N. (1991). Early Induan ammonoids of Spitsbergen. In *Problems of Permian and Triassic Biostratigraphy of East USSR*, ed. Yu. D. Zakharov & Yu. I. Onoprienko. Vladivostok: DVNC Akad. Nauk USSR. (In Russian.) (In press.)

Kotljar, G. V., Chedija, I. O., Kropatcheva, G. S. & Vuks, G. P. (1983). Concerning the *Codonofusiella* and *Araxilevis* beds in the Upper Permian of Trans-Caucasus. *Dokl. Akad. Nauk USSR*, **270**(1): 190–5. (In Russian.)

Kotljar, G. V., Komissarova, R. A., Khramov, A. N. & Chedia, I. O. (1984). Paleomagnetics of the Upper Permian of Trans-Caucasus. *Dokl. Acad. Nauk USSR*, **276**(3): 669–74. (In Russian.)

Kotljar, G. V., Vuks, G. P., Kropatcheva, G. S. & Kushnar, L. V. (1991). The Nakhodka Reef and the position of the Lyudyansa horizon of southern Primorye in the stage scale of the Permian deposits of the Tethys. In *Problems of the Permian and Triassic Biostratigraphy of East USSR*, ed. Yu. D. Zakharov & Yu. I. Onoprienko. Vladivostok: DVNC Akad. Nauk USSR. (In Russian). (In press.)

Kotljar, G. V., Zakharov, Yu. D., Koczyrkevicz, B. V., Kropatcheva, G. S., Rostovcev, K. O., Chediya, I. O., Vuks, G. P. & Guseva, E. A. (1983). *Evolution of the Latest Permian Biota. Dzhulfian and Dorashamian Regional Stages in the USSR.* Leningrad: 'Nauka'. (In Russian.)

Kotljar, G. V., Zakharov, Yu. D., Vuks, G. P., Kropatcheva, G. S., Chedija, I. O., Burago, V. I. & Guseva, E. A. (1991). *Evolution of the Latest Permian Biota. Midian Stage in the USSR.* Leningrad: 'Nauka'. (In Russian.) (In press.)

Kozur, H. (1986). The problem of the Permian–Triassic boundary [Abstract]. In *Field Conference on Permian and Permian–Triassic Boundary in the South-Alpine Segment of the Western Tethys, and Aditional Regional Reports*, Abstracts vol., pp. 32–5. Pavia: Soc. Geol. Ital. and IGCP Project 203.

Kozur, H., Leven, E. Ya., Lozovskiy, V. R. & Pjatakova, M. V. (1978). Division of the Permian–Triassic boundary beds of Trans-Caucasus on the basis of conodonts. *Bull. Mosk. obstsh. ispyt. prirody, otd. geol.*, **53**(5): 15–24. (In Russian.)

Kozur, H. & Mostler, H. (1976). Neue Conodonten aus dem Jungpaläozoikum und Trias. *Geol. Paläont. Mitt. Innsbruck*, **6**(3): 1–33.

Krystyn, L. (1974). Die *Tirolites*-Fauna (Ammonoidea) der untertriassischen Werfener Schichten Europas und ihre stratigraphische Bedeutung. *Sitzungsb. Österr. Akad. Wiss., Mat.-Naturwiss. Kl., Abt. 1*, **183**(1–3): 29–50.

Kulikov, M. V. & Tkachuk, G. A. (1979). Concerning the discovery of *Claraia* (Bivalvia) in the Upper Permian of North Caucasus. *Dokl. Akad. Nauk USSR*, **245**(4): 905–8. (In Russian.)

Kushlin, B. K. (1973). The Pamir geosyncline. In *Stratigraphy of USSR. Triassic System*, pp. 374–94. Moskva. (In Russian.)

Leven, E. Ya. (1980). *Explanatory Note to Stratigraphic Scale of the Permian in the Tethys.* Leningrad. (In Russian.)

Liang Xi-luo (1983). New material of Permian ammonoids with discussion on the origin, migration of Araxoceratidae and the horizon of *Paratirolites*. *Acta palaeont. sinica*, **22**(6): 606–15.

Miklukho-Maklai, K. V. (1954). *Upper Permian foraminifers of North Caucasus.* Moskva: VSEGEI. (In Russian.)

Miklukho-Maklai, K. V. (1963). *Upper Paleozoic of Middle Asia.* Leningrad: Leningrad State Univ. Press. (In Russian.)

Mojsisovics, E. (1879). Zur Altersbestimmung der Sedimentar-Formationen der Araxes-Enge bei Djoulfa in Armenien. *Verhandl. Geol. Reichsanst. Wien*, **8**:171–173.

Mojsisovics, E. (1892). Vorläufige Bemerkungen über die Cephalopodenfaunen der Himalaya-Trias. *Sitzungsber. Akad. Wiss. Wien, Mat.-Natur. Kl.*, **101**(1): 372–8.

Newell, N. D. (1986). The boundary between the Paleozoic and Mesozoic erathems? [Abstract]. In *Field Conference on Permian and Permian–Triassic Boundary in the South-Alpine Segment of the Western Tethys, and Additional Regional Reports*, Abstracts vol., p. 41. Pavia: Soc. Geol. Ital. and IGCP Project 203.

Popov, Yu. N. (1963). New genus *Daubichites* of the family Paragastrioceratidae. *Paleont. zhurn.*, **1963**(2): 148–50.

Rostovcev, K. O. & Azaryan, N. R. (1974). Paleozoic and Mesozoic boundary in Trans-Caucasus and new Upper Permian Stage. *Sovetskaja geol.*, **1974**(4): 70–82. (In Russian.)

Ruzhencev, V. E. (1959). Classification of superfamily Otocerataceae. *Paleont. zhurn.*, **1959**(2): 56–7. (In Russian.)

Ruzhencev, V. E. (1962). Classification of the family Araxoceratidae. *Paleont. zhurn.*, **1962**(4): 88–103.

Ruzhencev, V. E. (1976). Late Permian ammonoids in Far East. *Paleont. zhurn.*, **1976**(3): 36–50. (In Russian.)

Ruzhencev, V. E., Sarytcheva, T. G. & Shevyrev, A. A. (1965). Biostratigraphic conclusions. *Trudy Paleont. Inst.*, **108**: 93–116. (In Russian.)

Schenck, H. G., Childs. T. S. *et al.* (1941). Stratigraphic nomenclature. *Bull. Am. Assoc. Petrol. Geol.*, **25**(12): 2195–202.

Shevyrev, A. A. (1968). Triassic ammonoids. *Trudy Paleont. Inst. Akad. Nauk USSR*, **217**: 1–184. (In Russian.)

Steiner, M., Ogg, J., Zhang, Z. & Sun, S. (1989). The Late Permian/Early Triassic magnetic polarity time scale and plate motions of South China. *J. Geophys. Res.*, **94**(B6): 7343–63.

Stoyanow, A. A. (1910). On the character of the boundary of Palaeozoic and Mesozoic near Djulfa. *Zap. Peterb. Mineral. obstch., ser 2*, **47**(1): 61–135.

Spath, L. F. (1930). The Eotriassic invertebrate fauna of East Greenland. *Medd. Grøland*, **83**(1): 1–90.

Sun Yiyin *et al.* (1984). The discovery of iridium anomaly in the Permian–Triassic boundary clay in Changxing, Zhejiang, China and its significance. *Developments in Geoscience. Contrib. to 27th Internat. Geol. Cong.*, Preprint, pp. 235–45. Beijing: Science Press.

Tozer, E. T. (1961). Triassic stratigraphy and faunas, Queen Elizabeth Islands, Arctic Archipelago. *Mem. Geol. Surv. Canada*, **316**: 1–116.

Tozer, E. T. (1967). A Standard for Triassic Time. *Bull. Geol. Surv. Canada*, **156**: 1–103.

Waterhouse, J. B. & Gupta, V. J. (1985). *Cyclolobus* and *Otoceras* from Spiti, northwest India. *Contrib. Himalayan Geol.*, **3**: 219–24.

Yang Zheng, Xu Dacyi, Ye Lianfeng, Zhang Zinwen & Sun Yiyin (1987). Carbon isotope anomaly events near the P/Tr boundary: current results and possible causes [Abstract]. In *The Final Conference on Permo-Triassic Events of East Tethys Region and Their Intercontinental Correlations*, Abstracts vol., p. 27. Beijing: IGCP Project 203.

Zakharov, Yu. D. (1968). *Biostratigraphy and ammonoids of the Lower Triassic of southern Primorye*. Moskva: 'Nauka'. (In Russian.)

Zakharov, Yu. D. (1973). New stage and zone subdivision of the Lower Triassic. *Geologija i Geophizica*, **1973**(7): 51–8 (In Russian.)

Zakharov, Yu. D. (1978). *Early Triassic ammonoids of East USSR*. Moskva: 'Nauka'. (In Russian.)

Zakharov, Yu. D. (1983). New Permian cyclolobids (Goniatitida) of South USSR. *Paleont. zhurn.*, **1983**(2): 126–30. (In Russian.)

Zakharov, Yu. D. (1985). To the problem of type of the Permian–Triassic boundary. *Bull. Mosk. obstch. ispyt. prirody, otd. geol.*, **60**(50): 59–70. (In Russian.)

Zakharov, Yu. D. (1991). The problem of ammonoid standards for the Permian–Triasic in the Tethys. In *Internat. Sympos. Zechstein*. Hanover. (In press.)

Zakharov, Yu. D., Gupta, V. J., Rybalka, S. V., Sokarev, A. N. & Starkov, B. P. (1991). Comparison of the Upper Permian of Trans-Caucasus, South Primorye and Himalayas. In *Aspects of the Problem of the Permian–Triassic Boundary*. Contrib. Himalayan Geol. (In press.)

Zakharov, Yu. D. & Pavlov, A. M. (1986a). The first find of araxoceratid ammonoids in the Permian of the eastern USSR. In *Permian–Triassic Events During Evolution of the Northeast Asia Biota.*, eds. Yu. D. Zakharov & Yu. I. Onoprienko, pp. 74–85. Vladivostok: DVNC Acad. Nauk USSR. (In Russian.)

Zakharov, Yu. D. & Pavlov, A. M. (1986b). Permian cephalopods of Primorye and the problem of the Permian zonal stratification in the Tethys. In *Correlation of Permo-Triassic Sediments of East USSR*, ed. Yu. D. Zakharov & Yu. I. Onoprienko, pp. 74–85. Vladivostok: DVNC Acad. Nauk USSR. (In Russian.)

Zakharov, Yu. D. & Rybalka, S. V. (1991). A standard for the Permian–Triassic in the Tethys. In *Problems of Permian and Triassic Biostratigraphy of the East USSR*, eds. Yu. D. Zakharov & Yu. I. Onoprienko. Vladivostok: DVNC Acad. Nauk USSR. (In Russian.) (In press.)

Zhang Zhengkun (1987). Magnetostratigraphy of the Permian–Triassic boundary in China [Abstract]. In *The Final Conference on Permo-Triassic Events of East Tethys Region and Their Intercontinental Correlations*, Abstracts vol., pp. 42–3. Beijing: IGCP Project 203.

Zhao Jinko, Liang Xi-luo & Zheng Zhuogan (1978). On the stratigraphic position of the Dalong Formation. *J. Stratig.*, **2**(1): 46–52. (In Chinese.)

Zharnikova, N. K. (1985). Ammonoids from the *Columbites* Zone of the Lower Triassic in South Primorye. *Paleont. zhurn.*, **1985**(13): 50–9. (In Russian.)

6 Classification and correlation of nonmarine Permo-Triassic boundary in China

YANG JIDUAN, QU LIFAN, ZHOU HUIQIN, CHENG ZHENGWU, ZHOU TONGSHUN,

HOU JINGPENG, LI PEIXIAN, SUN SHUYING, WU SHAOZU, LI DAIYUN, AND LONG

JIARONG

Introduction

In China, nonmarine Permian and Triassic strata are distributed mainly in the vast area north of a line formed by the Kunlun, Qinling, and Dabie mountains. They are particularly well developed in the Junggar, Turpan, and Shaanxi-Gansu-Ningxia basins. South of the line, nonmarine Permo-Triassic sediments occur mostly in eastern Yunnan and western Guizhou. In addition, alternating marine and continental Permian and Triassic rocks are also developed in a few areas in both South and North China. In the past decade or so great progress has been made in studies of both the biological and nonbiological features of the nonmarine Permo-Triassic boundary. Existing data reveal distinct changes across the boundary in biologic groups such as vertebrates, bivalves, ostracodes, spores, pollen, and other flora, and in nonbiologic features such as grain size and color of sediments, sedimentary environments and climate. These changes suggest an event between the Permian and Triassic and can be applied to correlation of the nonmarine Permo-Triassic boundary.

Biotic changes

Determination of the Permo-Triassic boundary in nonmarine sequences is made at present mainly on the basis of vertebrates and spore-pollen assemblages, and to a lesser extent, on the basis of ostracodes, bivalves, and other elements of the flora.

Vertebrate fauna

The vertebrate fauna changes markedly from Permian to Triassic. Dicynodonts such as *Kunpania scopulusa*, *Striodon magnus*, *Jimusaria sinkiangensis*, *Dicynodon tienshanensis*, *Jimusaria taoshuyanensis*, *Turfandon bogdaensis* and *Kansuodon* sp. are represented in Upper Permian strata in the Junggar and Turpan basins of Xinjiang and Gansu provinces, northwest China (Zhao, 1980; Yang et al., 1986). A *Pareiasaurus* fauna, represented by Tapinocephalidae and *Shihtienfenia permica*, and *Shansisaurus xuecunensis*, has been reported from equivalent strata in Shanxi and Shaanxi provinces, North China. In recent years, dicynodonts have also been excavated in Hebei and Henan provinces and, particularly in the Daqingshan area,

Inner Mongolia, where abundant dicynodonts are mixed with *Pareiasaurus* (Yang et al., 1988).

In rocks of Early Triassic age, there are generally no dicynodonts or representatives of the *Pareiasaurus* fauna. Instead, these rocks yield a *Lystrosaurus* fauna, which is represented in the Xinjiang region by *L. hidini*, *L. weidenreichi*, *L. broomi*, *L. robustus*, *L. youngi*, *L. shichanggouensis*, *L. latifrons*, *Chasmatosaurus yuani*, *Prolacertoides jimusarensis*, and representatives of the Capitosauridae.

Significant faunal changes of this sort have also been reported from equivalent sequences in the USSR (Lapkin et al., 1973), South Africa (Keyser & Smith, 1979), and India (Dutta, 1987). It should be pointed out that individual Late Permian dicynodont species are mixed with members of the *Lystrosaurus* fauna at the base of the Lower Triassic in Xinjiang, China and in South Africa. This mixture suggests that there was no long hiatus between the Late Permian and Early Triassic in those regions.

Spore and pollen assemblages

Upper Permian–Lower Triassic nonmarine successions in China yield abundant spore and pollen assemblages, which show great changes and clear transition features.

In North China, Upper Permian spore-pollen assemblages differ from Lower Triassic ones in being dominated by gymnospermous pollen, whereas the latter are dominated by pteridophyte spores. For example, Upper Permian spore-pollen assemblages in the Junggar Basin, Xinjiang, contain up to 80% gymnospermous pollen, chiefly referable to *Cordaitina* spp., *Lueckisporites virkkiae*, *Protohaploxypinus*, *Striatoabeites*, *Striatopodocarpites*, *Klausipollenites* and *Vittatina*. A minor contingent of pteridophyte spores represents *Lophotriletes*, *Apiculatisporites*, *Limatulasporites*, and *Kraeuselisporites*.

In Shanxi and Shaanxi (Qu et al., 1983; Yang et al., 1986), Upper Permian spore-pollen assemblages are also dominated by gymnospermous pollen representing *Lueckisporites virkkiae*, *Protohaploxypinus*, *Striatoabeites*, *Taeniaesporites*, *Illinites*, *Jugasporites*, *Vittatina*, and *Cordaitina*. These assemblages contain only a few pteridophyte spores, which represent genera such as *Punctatisporites*, *Kraeuselisporites*, and *Cyclogranisporites*.

Lower Triassic spore-pollen assemblages are characterized by a significant increase in spores, both in quantity and in species. For example, in the Junggar Basin the average content of

Limatulasporites increases from 13% in Upper Permian rocks to more than 30% in lowermost Triassic strata. *Lundbladispora* is represented in uppermost Permian rocks by only a few *L. subornata* spores, but in Lower Triassic strata makes up 14.35% of the spore-pollen assemblages and is represented by a greater diversity of species, such as *L. nejburgii*, *L. subornata*, *L. watangensis*, *L. willmatti*, and *L. iphilegna*. In addition, a few specimens of *Aratrisporites* also appear in Lower Triassic samples. By contrast, gymnospermous pollen, dominant in late Paleozoic rocks, declines rapidly in abundance from Upper Permian to Lower Triassic. For example, *Protohaploxypinus* decreases from 4.1–9.7% to 0.6–5.9%; *Striatopodocarpites* from 0.4–4.4% to 1.0%; *Lueckisporites virkkiae* from 0.4–26% to 0–0.1%; *Cordaitina* from 2.1% to 0.1%; and *Vittatina* becomes completely extinct. These data also indicate that there was a gradual transition within many genera and species from Late Permian to Early Triassic.

In South China, Upper Permian and lowermost Triassic spore-pollen assemblages are all composed predominantly of spores and, to a lesser extent, of gymnospermous pollen. For example, in Fuyuan, Yunnan Province (Ouyang, 1986), the Upper Permian assemblage is characterized by *Yunnanospora* and *Gardenasporites*, whereas the Lower Triassic assemblage is distinguished by *Aratrisporites* and *Lundbladispora*. The two assemblages thus show some differences, but there is also a clear transitional relationship between them. Of the 135 species recognized in rocks of Late Permian age, 69 survived into the Early Triassic.

It should be pointed out that many features of the above-mentioned spore-pollen assemblages have also been reported from equivalent successions in other countries and areas of the world, such as east Greenland (Balme, 1979), Poland (Orlowska-Zwolinska, 1984), Hungary (Barabas, 1981), USSR, Pakistan (Balme, 1970), Madagascar (Wright & Askin, 1987), and Australia (Foster, 1982). Thus the nature of spore-pollen assemblages provides a key to regional or intercontinental correlation of the Permo-Triassic boundary.

In addition to the microspores mentioned above, megaspores have also been recovered in recent years from Upper Permian–Lower Triassic successions in Xinjiang, Gansu, and Guizhou (Yang & Sun, 1987). Upper Permian megaspore assemblages include representatives of *Triangulatisporites junggarensis*, *T. microspinosus*, *T.* sp. cf. *T. triangulatus*, while the lower Triassic assemblage is dominated by representatives of *T.* sp. cf. *T. trangulatus*, *Pusulosorites inflatus*, *Trileites* spp., and *Maexisporites* spp. Similar megaspore assemblages have also been reported from Poland (Fuglewicz, 1980), hence features of the megaspore assemblage are also important in the determination and correlation of the Permo-Triassic boundary.

Bivalve fauna

In Xinjiang and northeast China (Yang *et al.*, 1984, 1986), Upper Permian strata contain abundant *Palaeomutela* and *Palaeoanodonta*. The Sunjiagou Formation of North China also yields *Abiella* and *Microdontella*, which were all extinct before the Triassic. Such an extinction of the bivalve fauna has also been reported from the USSR (Lapkin *et al.*, 1973) and South Africa (Keyser & Smith, 1979).

Ostracodes

Permian and Triassic ostracode faunas differ greatly from one another in generic and specific composition, and in size, which is particularly significant in the Junggar Basin, Xinjiang (Yang *et al.*, 1984, 1986; Pang, 1985). Upper Permian rocks yield large-sized *Darwinula* and representatives of the *Vymella–Panxiania–Darwinuloides* assemblage, such as *Vymella subglobosa*, *Panxiania opima*, *Darwinuloides dobrinkaensis*, *D. buguruslanica*, *Suchonella posinoda*, *S. arcuaia*, *Darwinula parafragiliformis*, *D. subelongata*, *D. zhichangensis*, *D. inaffectata*, *D. inassueta*, *D. parallela*, *D. macra*, and *D. elongata*. Lower Triassic rocks, on the other hand, yield only a monogeneric assemblage of small-sized *Darwinula* representing *D. elongata*, *D. spizharskyi*, *D. ingrata*, *D. adducta*, *D. subpromissa*, *D. triossiana*, *D. rotundata*, *D. arta*, *D. accepta*, and *D. designata*.

Flora

In the Late Permian and Early Triassic, the Chinese flora experienced a history that ranged from thriving to extinction, followed by revival and prosperity (Zhou & Zhou, 1983; Yang *et al.*, 1986). In the Late Permian, the Angara *Callipteris–Iniopteris–Comia* flora flourished north of the Tianshan-Xing'an, and the Cathaysian *Gigantopteris–Lobatannularia* flora was developed south of the Kunlun-Qinling. Between these two areas, Angaran floral elements are mixed with those of the Cathaysian flora.

In the broad area north of Kunlun-Qinling, the flora became completely extinct in the latest Permian and only a few scattered plant fragments are found in lowermost Triassic rocks. Only late in the Early Triassic did a flora characterized by *Pleuromeia* and *Volzia* revive and prevail.

In South China, only a few Late Permian genera survived into the earliest Triassic, when the flora was also rather poor. It is worth noting that such an extinction and paucity of plants near the Permo-Triassic boundary also exists in Europe, Japan, and India, and is thus of worldwide significance.

Conchostracans

In the continuous Permo-Triassic sections in Dalongkou, Jimsar, Xinjiang, conchostracan assemblages are characterized by the fact that, of the seven genera that appear in the Upper Permian (i.e. *Polograpta*, *Cornia*, *Trinodus*, *Huanghestheria*, *Beijianglimnadia*, *Flasisca*, and *Cyclounquzetes*) only the latter three are represented in the Lower Triassic (Liu, 1987).

Nonbiotic changes

Significant nonbiotic changes in sedimentary materials, grain size, and color, indicating changes in environment and paleoclimate, have been observed on different sides of the Permo-Triassic boundary in China (Fig. 6.1).

In the Junggar Basin, Xinjiang, the lower and middle members of the Upper Permian Guodikeng Formation are composed, in ascending order, of dark grey, and greyish black siltstone intercalated with minor fine sandstone and lenticular muddy limestone. These features suggest a lacustrine sedimen-

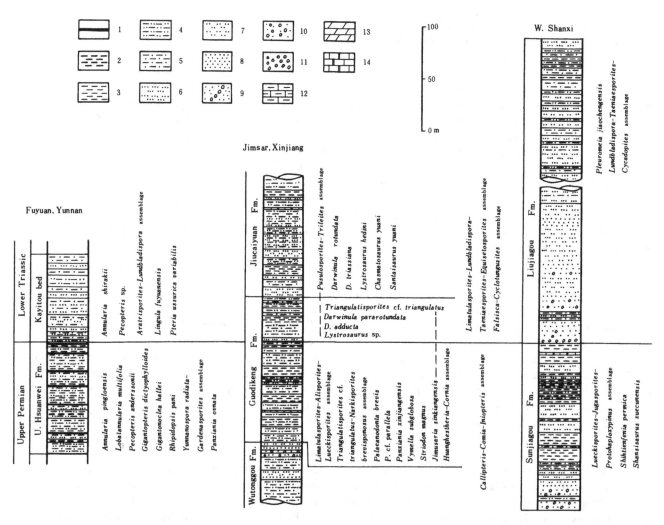

Fig. 6.1. Stratigraphic correlation of three Chinese sections composed of nonmarine strata in the Permo-Triassic boundary interval.

1, coal seam; 2, carbonaceous mudstone; 3, mudstone; 4, fine siltstone; 5, coarse siltstone; 6, fine sandstone; 7, medium-grained sandstone; 8, coarse sandstone; 9, pebbly sandstone; 10, sandy conglomerate; 11, conglomerate; 12, argillaceous limestone; 13, marl; 14, dolomitic limestone.

tary environment and a damp, warm climate. Additional siltstone and red mudstone in the Lower Triassic upper member of the Guodikeng Formation, and grey fine sandstone, purplish grey siltstone and purplish red massive mudstone in the Jiucaiyuan Formation suggest a shallow-water lake-shore environment and a hot, dry climate.

In North China, the upper member of the Upper Permian Sunjiagou Formation consists of interbedded dark purplish red mudstone and greyish green, yellowish green, and greyish purple sandstone intercalated with dolomitic limestone. These features indicate a lacustrine environment and a drier, hotter climate than prevailed in the Junggar Basin. The Lower Triassic Liujiagou Formation, on the other hand, is made up of thick-bedded, greyish white, purplish red, and greyish purple sandstone, locally intercalated with conglomerate, which is thought to represent a fluvial facies and an even drier and hotter climate.

At the junction between Yunnan, Guizhou, and Sichuan provinces, the Upper Permian Xuanwei Formation is composed chiefly of yellowish grey, greyish green, and grey fine-grained sandstone, siltstone, and mudstone of littoral-swamp facies, interbedded with carbonaceous mudstone and coal seams

(Ouyang, 1986). The overlying Lower Triassic consists, in the lower part (Kayitou Bed) of grey and greyish green siltstone, fine-grained sandstone, and silty mudstone, with intercalations mainly of mudstone, which become more variegated upward and are replaced by sediments of mainly dark purple and greyish purple colors. These represent a littoral-lake environment. The upper part of the Lower Triassic succession consists of red clastic sediments of fluvial facies. This sequence of facies, in which grain size coarsens upward and colors change from dark to red, suggests that the water body became shallower and the climate changed from damp and warm (or damp and hot) to drier and hotter. Such a sequence of changes is represented countrywide in China and also characterizes equivalent beds of the Zechstein and Buntsandstein in Germany; the Tatarian and Vetlugian in the USSR; the Kovagoszolos and Jakabhegy formations in Hungary (Barabas, 1981); the Kamthi Formation (or Raniganj and Panchet formations) in India (Dutta, 1987); the Chhidru and Mianwali formations in Pakistan; the Balfour and Katberg formations in South Africa; and the Blackwater Group and Rewan Formation in Australia.

In summary, both biotic and nonbiotic features of nonmarine

rocks near the Permo-Triassic boundary in China indicate that significant changes occurred in terrestrial environments. Recognition of these may help in determination and correlation of the boundary in nonmarine sequences.

References

Balme, B. E. (1970). Palynology of Permian and Triassic strata in the Salt Range and Surghar Range, West Pakistan. In *Stratigraphic Boundary Problems: Permian and Triassic of West Pakistan*, ed. B. Kummel & C. Teichert, pp. 306-453. (Dept. Geol., Univ. Kansas, Spec. Pub. 4). Lawrence: Univ. Kansas Press.

Balme, B. E. (1979). Palynology of Permian–Triassic boundary beds at Kap Stosch, East Greenland. *Medd. om Grønland*, **200**(6): 1–37.

Barabas, S. A. (1981). Microflora of the Permian and Lower Triassic sediments of the Mecsek Mountains (South Hungary). *Acta Geologica Academiae Scientiarum Hungaricae*, **24**(1): 49–97.

Dutta, P. K. (1987). Upper Kamthi: a riddle in the Gondwana stratigraphy of India. *Gondwana Six: Stratigraphy, Sedimentology, Geophysical Monograph*, **41**: 229–38.

Foster, C. G. (1982). Spore-pollen assemblages of the Bowen Basin, Queensland (Australia): their relationship to the Permian-Triassic boundary. *Rev. Palaeobot. Palynol.*, **36**: 165–83.

Fuglewicz, R. (1980). Stratigraphy and palaeogeography of Lower Triassic in Poland on the basis of megaspores. *Acta Geologica Polonica*, **30**(4): 405–46.

Keyser, A. W. & Smith, R. M. S. (1979). Vertebrate biozonation of the Beaufort Group with special reference to the western Karroo Basin. *Ann. Geol. Surv. S. Afr.*, **12**: 1–35.

Lapkin, I. J., Blom, G. I., Girgoryev, N. V., Entsova, F. L., Zamerenov, A. K., Kalantar, I. Z., Kisnerius, Y. L., Kuchtinov, D. A., Lutkevich, E. M., Movshovich, E. V., Sokolova, E. I., Sterlin, B. P., Suveizdis, P. I. & Tverdochlebov, V. F. (1973). The Permian–Triassic boundary on the Russian Platform. *Mem. Can. Soc. Petrol. Geol.*, **2**: 150–7.

Liu Shuwen (1987). Some Permian–Triassic conchostracans and their significance of geological age from the middle area Tianshan Mountains. *Prof. Paper Stratig. Paleont.*, **18**: 92–116. (In Chinese, with English abstract.)

Orlowska-Zwolinska, T. (1984). Palynostratigraphy of the Buntsandstein in sections of western Poland. *Acta Palaeont. Polonica*, **29**(3–4): 161–94.

Ouyang Shu (1986). Palynology of Upper Permian and Lower Triassic strata of Fuyuan District, eastern Yunnan. *Palaeont. Sinica*, **169**(n.ser. 9): 1–122. (In Chinese, with English summary.)

Pang Qiqing (1985). Preliminary discussion on the continental Permian–Triassic boundary in the northern foothills of Tianshan Mountains, Xinjiang. *Xinjiang Geol.*, **3**(4): 93–8. (In Chinese, with English abstract.)

Qu Lifan, Yang Jiduan, Bai Yunhong & Zhang Zhenlai (1983). A preliminary discussion on the characteristics and stratigraphic divisions of Triassic spores and pollen in China. *Bull. Chinese Acad. Geosci.*, **4**(5): 81–94. (In Chinese, with English abstract.)

Wright, R. P. & Askin, R. A. (1987). The Permian–Triassic boundary in the southern Morondava Basin of Madagascar as defined by plant microfossils. *Gondwana Six: Stratigraphy, Sedimentology, and Paleontology, Geophysical Monograph*, **41**: 151–66.

Yang Jiduan, Qu Lifan, Zhou Huiqin, Cheng Zhengwu, Zhou Tongshun, Hou Jingpeng, Li Peixian & Sun Suying (1984). Late Permian and Early Triassic continental strata and fossil assemblages in northern China. *Sci. Pap. Geol. Inter. Exch.*, **1**: 87–99. (In Chinese, with English abstract.)

Yang Jiduan, Qu Lifan, Zhou Huiqin, Cheng Zhengwu, Zhou Tongshun, Hou Jingpeng, Li Peixian, Sun Suying, Li Yougui, Zhang Yuxiu, Wu Xiaozu, Zhang Zhimin & Wang Zhi (1986). Permian and Triassic strata and fossil assemblages in the Dalongkou area of Jimsar, Xinjiang. *Geol. Mem.*, ser. 3: 1–235. (In Chinese, with English summary.)

Yang Jiduan, Qu Lifan, Zhou Huiqin, Cheng Zhengwu, Zhou Tongshun, Hou Jingpeng, Li Peixian & Sun Wuying (1988). Continental Permian–Triassic boundary and event. *Geoscience*, **2**(3): 366–74 (In Chinese, with English abstract.)

Yang Jiduan & Sun Suying (1987). Recent advances of megaspore studies in China. *Prof. Paper Stratig. Paleont.*, **18**: 366–76. (In Chinese, with English abstract.)

Zhao Xijin (1980). Mesozoic Vertebrata and strata in northern Xinjiang. *Bull. Inst. Paleont., Paleoanthropol., Acad. Sinica A*, **15**: 1–119. (In Chinese.)

Zhou Tongshun & Zhou Huiqin (1983). Triassic nonmarine strata and flora of China. *Bull. A. O. G. S.* **5**: 95–110. (In Chinese, with English abstract.)

7 Permian and Triassic events in the continental domains of Mediterranean Europe

GIUSEPPE CASSINIS, NADÈGE TOUTIN-MORIN, AND CARMINA VIRGILI

Introduction

This contribution is a synthesis of knowledge about late Paleozoic and early Mesozoic events in the primarily terrestrial domains now represented in various parts of Mediterranean Europe. Inclusion in this volume is justified by the fact that the areas discussed show the westernmost patterns of the ancient Tethys. Moreover, because a detailed evaluation of terrestrial history is generally more difficult for late Paleozoic and early Mesozoic time than is the reconstruction of history in marine domains, interest is aroused by recognition of major pre-Jurassic events, which are still not completely known. Thus this work leads to a tentative reconstruction of the still-debated geological history of the interval between the end of the Hercynian orogenic cycle and the beginning of the Alpine sedimentary cycle.

Data and interpretations vary from one region of terrestrial rocks to another. Despite this, we have tried to establish the most typical events. Although the correlation and nature of some are still in doubt, the effort of reconstructing them is significant. Validity of some of our conclusions, however, stems mainly from their widespread importance.

Each of us has been concerned with various aspects of the national geological situations considered. However, we have tried not to modify the content of our individual contributions in order to maintain each of our scientific attitudes, opinions and expressions.

Former continental domains

Former late Paleozoic and early Mesozoic continental domains of Mediterranean Europe west of the oldest Tethys include areas now in Italy, France, and Spain. We consider these areas in that order.

Italy

In Italy, Paleozoic and Mesozoic stratigraphic sequences are discontinuous and, in different places, have different characters and thickness. Thus a region-by-region description is indicated. From north to south, outcrops are found in the following regions:

Maritime Alps

In this region, mainly in the western Ligurian Briançonnais, the late Hercynian cover consists of metasedimentary and metavolcanic units. In the external sector of this domain (Fig. 7.1, col. 1) the earliest deposits are terrigenous continental rocks intercalated with, or surrounded by rare, essentially rhyolitic products and regarded as equivalents of the classic French 'Houiller productif'. In the internal sector, on the other hand, fine-grained sedimentary rocks rich in andesitic products prevail.

Above the rocks just described are conspicuous volumes of mostly Early Permian calcalkaline rhyolitic ignimbrites. Locally (Ormea-Viozene), the uppermost part of these is generally subalkaline and potassic in nature (Cortesogno et al., 1988).

Throughout this region of the Alps, and in adjacent areas, the Permian sequence ends with transgressive red-violet conglomerates and sandstones, the so-called 'Verrucano Brianzonese'. The unconformity at the base of this unit presumably marks an important regional uplift.

Above the Verrucano Brianzonese follow typical alluvial-deltaic 'Werfenian quartzites'. These are succeeded locally by a primarily terrigenous and carbonate sequence, which presumably represents a transition to the marine environment and is normally regarded as Scythian and earliest Anisian in age. Marine conditions are clearly documented by immediately overlying Middle Triassic carbonates, which indicate the end of an interval of continental deposition in the Ligurian Alps that had prevailed since the Middle Carboniferous (Namurian).

Northern Apennines

Punta Bianca Near La Spezia the Triassic sequence rests unconformably on a basement of Paleozoic metamorphic rocks that lacks a Permo-Carboniferous cover (Fig. 7.2, col. 2). The entire Punta Bianca succession, about 220 m thick, has been divided into two main sedimentary cycles, each characterized by a continental to marine depositional trend (Rau, Tongiorgi & Martini, 1988). The lower cycle is generally thought to be Anisian–Ladinian in age because it includes some fossiliferous beds with algae (Diploporae), bivalves, and ammonites, as well as conodonts of the Fassanian *Neogondolella transita* Assemblage Zone. Submarine basaltic lava flows, of intracratonic type, which also occur in the lower cycle (Ricci & Serri, 1975), are important for geodynamic interpretation.

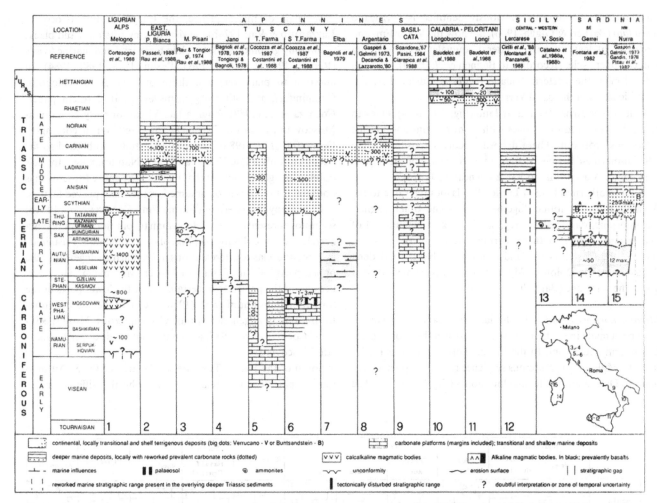

Fig. 7.1. Late Paleozoic–Early Mesozoic stratigraphic sequences in Mediterranean Italy. In section 5, the left sequence is compiled from data in Costantini *et al.* (1988); the right one is from Cocozza *et al.* (1987). In section 13, carbonate rocks with Wordian ammonoids are shown separately in the left column. Numbers in sections indicate thicknesses of some units. Sections are located on the map insert. Geologic time scale from Van Eysinga (1975) and Harland *et al.* (1982).

Due to rapid emergence, the upper cycle begins with terrigenous deposits of Verrucano type. According to Rau *et al.* (1988), this unit is completely similar to the typical Verrucano *sensu stricto* of Pisan Mts. and although it might have had its roots in the latest Ladinian it continues into the Carnian.

Pisan Mts. and nearby areas   At most places in these famous Tuscan mountains, which gave birth to the name 'Verrucano', strata unconformably above the Hercynian crystalline substrate (Fig. 7.1, col. 3) include the Verrucano Fm. *sensu stricto* and the superjacent Mt. Serra Quartzites, both 700 m thick. Pelecypods and vertebrate footprints of Carnian age have been discovered and described in the upper part of this unit (Rau & Tongiorgi, 1966). The typical basal conglomeratic formation, although nonfossiliferous, has generally been considered to be Middle Triassic or slightly older (Tongiorgi, Rau & Martini, 1977; Gandin *et al.*, 1982). However, even if there is no reason to presume a different age, data from Punta Bianca suggest that Verrucano deposition began in the late Ladinian (Rau *et al.*, 1988).

The entire group, which has the characters of a fluvio-deltaic system, is overlain by Norian carbonates and other sediments.

Locally the beds in question lie on a reduced red unit, the Asciano breccias and conglomerates, the exact age of which is as yet undefined. These clastics, considered to be scarp-foot deposits that formed in a subarid environment, rest either on the Permo-Carboniferous S. Lorenzo Shale or directly on the older metamorphic basement.

The presence of red rhyolitic rock fragments in basal Verrucano conglomerates of the Pisan Mts., a feature not previously observed in the Asciano Fm., suggests that volcanism was active in Early Permian or slightly more recent times, after deposition of the Asciano but before that of the Triassic Verrucano.

The S. Lorenzo Schists crop out in the northeastern Pisan Mts. (Fig. 7.1, col. 3). They are composed essentially of continental black, silty shales, rich in plant fossils that range in age from Westphalian D (?) to Early Permian (Remy, in Rau & Tongiorgi, 1974). During deposition of this unit the climate was probably humid intertropical. Below this unit the crystalline basement should be found.

Farther north, in the Apuane metamorphic area, the 'Grezzoni' Fm., represented by Norian–Rhaetian carbonate rocks, rests in most places on a Paleozoic crystalline substrate. However, lenses of massive quartzite are locally present beneath the carbonates. These deposits, previously identified as Verrucano *sensu lato* (Barberi & Giglia, 1965), are now the Vinca Fm. and represent, at least in part, a lateral equivalent of the Punta Bianca upper cycle. According to Bagnoli *et al.* (1979), Permian volcanism in the western Apuane region is indicated by red rhyolitic clasts in the basal Verrucano.

South of the Arno River, near Jano (Fig. 7.1, col. 4), black to greenish shales and sandstones with plant remains and marine organisms crop out. These deposits record deposition in an alluvial-marine-deltaic environment, and the age of the flora is Stephanian A according to Vai & Francavilla (1974).

In the geothermal field of Larderello clastics referable to the Asciano Fm. are found together with some other slightly metamorphic, dark red or grey, sandy rock fragments full of quartz and, here and there, Permian volcanic rocks. These clastics, the Castelnuovo Sandstones, resemble the Gardena Sandstone of the Dolomite Alps and thus are presumed to be of about the same age (Bagnoli *et al.*, 1978).

Monticiano-Roccastrada   In the last few years many studies have been made in southern Tuscany, between the Montagnola Senese and Mt. Leoni. In the Farma Torrente area, a Lower (*p.p.*) to Middle Carboniferous sequence rests on Upper Devonian? to Dinantian? lydites and limestones (Fig. 7.1, col. 5). The

lower part of this sequence, the Carpineta Fm. (Cocozza, Lazzarotto & Vai, 1974b) consists of grey siltstones and black, commonly graphitic shales, rich in terrestrial and marine fossils (Redini, 1941; Cocozza, 1965; Pasini, 1978, 1980, 1984). In addition to plant remains, those of crinoids, brachiopods, foraminifers, and other organisms occur. In a recent paper, Cocozza *et al.* (1987) assign a late Visean or early Namurian-Moscovian age to the Carpineta Fm. However, according to Costantini *et al.* (1988) paleontologic studies published so far document only a late Visean and early Namurian age. In their opinion an age assignment that includes the entire Moscovian is doubtful for it is based on assumed, rather than proved factors.

In Tuscany, the overlying Farma Fm. (Cocozza *et al.*, 1974b) seems to represent a unique Hercynian turbiditic deposit that is a correlative of the Hochwipfel Fm. of the Carnic Alps. Fusulinids in carbonate clastic intercalations in the Farma suggest that the upper parts of this flysch are upper Moscovian. However, because of the presently uncertain age of the lower parts of the formation, Costantini *et al.* (1988) prefer to include the entire Farma Fm. in a 'generic' Middle Carboniferous. More specifically, Cocozza *et al.* (1987) assign the unit to the upper Bashkirian–Moscovian.

South of the Farma Torrente, in the area of the S. Antonio mine, a paleotectonically and stratigraphically different situa-

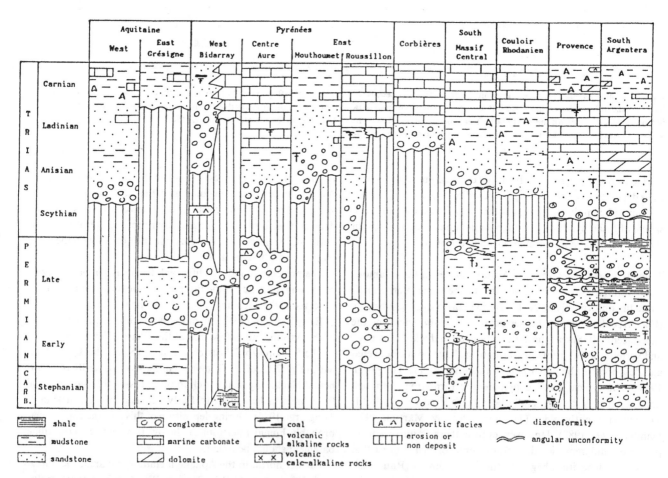

Fig. 7.2. Relationships between the Permian and Triassic in the south of France. (From Toutin-Morin, 1988.) F_0 = Stephanian flora; F_1 = Autunian flora; F_2 = Saxonian flora; F_3 = Thuringian flora.

tion exists (Fig. 7.1, col. 6). The rock sequence in this area begins with the Carpineta Fm. (without olistoliths), which is followed upward by the S. Antonio Limestone (Cocozza, 1965). The latter consists of shallow-water dark and black bioclastic limestones with early Moscovian foraminifers (Pasini, 1978, 1980). The so-called dark grey and green *Spirifer* Shales above the S. Antonio Limestone are full of brachiopods and fusulinids of Moscovian to early 'Cantabrian' age.

The sequences just described suggest the existence of an irregular sea-floor topography, which was expressed by a structural high to the south and a low to the north that have been disconnected by faults activated since the Bashkirian.

The Triassic Verrucano Group, about 500 m thick, lies unconformably on older stratigraphic sequences (Fig. 7.1, cols. 5, 6). Locally, the middle part of the group (Mt. Quoio Fm.) has yielded some calcareous pebbles with *Meandrospira pusilla* that suggest attribution of these samples to the Scythian or, at the outside, to the basal Anisian (Cocozza, Lazzarotto & Pasini, 1975). Furthermore, fusulinid-bearing clasts of end-Carboniferous to earliest Permian age have been found in the same unit (Engelbrecht, 1984).

Above the Verrucano generally are varicolored pelites and intercalated yellow carbonate beds and breccias (Tocchi Fm.) of Carnian (?) age. These represent the beginning of the Alpine marine cycle.

Island of Elba   Permo-Carboniferous dark grey clastic deposits crop out near Rio Marina, along the east coast of Elba (Fig. 7.1, col. 7). These rocks yield pelecypods, brachiopods, crinoids, echinoids, cephalopods (De Stefani, 1894), foraminifers referable to *Parafusulina* (Bodechtel, 1964; Kahler & Kahler, 1969) and plant remains. These fossils indicate the rocks range in age from near the Westphalian–Stephanian boundary to Early Permian (Vai, 1978). The Rio Marina Fm. has been interpreted as a prograding deltaic deposit.

Maremma Toscana   Verrucano, like that of Argentario, crops out again in the Uccellina Mts. (Gasperi & Gelmini, 1973). Farther south, in the eastern Argentario promontory, the same unit is about 300 m thick and is composed basically of two coarse- to fine-grained quartz-terrigenous units (Fig. 7.1, col. 8). The age of these deposits is thought to be Ladinian because of lithologic and stratigraphic affinities with other Tuscan Verrucanoes. However, considering the data from Punta Bianca, which indicate a younger age, this interpretation should be re-examined.

Sandy, phyllitic beds are also widespread beneath the Argentario Verrucano in the nearby Romani Mts. (Cocozza *et al.*, 1974a). These units, generally referred to the Carboniferous and/or Permian, have more recently been involved by Costantini *et al.* (1988) in the Hercynian Sudetian phase. They are not evaluated here. The Verrucano above, about 300 m thick, exhibits marine influences to a greater or lesser degree.

Sedimentologic study (Uncini, 1988) indicates that the Verrucano environment was a vast alluvial plain with braided rivers and, at its more distal margin, a coastal basin. Later, flattening of the source area and minor subsidence favored progradation of shallow-marine sediments and initiation of a continental-marine transition environment. Such a reconstruction is not unlike that advanced for the typical Verrucano of the Pisan Mts. Moreover, the climate was probably semi-arid and regulated by coastal rainfall.

Worthy of note is the presence of limestones with foraminifers of Middle to earliest Late Triassic age within a thick tectonic slice on the western side of Argentario (Gelmini & Mantovani, 1980). These beds seem to have been deposited more or less contemporaneously with the Verrucano.

Southern Apennines

Basilicata   In this region, and in Sicily, Late Paleozoic and Early Mesozoic stratigraphic sequences differ greatly from those just described. Some basins (Lagonegro and Sicanian), which were affected by intense synsedimentary tectonics, alkali-basaltic volcanism, crustal thinning, and consequently high subsidence rates, were filled with marine sediments indicating relatively deeper water than on surrounding carbonate platforms. Within these troughs there are, in addition to local Early?–Middle Triassic basinal deposits, large volumes of Permian rock, in many cases with obvious indications of pronounced resedimentation. On the basis of paleontologic evidence, these Permian rocks formed on, or on the margins of, carbonate platforms and in other environments.

In Basilicata, the Mt. Facito Fm., deposited in an intraplatform basin, constitutes the Triassic basal unit of the so-called Lagonegro Series (Scandone, 1968, 1972). Subsequent Alpine tectonics have fragmented outcrops so the original stratigraphic succession is doubtful. However, recent studies have enabled Ciarapica *et al.* (1988b) to recognize four main lithic associations within the Facito Fm. (Fig. 7.1, col. 9). In the lowermost of these, which locally includes basaltic rocks (Montanari & Panzanelli-Fratoni, 1988), there are palynomorphs (*Endosporites papillatus, Densoisporites* sp. div., *Lundbladispora* sp. div.), acritarchs, and foraminifers (*Meandrospira pusilla* and forms transitional to *M. dinarica*) that indicate an Early Triassic and perhaps also an earliest Middle Triassic age.

In the lowermost unit there are also thin breccias that contain reworked Permian fusulinids and other foraminifers, which have been noted in the past but are only now the objects of careful revision. In fact, according to Ciarapica *et al.* (1988a) it has been possible to distinguish at least two late Permian biozones, the *Neoschwagerina craticulifera* Zone (early Murghabian *sensu* Altiner, 1981) and the *Yabeina* Zone (early Dzhulfian *sensu lato* or 'Capitanian' = Midian of Altiner, 1981). The larger fusulinids disappear above the latter zone. In the opinion of Ciarapica *et al.* (1988a) these larger forms are replaced in the younger Permian (? late Dzhulfian *sensu lato*) by microfaunas with smaller foraminifers (*Crescentia vertebralis, Dagmarita chanakchiensis, Kamurana bronnimanni*, etc.). However, a continental regime at the end of the Permian (Dorashamian) may be presumed from the lack of data.

Redeposited specimens of *Meandrospira pusilla* indicate the presence of Lower Triassic rocks at some levels in the lower part of the Mt. Facito Fm. Moreover, the presence of faunal and floral elements no younger than Early Triassic seems to indicate that the reworking also took place in the Early Triassic. Pasini

(1984) emphasizes that, at Mt. Facito, this redeposition also affected the Lower Permian, from the *Zellia* Zone to the *Pseudofusulina vulgaris* Zone.

The Basilicata microfaunas show affinity with those of the Sosio Valley in Sicily (Miller, 1933; Skinner & Wilde, 1966; Kahler, 1974; Pasini, 1984) and with those of Djebel Tebaga in Tunisia (Skinner & Wilde, 1967). They are also similar to the 'Middle' (Murghabian) and Late Permian ('Capitanian', Dzhulfian, Changxingian) microfaunas reworked into the Ladinian conglomerates of the Julian Alps (Ramovš & Kochansky-Devidé, 1979), and to the Middle Permian microfaunas with *Neoschwagerina craticulifera*, *Afghanella*, *Chusenella*, *Sumatrina*, etc. mentioned from the Carnic Alps (Flügel, Kahler & Kahler, 1978). In fact, these fusulinids may be correlated with those of the '*Neoschwagerina* beds' of the Julian Alps (Kochansky-Devidé, 1967).

The 'Calabro-Peloritano Arc' This Alpine fragment, superimposed upon the Apenninic-Maghrebian chain, interrupts the western Tethyan Permo-Triassic marine facies of Basilicata and reintroduces continental domains.

In the Longobucco sequence (Sila Grande) the base of the Mesozoic consists essentially of red Verrucano-type clastics that rest unconformably on a Paleozoic metamorphic basement intruded by late Hercynian granitoids (Fig. 7.1, col. 10). Early Hettangian pollen occurs in pelites intercalated in an overlying yellowish grey detrital unit, and this leads us to place this Verrucano at the top of the Triassic (Baudelot *et al.*, 1988). Marine sedimentation began here in the Early Jurassic with formation of dark calcareous layers (Young, Teale & Brown, 1986).

Conglomerates and red sandstones of Verrucano type, a few meters to almost 300 m thick, also crop out in the so-called 'Calcareous Chain' that forms the sedimentary cover of the Peloritani metamorphic substrate. Immediately above these clastics, near Longi (Fig. 7.1, col. 11), a Hettangian palynoflora has also been found (Baudelot *et al.*, 1988). Thus it appears that the Verrucano at this locality (Longi Sandstone) may also be Triassic, an attribution that is supported by the Norian–Rhaetian age of alluvial grey sandstones directly below the red Verrucano in internal Peloritani areas.

Local continuity of marine sedimentation is early Liassic (Colacicchi & Filippello, 1966), a conclusion proved by discovery within the overlying fossiliferous dark limestones of two ammonites, *Arnioceras speciosum* and *Epophioceras carinatum*, of late Sinemurian–Lotharingian age (Lentini, 1975). According to Baudelot *et al.* (1988), these fossils, as well as the Rhynchonellinae in the upper 'massive limestones', characterize external units of the 'Calcareous Chains' of Kabilie, Rif and the Betic Cordillera, which represent the westernmost equivalent of the Peloritani 'Calcareous Chain'.

Sicily

In the central-western sector of this island (Sicanian Basin), a marine situation more or less comparable to that observed in Basilicata was present in the late Paleozoic (Fig. 7.1, cols. 12, 13). Permian material, here and there obviously reworked, lies tectonically alongside, and is also intercalated in Triassic basinal strata (Lercara Fm.?), in which basalts are found locally. Because of the great abundance and variety of Late Paleozoic fossils (e.g., ammonoids, brachiopods, trilobites, and other forms) this series has been intensively studied. Different opinions on events that determined accumulation of Paleozoic materials have until now impeded a clear lithostratigraphic reconstruction. However, studies undertaken by Catalano, Di Stefano & Kozur (1988a, 1988b) and Cirilli, Montanari & Panzanelli-Fratoni (1988) show that floras and faunas enable recognition of Early (Autunian, Artinskian, Kungurian–Chihsian) and Late Permian (Kubergandinian, Wordian, Capitanian, Abadehan *sensu* Kozur in Catalano *et al.*, 1988b) strata, as well as those of generic early Triassic age.

The Permian part of the Sicilian sequence is characterized by carbonate-platform and marginal (reef-slope) deposits, with some faunal elements like those found in Basilicata and other regions along the northerly Dinaric branch of the Tethys. The presence of several analogous forms in Tunisia leads us to hypothesize that Sicily and surrounding areas now occupied by the Tyrrhenian Sea, were part of a South Tethyan branch between a northern microcontinent (Apulia?) and the more southerly Gondwanian lands. In this furrow, basinal conditions prevailed not only in the Triassic, but also in the Permian, for radiolarians, psychrospheric ostracodes, and other organisms have been recognized in exposures of Permian rock in the San Calogero Torrente (Sosio Valley) and in the Roccapalumba-Lercara area (Catalano *et al.*, 1988a, 1988b; Di Stefano, 1988). The radiolarians (Albaillellacea), which have a circum-Pacific distribution, show that the paleogeography of Sicily was characterized in the Permian by a wide articulated basin connected with the Paleotethys.

The present lack of fossils representing the uppermost Permian, noted also for Basilicata, suggests once again that continental conditions may have prevailed at the end of this period and thus that there may have been emergence due to a regressive trend.

Sardinia

The Late Paleozoic–Early Mesozoic picture in Sardinia is similar in most aspects to that in nearby Corsica and is reminiscent also of conditions in Provence (France) and Spain, with which Sardinia and Corsica were united before beginning their drift toward the Tyrrhenian. This affinity is especially emphasized in central-northeast Sardinia by the widespread distribution of Hercynian intrusives (granites, diorites, etc.) with radiometric ages in Ogliastra and Gallura between 297 and 279 Ma (Del Moro *et al.*, 1974).

According to Gelmini (1986) and others, Sardinian deposits represent at least two main cycles (Fig. 7.1, cols. 14, 15). The first includes Stephanian and more widespread Autunian continental deposits, which range in thickness from a few dozen to some hundreds of meters, are normally rich in fossil plants and volcanic products, and rest unconformably on the Hercynian basement. Stephanian rocks occur in the S. Giorgio Basin of southwest Sardinia, whereas Autunian ones occur in Nurra and in central and southeastern Sardinia.

The Early Permian faultblock landscape, attributed previously to extensional movements, might have been formed in the transcurrent regime thought by many to have characterized the European Late Paleozoic climax. In fact, the presence in the northern and eastern areas of the Seui Basin of a thrust of metamorphic basement over the sedimentary cover has given rise to a hypothesis of late Hercynian compressional tectonics, which probably occurred between the end of terrigenous sedimentation in the basin and emplacement of the later Permian magmatic rocks (Sarria & Serri, 1987).

The first calcalkaline and intermediate volcanic cycle, which took place during the ?late Carboniferous and early Permian, resulted in widespread accumulation of rhyolites, rhyodacites, and andesites (tuffs, lavas, ignimbrites) a few hundreds of meters thick and similar to those in Corsica and in the Alps.

The second stratigraphic sequence probably began in Sardinia at about the end of the Permian (Fontana, Gelmini & Lombardi, 1982; Gelmini, 1986). Thus long tectonic inactivity in post-Autunian times would have been accompanied by reduced erosion of pre-existing rocks. This younger cycle, which initiates the Alpine era, is marked by the appearance of continental red clastics ('Verrucano Sardo') similar to the Buntsandstein, a few meters to about 250 m thick, and recording a prevalently alluvial environment. These clastics were generated contemporaneously with an interval of extensional tectonics that was probably stronger than the preceding one and gave rise to the formation of grabens (Gasperi & Gelmini, 1977, 1979).

While the Buntsandstein (or 'Verrucano Sardo') was being deposited, Sardinia experienced a second magmatic cycle, which, according to Gelmini (1986) '. . . seems similar in composition to the first one' and was confined largely to the early Triassic. Some studies (Lombardi et al., 1974) classify these more recent volcanics mainly as alkali–rhyolitic ignimbrites and pyroclastics.

In Sardinia, marine ingression began in the Middle Triassic. It is well recorded in Nurra by the typical Germanic facies (e.g., Oosterbaan, 1936; Gandin, Gasperi & Gelmini, 1977).

France

Southeast France and Corsica

In southeast France, including the southern border of Argentera, the Barrot Dome, eastern Provence, and Corsica, the Late Paleozoic and Early Mesozoic were characterized by continental, terrigenous sedimentation associated with complex magmatic activity (Debrand-Passard, Lienhardt & Courbouleix, 1984) (Figs 7.2, 7.3). Briefly, the history is as follows:

At the end of the Early Carboniferous, the southern branch of the Middle European Cordillera was formed in the Sudeten phase of the Hercynian orogenic cycle. This produced areas of marked relief such as Argentera, the Tanneron and Maures massifs, and the Corsican–Sardinian batholith. At the end of the Carboniferous and in the Early Permian (Autunian), sedimentary deposits accumulated in small, intermontane basins, which subsided along great displacements correlated with the Asturian phase of the Hercynian orogenic cycle (Aubouin & Mennessier, 1963). These deposits are dated by means of their plant remains, which are abundant in carbonaceous beds in Corsica or in black schists in Argentera. During this same interval there was calcalkaline magmatic activity (first cycle), connected with an interval

of subduction and collision (Bonin, 1987). Magmatic activity developed in Corsica between 330 and 290 Ma. In Provence notable plutonism occurred between 320 and 290 Ma, but volcanism is recorded only by pebbles derived from the Corsica–Sardinia block.

In Provence the Carboniferous was characterized by great NNE–SSW tectonic lineaments and was marked by a post-Stephanian compressional phase. During the Autunian, some WNW–ESE grabens (Avellan, Argentière) opened on the border of horsts oriented NNE–SSW (Toutin-Morin, Delfaud & Marocco, 1988). These represent the first manifestations of later extensional movements. An important tectonic climax tilted these early deposits so that, in Provence, the Autunian/'Saxono–Thuringian' boundary is locally marked by an angular unconformity attributed to the Saalian phase of the Hercynian cycle (Toutin-Morin, 1987) (Fig. 7.3, col. 3).

During the late Permian extensional grabens increased (Arthaud, Mégard & Séguret, 1977). Differences in tectonics at their borders determined the distribution of sedimentary bodies (as much as 2000 m thick), from marginal coarse conglomerates to distal fine sediments. At the same time, progressive erosion of Hercynian highs led to formation of molasse-like 'new red sandstones'.

In the latest Permian a vast peneplain characterized southeast areas of France (Toutin-Morin, 1987). At this time, the climate was generally tropical, with alternating dry and humid periods, and animal and plant life developed. The flora, represented by pollen, spores, and other plant remains, indicates a Thuringian age for the Late Permian formations of Provence and the Barrot Dome (Vinchon & Toutin-Morin, 1987). Throughout this period, the 'Saxono–Thuringian' of authors, intense volcanism took place (Fig. 7.3), especially in Corsica and Estérel. Volcanic rocks of acid, basic and, rarely, intermediate chemical composition, accumulated in the form of flows, dikes, sills and volcanoclastic deposits. At first these materials were erupted along the east–west marginal faults of grabens; later, eruption gave rise to small volcanic bodies and, locally, to calderas (Scandola, Sénino, Cinto in Corsica; Maurevieille in Provence). To the west (Toulon) and north (Barrot, Argentera) latest Permian volcanism is recorded mostly by pebbles included in sedimentary units. Plutonism is known only in Corsica, where it is represented by granitic intrusives.

The second magmatic cycle was of an alkaline nature. It was connected with an extensional regime of early rifting (Mermet et al., 1988) and took place between 278 and 224 Ma, with a particular climax around 270 Ma, even though it continued in some sectors during later Triassic times (Gondolo & Toutin-Morin, 1988).

The appearance of new relief at the end of the Permian, signalled by the formation of volcanic domes in the Estérel and in Corsica, by the uplift of new margins following successive collapse of grabens (Bas-Argens Basin, Provence), or by development of calderas (Corsica), renewed erosion and the spread of unaltered Permian detrital material within the almost filled basins (Toutin-Morin, 1988). Thus, Triassic sedimentation began with detrital deposits made of varicolored sandstones of Germanic 'Buntsandstein' facies, less than 100 m thick and attributed to the Anisian according to the latest palynological

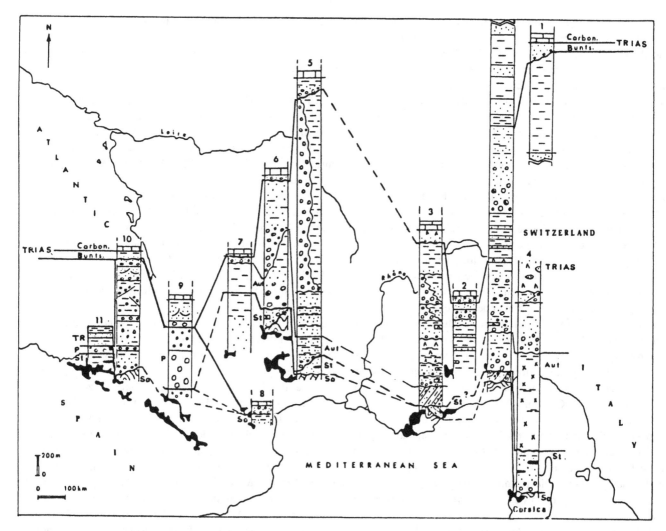

Fig. 7.3. Logs of stratigraphic sections in the south of France. 1, south Argentera; 2, Barrot; 3, Provence; 4, NW Corsica; 5, Lodève; 6, Rodez; 7, La Grésigne; 8, Roussillon; 9, Aure; 10, Bidarray; 11, Rhune. Carbon. = marine carbonates; Bunts. = Buntsandstein; P = Permian (black); Aut. = Autunian; St. = Stephanian; So. = Pre-Stephanian basement. Other symbols as in Fig. 7.2. (Toutin-Morin, 1988.)

research (Durand, 1988). Locally (Bas-Argens) eolian pebbles at the base of these deposits testify to a long arid phase (Toutin-Morin, 1987) that seems to correspond to part of the middle Buntsandstein interval (mid-Scythian?). Thus the infra-Triassic landscape was modelled during the Early Triassic.

In southeast France, Buntsandstein sedimentation may have commenced late in the Scythian (Fig. 7.2), but certainly in the early Middle Triassic, and the Permo-Triassic boundary is marked by a regional unconformity. Thus the first Mesozoic deposits lie on rocks of various Permian ages, on the Carboniferous, or on the basement. Locally, in the older Permian uplifts, the unconformity is angular, possibly as a result of the Palatine phase.

Westward from southeast France, the Triassic basin continues as far as the border of the Massif Central. In Languedoc, the sediments are at least 1000 m thick, but thicknesses decrease to the northeast, near the Delphino-Durancienne Ridge (Baudrimont & Dubois, 1977).

Only at the end of the Middle Triassic (Muschelkalk) did the sea, arriving from the east, reach the region just discussed. Initial marine deposits represent a lagoonal environment; later, during the Ladinian, a carbonate platform developed and marine sedimentation continued for a long time (Fig. 7.2).

In conclusion, the transition from a continental to a marine domain does not correspond in southeast France to the Paleozoic–Mesozoic boundary. That transition took place in the Triassic. The late Permian and early Triassic were 'hinge' intervals that saw the opening of the Tethys, and thus contrast with the Carboniferous and early Permian, which were linked to Hercynian orogenesis. The southernmost regions discussed here, Provence and Corsica, represent a weak zone in the Earth's crust, from which Corsican–Sardinian drift was to take place in the Tertiary.

South of the French Massif Central

Areas bordering the Massif Central include a number of small, continental basins filled with Carboniferous or Permian rock. Those with a history belonging to that of southern Europe include the Lodève, Gabian-Neffiès, Saint Affrique, Brousse-Broquiès, and Rodez basins (Odin, Doubinger & Conrad, 1986; Becq-Giraudon, Giot & González, 1987; Bourges, Rolando & Souquet, 1986).

During the Stephanian, filling of these basins was of intra-montane type. Initially, fluviolacustrine or fluviopalustrine deposits, rich in carbon-bearing layers, developed. Subsequently, however, the hot, humid climate changed from arid to semi-arid with very dry seasons alternating rapidly with more humid ones (Neffiès, western Saint Affrique). This is reflected in channelized alluvial or floodplain sediments (Becq-Giraudon et al., 1987).

During the transition from Stephanian to Autunian time, a compressional phase took place. In fact, the Carboniferous–Lower Permian contact is an angular unconformity. Moreover, the basal Permian appears to be absent in basins south of the Massif Central.

The alluvial-lacustrine lower Permian, which was deposited in extensional basins under a warm, tropical climate, is generally dated as late Autunian by its plant fossils (primarily *Potonieisporites*) and vertebrate footprints (especially in Lodève). Numerous volcanic-ash horizons also occur. South of Lodève and north of Saint Affrique, continental deposits lie directly on the basement.

Fluviolacustrine sedimentation continued through the 'Saxonian' of authors (see Visscher, 1973, for discussion of the term 'Saxonian'). Beginning in this interval, the basins began to subside rapidly and the appearance of a new flora in the dominantly red sediments points to a slight aridity of climate in the southeastern part (Lodève) of the region examined, where Doubinger, Odin & Conrad (1987) have recognized a 'Saxonian' flora. The plant assemblage is characterized by a decrease of *Potonieisporites*, by the presence of *Hamiapollenites saccatus*, and by the appearance of *Nuskoisporites klausii*. During the early Thuringian, a fluvio-lacustrine system developed and subsidence increased. A flora rich in bisaccates includes Thuringian forms such as *Lueckisporites globosus* (Odin et al., 1986). The basins correspond to grabens, whose faulted borders were active during a large part of the Permian (e.g. Rodez). In the 'Saxonian' and lower Thuringian parts of the section, cinerites, produced by explosive volcanism farther south, occur at various levels.

The next sedimentary cycle commonly begins with an alluvial conglomerate that rests on preceding formations or directly on the basement. This conglomerate was the product of an intra-Thuringian tectonic phase, which regenerated erosion on positive areas (Fig. 7.2). A basal unconformity is conspicuous, and is angular on basin borders (north of Rodez). The conglomerate passes laterally into fine materials (Fig. 7.3, col. 5), which accumulated in a semi-arid climate and locally (south of Lodève) to a thickness of more than 2000 m.

Thus, two sedimentary cycles are recognizable in the Permian. The first developed without interruption between the late Autunian and the early Thuringian, whereas the second cycle, which characterizes the same basins, is as yet undated (Rolando et al., 1988).

For the most part, a vast basin area, centered on Lodève, was present in the Permian south of the Massif Central.

As in southeastern France, the Triassic began with Buntsandstein deposits, normally 100–200 m thick, which rest unconformably on the Permian or locally on the Hercynian substrate. The base of these deposits is marked by a paleosurface distinguished by the formation of new minerals at the water table.

At Lodève, Triassic deposits are alluvial conglomerates and sandstones, the latter dated by pollen as middle Anisian (Doubinger & Adloff, 1981). Thus there is an Early Triassic gap in sedimentation, which is more extensive than in southeast France and is recorded by the absence or erosion of deposits. Later, with lowering of relief, clastic materials became finer and argillaceous and evaporitic sediments, dated as Anisian–Ladinian, were deposited in a playa environment (López, 1987). Rejuvenation of source areas through renewed tectonic activity caused the spread of new coarse material and essentially continuous uplift supplied detrital deposits from Anisian into the Ladinian, or even into the Rhaetian in the most northerly areas. The latter uplifts were more numerous in the southern part of the area considered, from the Cévennes to the Corbières regions (Fig. 7.2), where a minigraben was present. At the beginning of the Liassic this led to formation of the small tilted blocks characteristic of the rifting phase that introduced the wider opening of Tethys (Debrand-Passard et al., 1984). Invasion of the sea took place in the late Ladinian, a bit later than in southeastern France, and is recorded by carbonate, clay, and evaporite deposits. To the north, on the Cévennes-Ardèche border, transgression was still later, and clearly marine carbonates did not appear until the earliest Carnian. Along the entire southeast border of the Massif Central the seafloor remained more or less static during a large part of the Triassic. Detrital deposits were conspicuous along the margin of the shallow platform, which extended widely toward the east. Farther west, toward Aquitaine, transgression was evidently still later (Fig. 7.2), for detrital Triassic sediments extend upward into the Carnian northwest of Toulouse (La Grésigne).

The Pyrenees

In this chain, dominantly red continental terrigenous formations, the so-called 'grès rouge des Pyrénées' (Bixel & Lucas, 1983) crop out from the Mediterranean to the Atlantic coast in small basins in which the stratigraphic sequences, rarely complete, were deformed during the Alpine cycle. Through the Stephanian, NNW–SSE to N–S oriented basins developed. These accumulated grey lacustrine sandstones, shales and carbonates, which are locally rich in plant material and are of latest middle Stephanian age. During the Stephano–Autunian interval, calcalkaline magmatism took place in the east Pyrenees (Fig. 7.2), and in the Stephanian andesitic volcanism, connected with local caldera-type collapses (Ossau), affected the entire chain.

The first Permian sediments, dated by plant fossils, are Autunian. They are breccias and sandstones that were deposited along E–W and NW–SE drainage axes under a humid climate. At the same time, domes of acid extrusives formed along the northeast borders of basins in the eastern Pyrenees and in Catalonia (northeast Spain).

Following this, a tectonic episode rejuvenated relief and provoked spread of the 'lower red series', which consists of alluvial-fan conglomerates a few meters to 700 m thick and ephemeral lacustrine carbonates. Later, water played a more important role. This presumed Permian sequence concludes with andesites at Béarn, in Aragón, and in Catalonia.

A sedimentary discontinuity separates the Permian rocks just described from deposits above them. During the interval represented by this discontinuity volcanism expanded over the entire

Pyrenees. Alkaline basalts produced during this episode differ from older igneous products and announce Early Triassic extensional movements.

Above the rocks just described is the 'upper red series', which lies unconformably on older sedimentary rocks or, locally, on basalts. Deposition of this sequence of primarily fluvial conglomerates, as much as 400 m thick, was triggered by a second tectonic phase. Climate was characterized by rapidly varying seasons. Age of the lower part of this sequence is probably Anisian, but pollen and plant fossils indicate a Carnian–Norian age for the Bidarray sandstones (Fig. 7.2). These deposits include rocks as young as Rhaetian west of Bordeaux and even include Hettangian strata in the Catalan Pyrenees and in north Aquitaine. However, in the first area, Thuringian plants occur within the lowermost Buntsandstein (Broutin et al., 1988). Thus the plant fossils clearly show the diachronism of these clastic deposits, which everywhere rest on an altered surface (Lucas, 1977).

Both the lower and upper red series appear coarser toward the north and in the center of the Pyrenees chain, whereas they are generally finer westward (Bidarray, Aure) and toward Ossau. These Buntsandstein-facies deposits do not appear to be contemporaries of the ones in Aquitaine, which become progressively younger northward (Fig. 7.2).

In some parts of the Pyrenees, a transitional sequence of Middle Triassic lagoonal and marine sediments succeeds the upper red series. These deposits contain rare conodonts, bivalves, and echinoids, and mark the passage to the clearly marine environment that was represented at the same time farther west in the Catalonian Range (Lucas & Gisbert, 1981).

From a geographical point of view, two provinces are evident in the Pyrenees, an 'Atlantic Province' on the west and a 'Catalonian–Alpine Province' on the east. The former was characterized by NW–SE Stephanian basins and NNE–SSW to ENE–WSW Permian semigrabens, and by three volcanic episodes in the Stephanian, the Late Permian, and during the Permo-Triassic sedimentary discontinuity. The latter, on the other hand, was characterized by NE–SW Stephanian basins and E–W and NW–SE Permian semigrabens; by Triassic extension that began at the intersection of the latter two directions; and by five volcanic phases between the Stephanian and the earliest Triassic. Thus Paleozoic and Mesozoic tectonic regimes were quite different in the Pyrenees. The system of asymmetric early Permian basins evolved during the Triassic into anastomosing cruciform systems that permitted the first marine incursion in the Ladinian.

Unlike the situation in southeast France and along the southern border of the Massif Central, contrast was very marked in the Pyrenees between Paleozoic and Mesozoic tectonics (Fig. 7.3) and the differences were emphasized by magmatic events. The orogenic Stephano–Autunian calcalkaline volcanism, indicating subduction, was followed by anorogenic Early Triassic alkaline volcanism indicating extension, which corresponds with the tholeiitic volcanism connected with crustal thinning (Lucas, 1987). The latter preceded the Late Triassic ophiolitic magmatism that characterizes the Pyreneean Orogen (Lucas, 1977).

Spain

In Spain, the Permian as well as the Lower and much of the Middle Triassic are detrital continental redbeds, mainly alluvial-fan deposits. Thus there is great difficulty in establishing a boundary between the two systems (Virgili et al., 1983a, 1983b; Sopeña et al., 1983; Virgili, 1987).

The Alpine sedimentary cycle begins with the Buntsandstein, a red, detrital complex of alluvial and shallow littoral facies, which rests unconformably on all older deposits, whether they be early Permian in age or components of the older basement. This basal unconformity has traditionally been considered the Paleozoic–Mesozoic boundary. However, it has recently been discovered that the lowermost Buntsandstein contains a Thuringian microflora, similar to the one that is widespread in rocks of Late Permian age (Sopeña et al., 1988; Virgili & Ramos, 1983; Virgili et al., 1983a, 1983b).

Late Hercynian and Early Alpine tectonics and development of Permian and Triassic basins

The unconformity at the base of the Spanish Buntsandstein is related to different stages in the late Paleozoic and early Mesozoic evolution of the Iberian microplate (Arthaud & Matte, 1975; Alvaro, Capote & Vegas, 1979; Vegas & Banda, 1982; Fontboté & Virgili, 1983; Sopeña et al., 1988). The early stage, which began at the end of the Carboniferous and lasted most of the Permian, was characterized by intense Late Hercynian movements. An important NW–SE wrench system, with NE-striking faults developed, and a thick sequence of continental sediments and associated volcanic rocks accumulated in semigraben basins. These basins were of limited width, were subjected to moderate to rapid subsidence, and were filled by alluvial fans and lacustrine sediments that formed in an arid climate. Widespread andesitic-rhyolitic volcanism accompanied sedimentation, especially early in the evolution of the basins. The volcanism was of an intracontinental, calcalkaline type, and radiometric dating indicates it was contemporaneous with the last episodes of granitoid emplacement in the Variscan fold-belt (Ancochea, Hernán & Vegas, 1981; Hernando, 1977; Hernando et al., 1980; Gisbert, 1983; Muñoz et al. 1985; Navidad, 1983; Peña, Marfil & Ramos, 1979). Tectonic activity decreased with time, but gave rise to unconformities that may be useful in dividing these materials into different tectono-sedimentary units.

The end of the early deformational stage is marked by an important and widespread unconformity. Above it are quartz-detrital materials like those in the Buntsandstein of Western Europe. These represent the beginning of the Alpine cycle (Fig. 7.4) and include Triassic fossils in their upper parts (Sopeña et al., 1983; Visscher, Brugman & López-Gómez, 1982; López-Gómez, Arche & Doubinger, 1985).

On the Iberian microplate, the Buntsandstein announces a long stage of extensional tectonics. The older wrench-fault systems, reactivated as normal faults, developed a variety of rifts, in which the sedimentary fill begins characteristically with alluvial fans, grades into evaporites and red clays, and finally into shallow-marine carbonate deposits. In eastern Spain (Fig. 7.4, section 9', 10') there are alkaline volcanics.

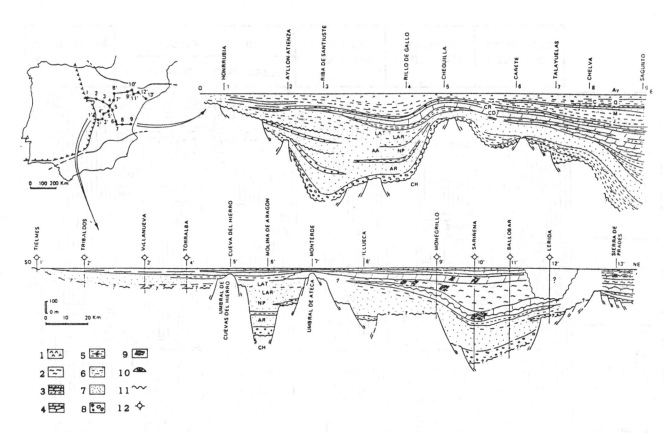

Fig. 7.4. Transverse sections showing geometry of basins and lateral changes from margins of the Iberian massif to the Mediterranean. Note the important highs that separate the basins. (After Sopeña et al., 1983, 1988.) CH and AR = Thuringian formations; AR, NP and higher units = Triassic formations. 1, gypsiferous mudstone and marl; 2, gypsum; 3, dolostone; 4, limestone; 5, mudstone; 6, sandy mudstone; 7, sandstone; 8, conglomerate; 9, anhydrite; 10, reefs; 11, unconformity; 12, well core.

The mature stage of tectonic subsidence, which began in the Middle Triassic, is recorded by shallow-marine siliciclastics followed by intertidal dolomites and marls. Marine sedimentation was clearly diachronous and advanced from Tethys to overlap the western continental margins (Gandin et al., 1982; Sopeña et al., 1988). The dominant process in this stage was thermal subsidence, which continued into the Jurassic.

Thuringian microflora in the Buntsandstein
In Iberia, the Permo-Triassic boundary has traditionally been placed at the unconformity beneath the Buntsandstein, for this level represents a change in both tectonic and paleogeographic evolution. However, with discovery of a Thuringian microflora above the unconformity, it can now be shown that Buntsandstein sedimentation began before the end of the Permian in many parts of Spain. This fact has already been pointed out in Ireland (Visscher, 1971) and in the North Sea (Geiger & Hopping, 1968).

The first discovery of a Permian microflora within the Spanish Buntsandstein was made by Boulouard & Viallard (1971) in Landete (Iberian Range). Interpretation was difficult, for the section is faulted. However, a comparable flora was later found near the basal conglomerate of the Buntsandstein in Molina de Aragón (Iberian Range), where this deposit clearly rests unconformably on older red Permian rocks (Ramos, 1979). More recently, Thuringian microfloras have been found in many parts of central, northern, and eastern Spain (Broutin et al., 1988;

Virgili, 1991; Sopeña et al., 1988; Boulouard & Viallard, 1971; Doubinger et al., 1978; López-Gómez et al., 1985; Ramos & Doubinger, 1979; Sopeña et al., 1983) (Fig. 7.5).

The most important discoveries of microfloras are as follows:

Landete-Talayuelas (Iberian Range) (Fig. 7.4, section 7). *Lueckisporites virkkiae* Potonié & Klaus, *Taeniaesporites albertae* Jansonius, *T. novialensis* Leschick, *Limitisporites* sp., *Nuskoisporites dulhuntyi* Potonié & Klaus, *Jugasporites delasaucei* Leschick, *Vesicaspora ovata* Hart, *Platysaccus umbrosus* Leschick.

Molina de Aragón (Iberian Range) (Fig. 7.4, section 6′; Fig. 7.5, col. IV). *Endosporites* sp., *Trizonaesporites grandis* Leschick, *Nuskoisporites dulhuntyi* Potonié & Klaus, *Lueckisporites virkkiae* Potonié & Klaus, *Paravesicasporites splendens* (Leschick), *Striatopodocarpidites* sp., *Jugasporites delasaucei* Leschick, *Protohaploxypinus microcorpus* Balme, *Gardenasporites heisseli* Klaus, *Falcisporites schaubergeri* Klaus, *Cycadopites* sp.

Palanca de Noves (East Pyrenees) (Fig. 7.5, col. II) (Broutin et al., 1988). *Nuskoisporites dulhuntyi*, *Falcisporites zapfei* Potonié & Klaus, *Limitisporites parvus* Klaus, *Gardenasporites* cf. *heisseli* Klaus, *Illinites* cf. *parvus* Klaus, *Vitreisporites pallidus* Nilsson, *Gigantoporites hallstatensis* Klaus, *Platysaccus papilionis* Potonié & Klaus, *Lunatisporites* cf *noviaulensis* Leschick, *Protohaploxypinus* sp., *Lueckisporites parvus* Klaus, *L. virkkiae*,

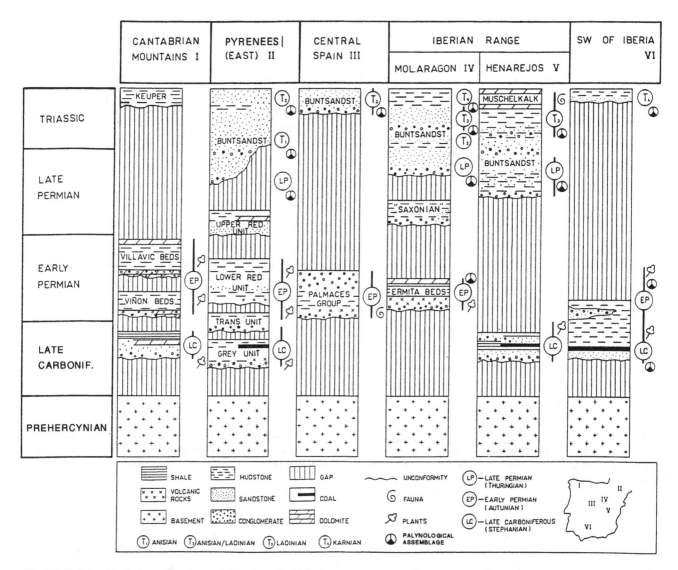

Fig. 7.5. Relationships between Permian and Triassic rocks in Spain. (From Virgili, 1991.)

Punctatisporites sp. aff. *P. fongosus* Balme, *Verrucosisporites morulatus* Smith & Butt, *Converrucosporites eggeri* Klaus, *Densoisporites playfordi* Balme, *Endosporites papillatus* Jansonius.

Regional Framework

The features just described apply to central and eastern Iberia. Those of the northern, southern, and western margins are somewhat different (Fig. 7.5, cols. III, IV, V).

In the western area, in Portugal, the Permian–Triassic gap is greatest. Locally, detrital late Triassic deposits overlap rocks of early Permian or late Carboniferous age (Palain, Doubinger & Adloff, 1977; Wagner & Martínez-García, 1982; Wagner, 1983; Wagner & Lemos de Sousa, 1983). On the southern border lithic features are similar (Bessems, 1981; Broutin, 1974, 1977) (Fig. 7.5, col. VI) and Upper Triassic rocks overlie either Lower Permian or Carboniferous strata, or the older basement. Farther south, in the Betic Range, there is no evidence of Permian, and paleogeographic connections between the Triassic and

coeval strata in the northern areas are not clear (Simon, 1987).

The northern part of Spain is more complex. In the Cantabrian area (Fig. 7.5, col. I) Upper Triassic or Lower Jurassic overlies the Lower Permian or Carboniferous (Martínez García, 1981, 1983). Paleogeographic relations between the Triassic of the Pyrenees and that of central Spain have not yet been totally solved, and the position of the lower systemic boundary is also questionable. There is a well-developed Buntsandstein, with an Anisian microflora in its uppermost parts and a Thuringian one here and there at its base (Gisbert, 1983; Broutin *et al.*, 1988; Lucas, 1977, 1987).

Permian and Triassic stratigraphy is also complex in the Pyrenees (Fig. 7.5, col. II), and involves different tectonosedimentary units (Gascón-Cuello & Gisbert-Aguilar, 1987), the study of which should be related to the situation in the south of France.

In the northeastern extremity of Spain, in the Catalonian Ranges, no Permian occurs. Detrital sediments of probable Scythian age are merely presumed because of the presence of

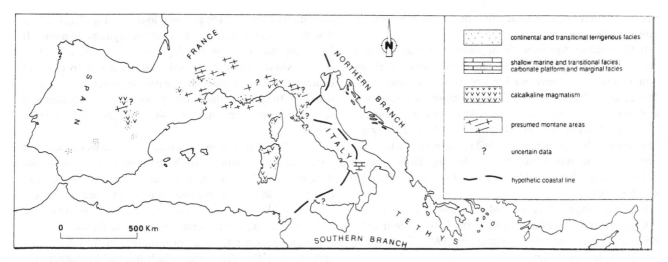

Fig. 7.6. General distribution in central and western Mediterranean Europe of deposits assumed to represent the Late Carboniferous–Early Permian interval. Betic area not considered in this figure or in Fig. 7.7, and intrusive bodies not shown in some sectors because radiometric dates are doubtful or it is difficult to determine the time of their emplacement.

early Anisian remains in the uppermost layers (see Anadón *et al.*, 1979; Calzada, 1987; Virgili, 1958). These strata overlap Carboniferous or older Paleozoic deposits (Fig. 7.4, col. 13').

In the Balearic Islands, the Permo-Triassic boundary is unknown. In Majorca, a Thuringian microflora is represented in the middle Buntsandstein (Ramos & Doubinger, 1991). In Menorca, an unconformity has been pointed out within the Buntsandstein. Rocks beneath the unconformity are attributed to the Upper Permian and a gap representing all the Scythian is postulated (Rossell *et al.*, 1988). However, no paleontologic or other chronostratigraphic evidence has been given.

Diachronism of basal Buntsandstein
Because the Buntsandstein was deposited in a tectonically active area with considerable paleorelief, accumulation of this deposit began at different times in different parts of Spain. Thickness is also quite variable, between 50 and 800 m (Fig. 7.4). However, as already noted, deposition of the Buntsandstein began in some places in the Late Permian (Ramos, 1979; Ramos & Doubinger, 1979; Arche, Ramos & Sopeña, 1983; Sopeña *et al.*, 1983; Virgili *et al.*, 1983b; López-Gómez *et al.*, 1985; López-Gómez & Arche, 1986). In other places it began in the Anisian or Ladinian (Sopeña, 1979; Visscher *et al.*, 1982; Pérez-Arlucea & Sopeña, 1983, 1985; Perez-Arlucea, 1987; Boulouard & Vaillard, 1981; Doubinger *et al.*, 1978). In other words, the Alpine sedimentary cycle began in some places at the end of the Permian and in others early in the Triassic. Thus at many places there is sedimentary continuity between the Upper Permian and Lower Triassic and the significant paleogeographic change linked to late Hercynian events is intra-Permian.

Whatever the age of the unconformity at the base of the Buntsandstein, it marks a level of important lithic change. Rocks above the surface are mostly quartz conglomerates, arkoses and quartz sandstones. Those below are arkose and greywacke sandstones and polymict conglomerates that consist of a mixture of volcanic, metamorphic, and sedimentary pebbles.

Whether the Buntsandstein rests on lower Permian or older Paleozoic rocks, there is between them a well-developed zone of paleoweathering. Reddish levels and ferruginous crusts are recognizable, as are kaolinitic alterations. This indicates that there was a rather significant hiatus before the advent of Buntsandstein sedimentation (Virgili *et al.*, 1974, 1983a). Locally this gap represents only part of the Late Permian or Early Triassic; however, in other sectors it evidently represents all the Permian and an important part of the Triassic.

General remarks on the Late Paleozoic–Early Mesozoic continental framework of Mediterranean Europe

In the Carboniferous through Triassic record of continental domains in Mediterranean Europe we recognize several sedimentary and/or tectonic cycles as well as other events.

The oldest cycle dealt with in this respect, presumably of Early *pro parte* to Middle Carboniferous age (Costantini *et al.*, 1988) seems to be represented only in the province of Siena (Torrente Farma and neighboring areas, Fig. 7.1), where it is documented by shallow- and deeper-water deposits. This cycle might represent a relic sea or it might announce the birth of the future Tethys. A stratigraphic sequence that is partly coeval, but not at all similar, occurs in the Carnic region of northeast Italy. At present, however, the sea-land pattern of this cycle is unknown.

A second main cycle, apparently more widespread (Figs. 7.1, 7.2, 7.5), began in the late Carboniferous and continued into various parts of the Permian. In Provence and Corsica it comprises only the Autunian *sensu stricto*, but in other continental regions it may also include the Thuringian. Structural highs and lows of this cycle formed in what most recent studies interpret as a tectonic transcurrent regime (see Arthaud & Matte, 1975; Rau & Tongiorgi, 1981, 1982; Vai *et al.*, 1984; Ziegler, 1984; Massari, 1988; Rau, 1988; and others). In depressions, continental sedimentation alternated with volcanic activity, mostly of calcalkaline nature (Fig. 7.6). To some authors (e.g. Bonin, 1987) this volcanism is linked to subduction-collision processes between crustal plates, accompanied by essentially compressional events. Such an interpretation clashes, however, with the commonly accepted one, which recognizes an extensional framework.

In the northern Apennines the appearance of marine intercalations in some Stephano–Autunian rocks (Jano, Elba) also demonstrates the relative proximity of the sea (see also Rau & Tongiorgi, 1974; Vai, 1978). Paleogeographic considerations lead us to suppose that these marine incursions came generally from the east, from an Italian-Dinaric furrow between Italy and Yugoslavia. These incursions reached north to the Carnic Alps, where they are represented by the Auernig Group, which includes marine deposits of Westphalian D to earliest Permian age (Fig. 7.6).

Existence of the Tethys in the Mediterranean region as early as Carboniferous is neither an unsupportable nor a new hypothesis (e.g. Vai, 1976; Rau & Tongiorgi, 1981; Tollmann & Krystan-Tollmann, 1985).

The Italian-Dinaric furrow, which probably formed at the end of the Hercynian orogenic cycle, clearly continued into the Permian without substantial modification in position. However, south of it there must have been one or more marine branches of currently uncertain position. Some Triassic sequences in the Lagonegro Basin (Mt. Facito Fm., Basilicata), and in the Sicanian Basin (Lercara Fm., Sicily) include well-documented reworked deposits of Early and Late Permian age, which confirm the presence of a pronounced penetration of Tethys toward Africa (Fig. 7.6). This penetration could not have extended much farther than Tunisia, for only continental Permian is noted in Morocco and the Betic cordillera (e.g. Michard, 1976; Mermet et al., 1988).

On the irregular sea-floor in the south-Tethyan sectors (as mentioned above) there accumulated shallow- to deeper-water sediments. Radiolarian and ostracode faunas from the latter in Sicily are also found in Japan and other circum-Pacific areas (Catalano et al., 1988a; 1988b). This leads us to suppose that during the Permian, the Paleotethys already had the physiographic characters that were to be more definitely outlined in the Mesozoic (Figs 7.7, 7.8). The separation of Hercynian Meso-Europa and the Adriatic microplate from Gondwana seems thus to have had its roots in these ancient times.

According to some authors (e.g., Irving in Rau & Tongiorgi, 1982) the opening of Tethys took place by means of important dextral shears situated approximately at the latitude of the present Mediterranean. Along these shears, from the last Hercynian pulses into the Late Triassic, Africa drifted so as to assume the classic position of Wegener's Pangea. Relative movements between Africa and Europe may also have begun the breakup of the Hercynian chain, thus giving rise to minor continental plates such as Atlas, Iberia, and so forth.

The sedimentary-volcanic sequences of the second cycle, which were involved in a very active tectonic regime, include numerous unconformities of varying importance. By convention most authors refer these events to Stille's 'Asturian' and 'Saalian' tectogenetic phases (or movements), which were widespread but have not yet been thoroughly investigated.

A regional tectonic event that affected Meso-Europa and the present Alpine and Mediterranean areas, occurred at a time thought to coincide in the Alps with the Early–Late Permian boundary or, in France and Spain, during the Late Permian and the Triassic. In all probability this event was accompanied by the formation of structural 'highs' and 'lows' (see Figs 7.1, 7.2, 7.5). The likely causes of this variation were geodynamic in nature. It is certain that the new physiography of present-day south Europe varied considerably, subjected as it was to constant demolition of relief associated with the above horst-graben situation. In any case, the landscape was progressively flattened so as to receive, even diachronously, sediments of the Triassic seas.

West of the Alps, and in the Sardinia–Corsica block, the event just mentioned is characterized by appearance of the 'Buntsandstein' (Figs 7.7, 7.8). This unit rests unconformably on a variety of metamorphic and nonmetamorphic rocks, the latter including some of latest Paleozoic age, and ranges in age from Late Permian (Thuringian) to various levels in the Triassic.

On the other hand, the event is marked in the Alps by the appearance of the 'Verrucano', which, like the 'Buntsandstein', occupies various stratigraphic positions and is thus heterochronous. The western Briançonnais Verrucano, however, which passes vertically and (?) laterally into so-called 'Werfenian quartzites' (Vanossi, 1980, Pl. II) overlain by terrigenous and carbonate rocks generally dated as Middle Triassic, certainly does not seem to represent a significant part of the Mesozoic.

In the evolution of the post-Hercynian sequences, the Verrucano and Buntsandstein thus document a crucial interval during which pre-existing paleogeographic situations changed into new ones over widespread areas. The unconformity at the base of these units also represents a gap of different, but still uncertain magnitude. Thus the appearance of these units may be considered to represent the beginning of the long Alpine sedimentary cycle.

Before and during deposition of the lower Buntsandstein, especially in southern Provence and Corsica, a widespread alkaline magmatism took place (Figs 7.2, 7.7). This second igneous episode, which started in the Early Permian and continued into the Triassic, generated in southern Provence one or several additional minor Permian cycles and was connected with the extensional regime that gave rise to the Buntsandstein and also perhaps to the Ligurian Verrucano. In France many authors suppose that this plutonic-volcanic activity initiated the Alpine cycle.

In Sardinia, as in Provence and Corsica, two magmatic cycles also occur (Figs 7.1, 7.6, 7.7). According to the literature, the first extrusives, of Early Permian age, are calcalkaline in composition, whereas the nature of rocks formed in the later cycle, probably of latest Permian and earliest Triassic age, is not completely known.

In the light of the above, the Permo-Triassic boundary, where it is not involved in a hiatus, must locally be sought within either the Buntsandstein or Alpine Verrucano. In the latter the boundary is most probably at the top, near the overlying Scythian terrigenous deposits. Considering the laterally variable alluvial nature of both units, however, placement of the boundary is not an easily solved problem.

In the northern Apennines, by contrast, the base of the Verrucano is probably Middle Triassic. In that area, Permo-Carboniferous taphrogenesis was apparently rarer, was accompanied by less intense volcanism than in areas farther west, and seems to have given rise to either a broad, general uplift or to relative structural rigidity of the region in the Late Permian and

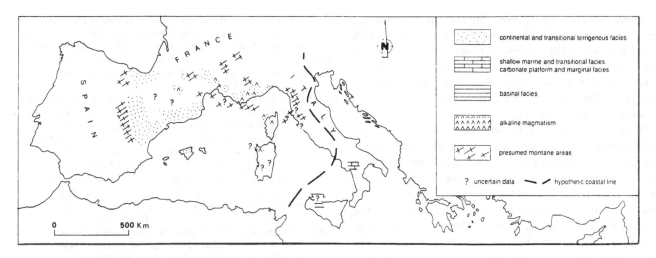

Fig. 7.7. General distribution in central and western Mediterranean Europe of deposits assumed to be of Late Permian age. Some Triassic rocks are also included in Spain because the Buntsandstein of Spain includes both Late Permian and Triassic strata.

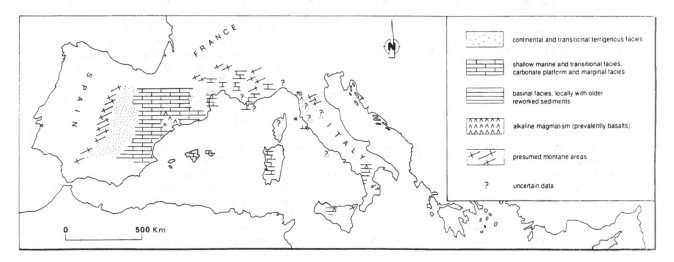

Fig. 7.8. General distribution in central and western Mediterranean Europe of rocks assumed to be of late Middle Triassic age.

Early Triassic (Fig. 7.7). Lithic sequences representing this time interval have not yet been discovered. The landmass in question, surrounded by areas that experienced different stratigraphic and structural histories and by seaways that have an articulation that is not easy to reconstruct today, seems to be represented entirely or in part by the Apulian continental microplate, which separated from Africa at the end of the Hercynian orogeny and assumed an important role in the Mesozoic.

Before becoming completely buried beneath the Verrucano, which initiated the Alpine sedimentary cycle, the Apenninic microcontinent experienced several marine incursions, which presumably were facilitated by rifting and eustatic fluctuations. Proof of this may be the discovery in the mid-portion of the Verrucano (Mt. Quoio Fm.) in the Siena region of calcareous pebbles with Scythian or early Anisian fossils (Cocozza et al., 1975), and of other fusulinid-bearing clasts of latest Carboniferous and Early Permian age (Engelbrecht, 1984).

On the other hand, the lower, pre-Verrucano, sequence of Punta Bianca, near La Spezia, with limestones and calcarenites yielding ammonites, algae, and conodonts of late Anisian and Ladinian age, is clear evidence of a marine event. Intercalated basaltic lavas of intermediate alkaline chemistry (Ricci & Serri, 1975) indicate that the local basin corresponded to an aborted continental rift that approximated the western margin of Apulia (e.g., Rau et al., 1988).

Appearance of the Apenninic Verrucano, which initiated an important new cycle, may generally be dated as Middle Triassic. Rau et al. (1988) tentatively fix the beginning of its deposition at Punta Bianca, and consequently in the Pisan Mts., as late Ladinian. As is true of the Buntsandstein and the Briançonnais Verrucano, however, the typical Tuscan Verrucano and similar deposits as far south as the Argentario were probably formed diachronously. Although the Apenninic Verrucano is not generally as young as Carnian, recent dating of the Longobucco and Longi Verrucanoes in the 'Calabro-Peloritan Arc' as latest Triassic (Baudelot et al., 1988) places an outside limit on diachronism of the deposit.

Erosional activity during formation of the Verrucano and

Buntsandstein caused progressive flattening of the landscape and this ultimately favored incursion of the sea (Fig. 7.8), which is generally recorded by a sedimentary sequence that may be rich in evaporites initially but is subsequently composed of open-water, shallow-marine sediments. Thus, the drowning of Apulia took place primarily in the Late Triassic and this resulted in a significant extension of Tethys.

The southern Apennines and Sicily, crossed by the Lagonegro and Sicanian marine basins, which assumed their identities during the latter part of Hercynian orogenesis, confirm continuity of the paleogeographic and structural situations. In the Middle Triassic (Fig. 7.8) tectonics repeatedly caused erosion and redeposition of Permian shelf carbonates in deeper-water settings. Volcanic products, mainly alkaline basalts, are locally present and, in all probability record activated rifting. The Permo-Triassic boundary is not documented faunally, hence we suppose that the transition between these two periods was characterized by emergence.

In conclusion, continental settings in the Late Paleozoic and Early Mesozoic of the central-western Mediterranean region show clear connections, which in all probability indicate a general heredity of events. This leads us to emphasize the importance assumed later by the structural alignments described in 'piloting' the course of Alpine evolution in those areas of the Mediterranean region we have examined.

References

Altiner, D. (1981). Recherches stratigraphiques et micropaléontologiques dans le Taurus oriental au NW de Pinarbasi (Turquie). *Thesis, Univ. Geneva*, 450 pp.

Álvaro, M., Capote, E. & Vegas, R. (1979). Un modelo de evolución geotectónica para la Cadena Celtibérica. *Acta Geol. Hispánica*, **14**: 172–7.

Anadón, P., Colombo, F., Esteban, M., Marzo, M., Robles, S., Santanach, P. & Solé, L. (1979). Evolución tectonoestratigráfica de los Catalánides. *Acta Geol. Hispánica*, **14**: 242–70.

Ancochea, E., Hernán, F. & Vegas, R. (1981). Un marco geotectónico para el vulcanismo de Atienza (Provincia de Guadalajara). *Cuad. Geol. Ibérica*, **7**: 421–30.

Arche, A., Ramos, A. & Sopeña, A. (1983). El Pérmico de la Cordillera Ibérica y bordes del Sistema Central. In *Carbonífero y Pérmico de España, X Congr. Int. Carbonífero*, pp. 421–38. Madrid: Inst. Geol. Minero España.

Arthaud, F. & Matte, P. (1975). Les décrochements tardi-hercyniens du Sud Ouest de l'Europe. *Tectonophysics*, **25**: 139–71.

Arthaud, F., Megard, F. & Seguret, M. (1977). Cadre tectonique de quelques bassins sédimentaires. *Bull. Cent. Rech. Expl.-Product. Elf-Aquitaine*, **1**: 147–88.

Aubouin, J. & Mennessier, G. (1963). Essai sur la structure de la Provence. In *Livre à la mémoire du Professeur Paul Fallot (1960 à 1963)*, Soc. Géol. France, 45–98.

Bagnoli, G., Gianelli, G., Puxeddu, M., Rau, A., Squarci, P. & Tongiorgi, M. (1978). The Tuscan Paleozoic: a critical review. In *Report on the Tuscan Paleozoic Basement*, ed. Tongiorgi, M., pp. 9–26. Pisa: C. N. R., Internal Rep. Progetto finalizzato Energetica-Sottoprogetto Energia Geotermica.

Bagnoli, G., Gianelli, G., Puxeddu, M., Rau, A., Squarci, P. & Tongiorgi, M. (1979). A tentative stratigraphic reconstruction of the Tuscan Paleozoic basement. *Mem. Soc. Geol. Ital.*, **20**: 99–116.

Barberi, F. & Giglia, G. (1965). La serie scistosa basale dell'Autoctono delle Alpi Apuane. *Boll. Soc. Geol. Ital.*, **84**: 41–92.

Baudelot, S., Bouillin, J.-P., Durand Delga, M., Giunta, G. & Olivier, P.

(1988). Datazioni palinologiche dell'Hettangiano alla base della trasgressione mesozoica del 'Verrucano' della Sila (Calabria) e dei Monti Peloritani (Sicilia). *Boll. Soc. Geol. Ital.*, **107**: 51–61.

Baudrimont, A. F. & Dubois, P. (1977). Un bassin mésogéen du domaine péri-alpin: le Sud-Est de la France. *Bull. Cent. Rech. Expl.-Product. Elf-Aquitaine*, **1**: 261–308.

Becq-Giraudon, J.-F., Giot, D. & González, G. (1987). Cycles sédimentaires du Carbonifère Supérieur (Stéphanien B) du Bassin de Neffiès (SE de la Montagne Noire, Hérault) et comparaison avec le houiller de Saint Rome-de-Tarn, bassin de Saint Affrique (Aveyron). *Géol. Alp. (Grenoble)*, *Mém. hors-sér.*, **13**: 5–17.

Bessems, R. E. (1981). Aspects of Middle and Late Triassic palynology. 1. Palynostratigraphical data from the Chiclana de Segura Formation of the Linares-Alcaraz region (southeastern Spain) and correlation with palynological assemblages from the Iberian Peninsula. *Rev. Paleobot. Palynol.*, **32**: 309–400.

Bixel, F. & Lucas, C. (1983). Magmatisme, tectonique et sédimentation dans les fossés stéphano-permiens des Pyrénées occidentales. *Rev. Géol. Dynam. Géogr. Phys., Paris*, **24**: 329–42.

Bodechtel, J. (1964). Stratigraphie und Tektonik der Schuppenzone Elbas. *Geol. Rdsch.*, **53**: 25–41.

Bonin, B. (1987). Réflexions à propos de la répartition des granitoïdes dans les massifs cristallins externes des Alpes françaises. *Géol. Alp. (Grenoble)*, **63**: 137–49.

Boulouard, Ch. & Vaillard, P. (1971). Identification du Permien dans la Chaîne Ibérique. *C. R. Acad. Sci. Paris*, **273**: 2441–4.

Boulouard, Ch. & Vaillard, P. (1981). Identification du Ladinien et du Carnien dans les marnes triasiques de la Serrania de Cuenca (Chaîne Ibérique sud occidentale, Espagne): considérations stratigraphiques et structurales. *Bull. Cent. Rech. Expl.-Product. Elf-Aquitaine*, **5**: 31–41.

Bourges, Ph., Rolando, J. P. & Souquet, P. (1986). Le Permien du détroit de Rodez: systèmes de dépôt, dynamique du bassin. *Ann. Soc. Géol. Nord*, **CVI**: 173–82.

Broutin, J. (1974). Découverte de l'Autunien dans le bassin de Guadalcanal (Nord de la province de Séville, Espagne du Sud). *C. R. Acad. Sci. Paris*, **278**(D): 1709–10.

Broutin, J. (1977). Nouvelles données sur la flore des bassins autuno-stéphaniens des environs de Guadalcanal (province de Séville-Espagne). *Cuad. Geol. Ibérica*, **4**: 91–8.

Broutin, J., Doubinger, J., Gisbert, J. & Satta Pasini, S. (1988). Premières datations palynologiques dans le faciès Buntsandstein des Pyrénées catalanes espagnoles. *C. R. Acad. Sci. Paris*, **301**(s. II): 159–63.

Calzada, S. (1987). Niveles fosilíferos de la facies Buntsandstein (Trias) en el sector norte de los Catalinedes. *Cuad. Geol. Ibérica*, **11**: 256–71.

Catalano, R., Di Stefano, P. & Kozur, H. (1988a). First evidence of Lower Permian Albaillellacea (Radiolaria) in the Tethyan Eurasia. *Atti 74th Cong. Soc. Geol. Ital.*, **A**: 119–23.

Catalano, R., Di Stefano, P. & Kozur, H. (1988b). New results in the Permian and Triassic stratigraphy of western Sicily with special reference to the section at Torrente San Calogero SW of the Pietra di Salomone (Sosio Valley). *Atti 74th Congr. Soc. Geol. Ital.*, **A**: 126–35.

Ciarapica, G., Cirilli, S., Martini, R., Panzanelli-Fratoni, R., Passeri, L., Salvini-Bonnard, G. & Zaninetti, L. (1988a). Le Fusuline rimaneggiate della Formazione triassica del Monte Facito, Appennino meridionale. *Atti 74th Congr. Soc. Geol. Ital.*, **B**: 117–24.

Ciarapica, G., Cirilli, S., Panzanelli-Fratoni, R., Passeri, L. & Zaninetti, L. (1988b). La Formazione di Monte Facito (Appennino meridionale). *Atti 74th Congr. Soc. Geol. Ital.*, **B**: 132–5.

Cirilli, S., Montanari, L. & Panzanelli-Fratoni, R. (1988). Palinomorfi nella Formazione Lercara: nuovi elementi di datazione. *Atti 74th Congr. Soc. Geol. Ital.*, **B**: 140–2.

Cocozza, T. (1965). Il Carbonifero nel Gruppo Monticiano-Roccastrada. *Ric. Sci.*, **35**(II–A): 1–38.

Cocozza, T., Decandia, F. A., Lazzarotto, A., Pasini, M. & Vai, G. B. (1987). The marine Carboniferous sequence in southern Tuscany: its bearing for Hercynian paleogeography and tectofacies.

In *Pre-Variscan and Variscan events in the Alpine-Mediterranean mountain belts*, ed. Flügel, H. W., Sassi, F. P. & Grecula, P., pp. 135–44. Bratislava: Mineralia slovaca Monogr.

Cocozza, T., Gasperi, G., Gelmini, R., & Lazzarotto, A. (1974a). Segnalazione di nuovi affioramenti paleozoici (Permo-Carbonifero?) a Boccheggiano e tra Capalbio e i Monti Romani (Toscana meridionale-Lazio settentrionale). *Boll. Soc. Geol. Ital.*, **93**: 47–60.

Cocozza, T., Lazzarotto, A. & Pasini, M. (1975). Segnalazione di una fauna triassica nel conglomerato di M. Quoio (Verrucano del T. Farma-Toscana meridionale). *Riv. Ital. Paleont.*, **81**: 425–36.

Cocozza, T., Lazzarotto, A. & Vai, G. B. (1974b). Flysch e molassa ercinici del Torrente Farma (Toscana). *Boll. Soc. Geol. Ital.*, **93**: 115–28.

Colacicchi, R. & Filippello, M. P. (1966). L'inizio del Mesozoico marino nella Sicilia nord-orientale. *Riv. Ital. Paleont.*, **72**: 755–94.

Cortesogno, L., Dallagiovanna, G., Vannucci, R. & Vanossi, M. (1988). Volcanisme, sédimentation et tectonique pendant le Permo-Carbonifére en Briançonnais ligure: Une revue. *Eclogae Geol. Helv.*, **81**: 487–510.

Costantini, A., Decandia, F. A., Lazzarotto, A. & Sandrelli, F. (1988). L'Unità di Monticiano-Roccastrada fra la Montagnola Senese e il Monte Leoni (Toscana meridionale). *Atti Ticin. Sci. Terra*, **31**(1987–88): 382–420.

Debrand-Passard, S., Lienhardt, M. T. & Courbouleix, S. (1984). Synthèse géologique du Sud-Est de la France. *Mém. Bur. Rech. Géol. Min.*, 125–126. Orléans: ed. B. R. G. M.

Decandia, F. A. & Lazzarotto, A. (1980). Le unità tettoniche di M. Argentario. *Mem. Soc. Geol. Ital.*, **21**: 385–93.

De Stefani, C. (1894). Découverte d'une faune paléozoique à l'Ile d'Elba. *Bull. Soc. Géol. France*, sér. 3, **22**: 30–3.

Del Moro, A., Di Simplicio, P., Ghezzo, C., Guasparri, G., Rita, F. & Sabatini, G. (1974). Radiometric data and intrusive sequence in the Sardinia batholith. *N. Jb. Mineral. Abh.*, **126**: 28–44.

Di Stefano, P. (1988). Il Trias della Sicilia e dell'Appennino meridionale: una rassegna. *Atti 74th Congr. Soc. Geol. Ital.*, **A**: 263–78.

Doubinger, J. & Adloff, M. C. (1981). Précisions palynologiques sur l'âge anisien moyen du gisement à plantes du Trias des Lavarèdes (SE de Lodève, Sud de la France). *Bull. Sci. Géol. (Strasbourg)*, **34**: 239–42.

Doubinger, J., Adloff, M. C., Ramos, A., Sopeña, A. & Hernando, S. (1978). Primeros estudios palinológicos en el Pérmico y Triásico de la Cordillera Ibérica y bordes del Sistema Central. *Palinol.*, num. extr., **1**: 27–33.

Doubinger, J., Odin, B. & Conrad, G. (1987). Les associations sporopolliniques du Permien continental du bassin de Lodève (Hérault, France); caractérisation de l'Autunien supérieur, du 'Saxonien' et du Thuringien. *Ann. Soc. Géol. Nord.*, **CVI**: 103–9.

Durand, M. (1988). Le Trias détritique du 'Bassin du Sud-Est'. Paléogéographie et environnements de dépôt. *Géol. Alp. (Grenoble)*, Mém., hors-sér., **114**: 69–78.

Engelbrecht, H. (1984). Bericht über die Kartierung und tectonische sowie sediment-petrographische Untersuchung eines Gebietes südlich Monticiano (Toscana). *Inst. Allgemeine und Aug. Geol. Univ. München, Diplomarbeit*, 108 pp.

Flügel, E., Kahler, F., & Kahler, G. (1978). Nachweis von marinem Mittelperm bei Forni Avoltri (Carnia, Südalpen). *N. Jb. Geol. Paläont. Mh.*, **8**: 449–58.

Fontana, D., Gelmini, R. & Lombardi, G. (1982). Le successioni sedimentarie e vulcaniche carbonifere e permo-triassiche della Sardegna. In *Guida alla Geologia del Paleozoico Sardo, Guide Geologiche Regionali*, ed. Carmignani, L. *et al.*, pp. 183–92. Mem. Soc. Geol. Ital., **24**: (Suppl. B.)

Fontboté, J. M., & Virgili, C. (1983). Evolución tardihercínica y ciclo alpino. In *Geologia de España (Libro Jubilar J. M. Rios)*, vol. 2, pp. 3–16. Madrid: Inst. Geol. Minero España.

Gandin, A. (1978). Il Trias medio di Punta del Lavatoio (Alghero, Sardegna NW). *Mem. Soc. Geol. Ital.*, **18**: 3–13.

Gandin, A., Gasperi, G. & Gelmini, R. (1977). Il passaggio Permo-Trias in Sardegna. In *Escursione in Sardegna: risultati e commenti*, a

cura di Vai, G. B., 2 (suppl.), pp. 35–7. Parma: Gruppo di Lavoro sul Paleozoico, CNR.

Gandin, A., Tongiorgi, M., Rau, A. & Virgili, C. (1982). Some examples of the Middle Triassic marine transgression in South-Western Mediterranean Europe. *Geol. Rdsch.*, **71**: 881–94.

Gascón-Cuello, J. & Gisbert-Aguilar, J. (1987). La evolución climática del Stephaniense, Pérmico y Buntsandstein del Pirineo Catalán en base al estudio de paléosuelos. *Cuad. Geol. Ibérica*, **11**: 83–96.

Gasperi, G. & Gelmini, R. (1973). Ricerche sul Verrucano. 1. Il Verrucano del M. Argentario e dei Monti dell'Uccellina in Toscana. *Boll. Soc. Geol. Ital.*, **92**: 115–40.

Gasperi, G. & Gelmini, R. (1977). I bacini permo-carboniferi della Sardegna. In *Escursione in Sardegna, risultati e commenti*, a cura di Vai, G. B., 2 (suppl.), pp. 39–40. Parma: Boll. Gruppo di Lavoro sul Paleozoico, CNR.

Gasperi, G. & Gelmini, R. (1979). Ricerche sul Verrucano. 4. Il Verrucano della Nurra (Sardegna nord-occidentale). *Mem. Soc. Geol. Ital.*, **20**: 215–31.

Geiger, M. E. & Hopping, C. A. (1968). Triassic stratigraphy of the southern North Sea basin. *Phil. Trans. Roy. Soc. London*, ser. B, **254**(790): 1–36.

Gelmini, R. (1986). A paleogeographical reconstruction of the Permo-Triassic circumtyrrhenian area. *Boll. Soc. Geol. Ital.*, **104**(1985): 561–74.

Gelmini, R. & Mantovani, M. P. (1980). Ritrovamento di fossili triassici nel calcare cristallino stratificato nero dell'Argentario (Toscana meridionale). *Mem. Soc. Geol. Ital.*, **21**: 427–30.

Gisbert, J. (1983). El Pérmico de los Pirineos españoles. In *Carbonifero y Pérmico de España*, ed. IGME, pp. 403–420. Madrid: X Congr. Int. Carbonifero.

Gondolo, A. & Toutin–Morin, N. (1988). Relation entre les données de la Géochronologie et les coupures stratigraphiques en Provence orientale et en Corse. *Assoc. Géol. Permien, 3è Jour. thématique, résumés*. Paris: Ecole des Mines.

Harland, W. B., Cox, A. V., Llewellyn, P. G., Pickton, C. A. G., Smith, A. G. & Walters, R. (1982). *A Geologic Time Scale*. Cambridge: Cambridge Univ. Press.

Hernando, S. (1977). Pérmico y Triásico de la región Ayllón-Atienza (Provincias de Segovia, Soria y Guadalajara). *Tesis Doct. Univ. Complutense Madrid (1975)*. Semin. Estratigrafia. Ser. monogr., **2**: 1–408.

Hernando, S., Schott, J. J., Thuizat, R. & Montigny, R. (1980). Ages des andésites et des sédiments interstratifiés de la région d'Atienza (Espagne): Étude stratigraphique géochronologique et paléomagnetique. *Bull. Sci. Géol. (Strasbourg)*, **33**: 119–28.

Kahler, F. (1974). Fusuliniden aus T'ien-chan und Tibet. *The Sino-Swedish Expedition*, 52(V), Invertebrate Pal., **4**: 1–148.

Kahler, F. & Kahler, G. (1969). Einige südeuropäische Vorkommen von Fusuliniden. *Mitt. Geol. Ges. Wien*, **61**: 40–60.

Kochansky-Devidé, V. (1967). Neoschwagerinenschichten einer Tiefbohrung in Istrien (Jugoslawien). *N. Jb. Geol. Paläont. Abh.*, **128**: 201–4.

Lentini, F. (1975). Le successioni Mesozoiche-Terziarie dell'Unita' di Longi (Complesso Calabride) nei Peloritani occidentali (Sicilia). *Boll. Soc. Geol. Ital.*, **94**: 1477–503.

Lombardi, G., Cozzupoli, D. & Nicoletti, M. (1974). Notizie geopetrografiche e dati sulla cronologia K-Ar del vulcanesimo tardopaleozoico sardo. *Period. Mineral.*, **43**: 221–312.

López, M. (1987). Caractérisation de la sédimentation du faciès Buntsandstein (formation inférieure) du Trias de Lodève, Sud de la France. *Géol. Alp. (Grenoble)*, Mém., hors-sér. **113**: 91–101.

López-Gómez, J. & Arche, A. (1986). Estratigrafia del Pérmico y Triásico en facies Buntsandstein y Muschelkalk en el sector sureste de la Rama Castellana de la Cordillera Ibérica (Provincias de Cuenca y Valencia). *Estud. Geol.*, **42**: 259–70.

López-Gómez, J., Arche, A. & Doubinger, J. (1985). Las facies Buntsandstein entre Canete y Talayuelas (Prov. de Cuenca): Características sedimentológicas y asociaciones palinólogicas. *Rev. Españ. Micropaleont.*, **17**: 93–112.

Lucas, Cl. (1977). Le Trias des Pyrénées, corrélations stratigraphiques et paléogéographiques. *Bull. Bur. Rech. Géol. Min.*, 2d ser., **IV**: 225–31.

Lucas, Cl. (1987). Le passage du Permien au Trias dans les Pyrénées septentrionales. *Assoc. Géol. Permien, 2è journee thématique, résumés*. Paris.

Lucas, Cl., & Gisbert, J. (1981). Eléments nouveaux pour l'évolution des paysages du grès rouge pyrénéen, du Permien au Trias. *106 Cong. Nat. Soc. Savantes*, Paris, Bibl. Nat., sect. Sci., **III**: 351–62.

Martínez-García, E. C. (1981). El Paleozoico de la zona cantábrica oriental (Nordeste de España). *Trab. Geol. Univ. Oviedo*, **11**: 95–127.

Martínez-García, E. C. (1983). El Pérmico de la Cordillera Cantábrica. In *Carbonifero y Pérmico de España*, ed. IGME, pp. 391–402. Madrid: X Congr. Int. Carbonífero.

Massari, F. (1988). Some thoughts on the Permo-Triassic evolution of the South-Alpine area (Italy). *Mem. Soc. Geol. Ital.*, **36**: 179–88.

Mermet, V., Perriaux, J., Tane, J. L. & Toutin-Morin, N. (1988). Le détritisme tardi et post-hercynien des bassins carbonifères, permiens et triasiques du Sud-Est de la France. *Géol. Alp. (Grenoble), Mém.*, hors-sér., **114**: 55–68.

Michard, A. (1976). Eléments de géologie marocaine. *Notes et Mém. Serv. Géol. Maroc*, **252**: 1–405.

Miller, A. K. (1933). Age of the Permian limestones of Sicily. *Am. J. Sci.*, **24**: 409–27.

Montanari, L. & Panzanelli-Fratoni, R. (1988). Confronti tra la Formazione Lercara e la Formazione di Monte Facito. *Atti 74th Congr. Soc. Geol. Ital.*, **B**: 325–8.

Muñoz, M., Ancochea, E., Sagredo, J., Peña, J. A., Hernán, F., Brandle, J. L. & Marfil, R. (1985). Vulcanismo Permo-Carbonífero de la Cordillera Ibérica. *C. R. X. Congr. Int. Strat. Geol. Carbonif., Madrid, 1983*, **3**: 27–52.

Navidad, M. (1983). El vulcanismo Permo-Carbonífero de la Península Ibérica. In *Carbonífero y Pérmico de España*, ed. IGME, pp. 471–82. Madrid: X Congr. Int. Strat. Geol. Carbonif.

Odin, B., Doubinger, J. & Conrad, G. (1986). Attribution des formations détritiques, rouges, du Permien du Sud de la France au Thuringien, d'après l'étude du bassin de Lodève. *C. R. Acad. Sci. Paris*, 302, II, **16**: 1015–20.

Oosterbaan, A. M. (1936). Etude géologique et paléontologique de la Nurra avec quelques notes sur le Permien et le Trias de la Sardaigne méridionale. *Univ. Utrecht*, 130 pp.

Palain, C., Doubinger, J. & Adloff, M. C. (1977). La base du Mésozoïque du Portugal et les problèmes posés par la Stratigraphie du Trias. *Cuad. Geol. Ibérica*, **4**: 269–80.

Pasini, M. (1978). Further paleontological records from the Upper Paleozoic outcrops of the Farma valley. In *Report on the Tuscan Paleozoic Basement*, a cura di Tongiorgi, M., pp. 71–6. Rapp. interno 'Programma Finalizzato Energetica, Sottoprogetto Energia Geotermica'.

Pasini, M. (1980). I Fusulinidi della Valle del Torrente Farma (Toscana Meridionale). *Mem. Soc. Geol. Ital*, **20**: 323–42.

Pasini, M. (1984). Fusulinidi permiani nel Trias medio dell'Appennino Meridonale (Formazione di M. Facito). *Mem. Soc. Geol. Ital.*, **24**: 169–82.

Passeri, L. (1988). Il Trias dell'Unita' di Punta Bianca. *Mem. Soc. Geol. Ital.*, **30**: 105–14.

Peña, J. A., Marfil, R. & Ramos, A. (1979). Desarrollo del magmatismo en el tránsito Paleozoico-Mesozoico de la Cordillera Ibérica: los basaltos de la zona de Ojos Negros (Guadalajara-Teruel). *Estud. Geol.*, **35**: 465–72.

Pérez-Arlucea, M. (1987). Distribución paleogeográfica de las unidades del Pérmico y el Triásico en el sector de Albarracín. *Cuad. Geol. Ibérica*, **11**: 607–22.

Pérez-Arlucea, M. & Sopeña, A. (1983). Estudio estratigráfico y sedimentológico de los materiales pérmicos y triásicos de la Sierra de Albarracín. *Estud. Geol.*, **39**: 329–43.

Pérez-Arlucea, M., & Sopeña, A. (1985). Estratigrafía del Pérmico y Triásico en el sector central de la Rama Castellana de la Cordillera Ibérica. *Estud. Geol.*, **41**: 207–22.

Pittau Demelia, P. & Flaviani, A. (1982). Aspects of the palynostratigra-

phy of the Triassic Sardinian sequences (Preliminary report). *Rev. Palaeobot. Palynol.*, **37**: 329–43.

Ramos, A. (1979). Estratigrafia y paleogeografia del Pérmico y Triásico al Oeste de Molina de Aragón. *Seminarios Estratigr.*, ser. monografias, **6**: 1–313.

Ramos, A. & Doubinger, J. (1979). Découverte d'une microflore thuringienne dans le Buntsandstein de la Cordillera Ibérique (Espagne). *C. R. Acad. Sci. Paris*, **289** (I): 525–8.

Ramos, A. & Doubinger, J. (1991). Datation palinostratigraphique du Buntsandstein (Permien supérieur-Anisien) de l'Ile de Majorque. *C. R. Acad. Sci. Paris*. (In press.)

Ramovš, A. & Kochansky-Devidé, V. (1979). Ladinijske Konglomeratne brece na Ursici in njin permijski ten Triasni microfosili. *Metal. Quart., Ljubljana*, **29**: 155–65.

Rau, A. (1988). Evoluzione del dominio toscano tra il Carbonifero superiore ed il Trias medio: una nuova ipotesi. *Atti 74th Congr. Soc. Geol. Ital.*, **B**: 341–5.

Rau, A. & Tongiorgi, M. (1966). I lamellibranchi triassici del Verrucano dei Monti Pisani. Nuova revisione. *Palaeontographica Ital.*, **61**: 187–234.

Rau, A. & Tongiorgi, M. (1974). Geologia dei Monti Pisani a Sud-Est della Valle del Guappero. *Mem. Soc. Geol. Ital.*, **13**: 227–408.

Rau, A. & Tongiorgi, M. (1981). Some problems regarding the Paleozoic paleogeography in Mediterranean western Europe. *J. Geol.*, **89**: 663–73.

Rau, A. & Tongiorgi, M. (1982). Alcune ipotesi sulla storia pre-Giurassica del futuro margine continentale nord-appenninico. *Mem. Soc. Geol. Ital.*, **21**: 23–31.

Rau, A., Tongiorgi, M. & Martini, I. P. (1988). La successione di Punta Bianca: un esempio di rift 'abortivo' del Trias medio del dominio Toscano. *Mem. Soc. Geol. Ital.*, **30**: 115–25.

Redini, R. (1941). Rinvenimento di Antracolitico nel Gruppo Monticiano-Roccastrada. *Boll. Soc. Geol. Ital.*, **60**: 331–4.

Ricci, C. A. & Serri, G. (1975). Evidenza geochimica sulla diversa affinitá petrogenetica delle rocce basiche comprese nelle serie a facies toscana. *Boll. Soc. Geol. Ital.*, **94**: 1187–98.

Rolando, J. P., Doubinger, J., Bourges, Ph. & Legrand, X. (1988). Identification de l'Autunien supérieur, du Saxonien et du Thuringien inférieur dans le bassin de Saint Affrique (Aveyron, France). Corrélations séquentielles et chronostratigraphiques avec les bassins de Lodève (Hérault) et de Rodez (Aveyron). *C. R. Acad. Sci. Paris*, **307**(II): 1459–64.

Rossell, J., Arribas, J., Elizaga, E. & Gómez, D. (1988). Caracterización sedimentológica y petrográfica de la serie roja Permo-Triásica de la isla de Menorca. *Boll. I. G. M. E.*, **49**: 71–82.

Sarria, E. & Serri, R. (1987). Tettonica compressiva tardo-paleozoica nel bacino antracitifero di Seui (Sardegna Centrale). *Rend. Soc. Geol. Ital.*, **9**: 7–10.

Scandone, P. (1968). Studi di geologia lucana: la serie calcareo-siliceo-marnosa ed i suoi rapporti con l'Appennino calcareo. *Boll. Soc. Nat. Napoli*, **76**: 301–469.

Scandone, P. (1972). Studi di geologia lucana: carta dei terreni della serie calcareo-siliceo-marnosa e note illustrative. *Boll. Soc. Nat. Napoli*, **81**: 225–300.

Simon, O. J. (1987). On the Triassic of the Betic Cordilleras (southern Spain). *Cuad. Geol. Ibérica*, **11**: 385–402.

Skinner, J. W. & Wilde, G. L. (1966). Permian fusulinids from Sicily. *Univ. Kansas Paleont. Contr.*, **8**: 1–16.

Skinner, J. W. & Wilde, G. L. (1967). Permian Foraminifera from Tunisia. *Univ. Kansas Paleont. Contr.*, **30**: 1–22.

Sopeña, A. (1979). Estratigrafía del Pérmico y Triásico del nordeste de la provincia de Guadalajara. *Seminarios Estratigrafia*, ser. monogr., **5**: 1–329.

Sopeña, A., López, J., Arche, A., Pérez-Arlucea, M., Ramos, A., Virgili, C. & Hernando, S. (1988). Permian and Triassic rift basins of the Iberian Peninsula. In *Triassic-Jurassic Rifting*, ed. Manspeizer, W. Amsterdam: Elsevier.

Sopeña, A., Virgili, C., Arche, A., Ramos, A. & Hernando, S. (1983). El Triásico. In *Geologia de España (Libro jubilar J. M. Rios)*, ed. IGME, **2**: 47–62.

Tollman, A. & Krystan-Tollmann, E. (1985). Paleogeography of the European Tethys from Paleozoic to Mesozoic and the Triassic relations of the eastern part of Tethys and Panthalassa. In *The Tethys, Her Paleogeography and Paleobiogeography from Paleozoic to Mesozoic*, eds. Nakazawa, K. & Dickins, J. M., pp. 3–22. Tokyo: Tokai Univ. Press.

Tongiorgi, M. & Bagnoli, G. (1981). Stratigraphie du socle paléozoïque de la bordure Continentale de l'Apennin septentrional (Italie centrale). *Bull. Soc. Géol. France*, (7), **23**: 319–23.

Tongiorgi, M., Rau, A. & Martini, I. P. (1977). Sedimentology of early Alpine, fluvio-marine clastic deposits (Verrucano, Triassic) in the Monti Pisani (Italy). *Sedimentary Geol.*, **17**: 311–52.

Toutin-Morin, N. (1987). Les bassins permiens provençaux, témoins de l'orogenèse hercynienne et de l'ouverture de la Téthys dans le Sud-Est de la France. *Ann. Soc. Géol. Nord.*, **CVI**: 183–7.

Toutin-Morin, N. (1988). Séquelles du cycle hercynien et prémices du cycle alpin dans la sédimentation du Sud-Est méditerranéen. *Géol. Alp. (Grenoble), Mém.*, hors-sér., **14**: 49–54.

Toutin-Morin, N., Delfaud, J. & Marocco, R. (1988). Dynamique des bassins permiens du Sud-Est de la France. Sédimentation, volcanisme, tectonique. *Géol. Alp. (Grenoble), Mém.*, hors-sér., **14**: 29–38.

Toutin-Morin, N., Durand, M., Vinchon, Ch. & Arril, G. (1987). Importance de la discontinuité entre dépôts permiens et triasiques dans le Sud-Est de la France. *Assoc. Géol. Permien, 2è Jour. thématique, résumés*, Paris.

Uncini, G. (1988). Il Verrucano del complesso basale del promontorio dell'Argentario. *Mem. Soc. Geol. Ital.*, **30**: 347–59.

Vai, G. B. (1976). Stratigrafia e paleogeografia ercinica delle Alpi. (Relazione ufficiale). *Mem. Soc. Geol. Ital.*, **13** (suppl. 1): 7–37.

Vai, G. B. (1978). Tentative correlation of Paleozoic rocks, Italian Peninsula and Islands. *Schriftenreihe Erdwiss. Komm. Österr. Akad. Wiss.*, **3**: 313–29.

Vai, G. B., Boriani, A., Rivalenti, G., & Sassi, F. P. (1984). Catena ercinica e Paleozoico nelle Alpi Meridionali. In *Cento anni di Geologia Italiana. Jubilee volume for first century S. G. I.*, pp. 133–54. Bologna.

Vai, G. B. & Francavilla, F. (1974). Nuovo rinvenimento di piante dello Stefaniano a Iano. *Boll. Soc. Geol. Ital.*, **93**: 73–80.

Van Eysinga, F. W. B. (1975). *Geological Time Table*, 3d ed. Amsterdam: Elsevier.

Vanossi, M. (1980). Les unités géologiques des Alpes Maritimes entre l'Ellero et la mer Ligure: un aperçu schématique. *Mem. Sci. Geol. (Padova)*, **34**: 101–42.

Vegas, R. & Banda, E. (1982). Tectonic framework and Alpine evolution of the Iberian Peninsula. *Earth Evol. Sci.*, **4**: 320–43.

Vinchon, Ch. & Toutin-Morin, N. (1987). Convergence de faciès et paléoenvironnements dans les bassins sédimentaires du Permien supérieur de l'Argentera-Barrot et de Provence orientale. *Géol. Alp. (Grenoble), Mém.*, hors-sér., **13**: 57–67.

Virgili, C. (1958). El Triásico de los Catalanides. *Boll. I. G. M. E.*, **69**: 1–856.

Virgili, C. (1987). Problemática del Trias y Pérmico superior del Bloque Ibérico. *Cuad. Geol. Ibérica*, **11**: 39–52.

Virgili, C. (1991). Permian subdivision in the Iberian microplate (western Tethys). *XI Internat. Congr. Strat. Geol. Carboniferous*, Beijing. (In press.)

Virgili, C., Paquet, H. & Millot, G. (1974). Altérations du soubassement de la couverture Permo-Triasique en Espagne. *Bull. Group Franç. Argiles*, **26**: 277–85.

Virgili, C. & Ramos, A. (1983). El Pérmico en España. In *Carbonífero y Pérmico en España*, ed. IGME, pp. 381–502. Madrid. X. Congr. Int. Estrat. Geol. Carbonífero.

Virgili, C., Sopeña, A., Arche, A., Ramos, A. & Hernando, S. (1983a). Some observations on the Triassic of the Iberian Peninsula. *Schriftenreihe Erdwiss. Komm. Österr. Akad. Wiss.*, **5**: 287–94.

Virgili, C., Sopeña, A., Ramos, A., Arche, A., & Hernando, S. (1983b). El relleno posthercínico y el comienzo de la sedimentación mesozoica. In *Geología de España (Libro jubilar J. M. Rios)*, ed. IGME, pp. **2**: 25–36.

Visscher, H. (1971). The Permian and Triassic of the Kingscourt outlier, Ireland. A palynological investigation related to regional stratigraphical problems in the Permian and Triassic of western Europe. *Geol. Surv. Ireland. Spec. Pap.*, **1**: 1–114.

Visscher, H. (1973). The Upper Permian of western Europe – A palynological approach to chronostratigraphy. *Can. Soc. Petrol. Geol., Mem.*, **2**: 200–19.

Visscher, H., Brugman, W. A. & López-Gómez, J. (1982). Nota sobre la presencia de una palinoflora triásica en el supuesto Pérmico del anticlinorio de Cueva del Hierro (Serranía de Cuenca), España. *Rev. Españ. Micropaleont.*, **14**: 315–22.

Wagner, R. H. (1983). The palaeogeographical and age relationships of the Portuguese Carboniferous floras with those of other parts of the western Iberian Peninsula. *Mem. Serv. Geol. Portugal*, **29**: 153–77.

Wagner, R. H. & Lemos De Sousa, M. J. (1983). The Carboniferous megafloras of Portugal. A revision of identifications and discussion of stratigraphic ages. *Mem. Serv. Geol. Portugal*, **29**: 127–52.

Wagner, R. H. & Martínez-García, E. (1982). Description of an Early Permian flora from Asturias and comments on similar occurrences in the Iberian Peninsula. *Trab. Geol. Univ. Oviedo*, **12**: 273–87.

Young, J. R., Teale, C. T. & Brown, P. R. (1986). Revision of the stratigraphy of the Longobucco Group (Liassic, southern Italy), based on new data from nannofossils and ammonites. *Eclogae Geol. Helv.*, **79**: 117–35.

Ziegler, P. A. (1984). Caledonian and Hercynian crustal consolidation of western and central Europe. A working hypothesis. *Geol. en Minjnb.*, pp. 93–108.

8 The Permo-Triassic boundary in the Southern Alps (Italy) and in adjacent Periadriatic regions ·

CARMELA BROGLIO LORIGA AND GIUSEPPE CASSINIS

Introduction

The purpose of this report is to illustrate the characters of the Permo-Triassic boundary in the Southern Alps. This was also the object of an international meeting in July, 1986, which was arranged cooperatively by the Geological Society of Italy and the leadership of IGCP Project 203. Most of the opinions expressed and conclusions reached at that meeting have been published (Cassinis, 1988). We have undertaken a further revision, however, in view of later reflections and subsequent data.

In the following report, the senior author considers the primarily marine deposits between the Adige Valley and Carnia, and compares them with deposits in adjacent Yugoslavia, whereas the junior author reviews Permian continental sequences of the South-Alpine region, mainly west of the sector discussed by Broglio Loriga, and considers their contact with overlying units.

The marine Upper Permian and the Permo-Triassic boundary

Eastern Southern Alps

The Middle(?) and Upper Permian of the Dolomites and Carnia consists of two lithostratigraphic units, the Val Gardena Sandstone and the Bellerophon Formation, which taken together may be regarded as a composite transgressive sequence. The Val Gardena Sandstone consists mainly of terrigenous clastics, in part typical red beds, which were deposited in a semi-arid setting. The Bellerophon Formation is composed mostly of chemical sediments (carbonates and sulphates) and its facies associations suggest a wide spectrum of depositional environments, ranging from a coastal sabkha to shallow-shelf settings. The Bellerophon Formation wedges out westward and disappears west of the Adige Valley by gradation into continental deposits (Figs 8.1, 8.2).

The Val Gardena Sandstone and Bellerophon Formation are partially intergradational laterally. That is, internal subdivision by second-order transgressive-regressive cycles results in a mutual intertonguing of terrigenous, carbonate, and evaporitic sulphate units.

In the central western Dolomites, the Val Gardena–Bellerophon sequence unconformably overlies Lower Permian volcanics (Bolzano volcanic complex) or, where these are lacking, it rests directly on the Hercynian crystalline basement (Figs 8.1, 8.2). In the northeastern Dolomites, the lower part of the sequence consists of thick fanglomerates (Sesto Conglomerate or Grodener Konglomerat), and in the Carnic Alps, it overlies the strongly deformed Hercynian sequence or lies unconformably on late-orogenic Permo-Carboniferous sediments, of which the youngest is the marine Trogkofel Limestone (Artinskian). In the Carnic Alps, the basal contact is everywhere sharp and marked by the Tarvisio Breccia, a chronologic equivalent of the Sesto Conglomerate.

The Permian Val Gardena–Bellerophon sequence is overlain by the shallow-marine Lower Triassic Werfen Formation, the basal deposits of which include two laterally equivalent facies, the oolitic Tesero Horizon in the central western Dolomites and the marly, micritic Mazzin Member in the Cadore-Comelico district and Carnia.

Val Gardena Sandstone

The Val Gardena Sandstone is absent in the Trento area, but thickens eastward to more than 500 m in the Comelico area. Irregular thickening and thinning over short distances suggests that sedimentation took place in a markedly differentiated topographic setting that was repeatedly rejuvenated by synsedimentary tectonics. A stratigraphic gap of varying magnitude may be recognized at the base of the formation (Figs 8.1, 8.2).

The Val Gardena Sandstone is dominated by red and gray conglomerates, sandstones and mudstones. Chemical sediments of Bellerophon facies separate a number of the main terrigenous bodies. Clastic facies represent former alluvial fans, braided stream deposits, point-bar sequences, overbank deposits, and the sandy terminal lobes of fans, and grade into coastal sabkha sequences. Continental sabkha episodes are recorded at some places in the lower part of the formation; and semi-arid climatic conditions are suggested by pedogenic calcretes, local pedogenic gypsum, and by the character of channel-fill sequences. Source terrains were the underlying Permian volcanic rocks, the pre-Permian crystalline basement, and granitic rocks.

From a paleontological point of view, the most interesting exposure of the Val Gardena Sandstone is in the Butterloch-Bletterbach Gorge of the western Dolomites (Fig. 8.3), where a marine bed a few meters thick occurs in the lower part of the sequence. That bed, which yields an Orthoconic Nautiloid Assemblage, is overlain by a continental red terrigenous unit in which the famous footprint beds occur. In addition, a more widely distributed marine fauna has been recognized in the upper Val Gardena and at the contact with the overlying Bellerophon Formation (Glomospirid Assemblage in silty dolomite). The latter horizon marks the slow transition from continental, through lagoonal to peritidal conditions (Fig. 8.3).

The Orthoconic Nautiloid Assemblage contains representatives of *Neocycloceras*, *Lopingoceras*, *Tainoceras*, ?*Metacoceras*, *Mojsvaroceras*, *Pleuronautilus*, *Thuringionautilus*, *Germanonautilus* or *Stearoceras*, ?*Permonautilus*, and ?*Parapronorites* (Figs 8.3, 8.4). The nautiloids, as well as the tetrapod footprints in beds very close to the marine bed, suggest a general Late Permian age. According to Pittau (in Italian Research Group, 1986) an early Dzhulfian age is indicated by palynomorphs of the *Playfordiaspora crenulata–Protohaploxypinus microcarpus* Zone (Fig. 8.3).

Glomospira-like foraminifers, *Ammodiscus*, *Agathammina*, ostracodes, and the bivalves *Permophorus* sp. and *Aviculopecten* sp. have been identified in collections from the Glomospirid Assemblage (Fig. 8.4). According to palynomorphs from the same beds, the age is Dzhulfian (Pittau, in Italian Research Group, 1986).

Bellerophon Formation

The Bellerophon Formation ranges in thickness from zero in the westernmost part of the eastern sector of the Southern Alps to 400–500 m at the depocenter in the Cadore area (Figs 8.1*B*, 8.2).

At its base the formation interfingers with underlying clastic, carbonate, and evaporitic rocks and, within the unit, a second-order cyclicity is recognizable. Each major cycle consists of a transgressive sequence of sulphate-evaporitic, dolomitic, and fossiliferous carbonate rocks. Some elements of this general sequence may be absent from individual cycles, reduced in thickness, or replaced by marginal facies, but the cyclicity is everywhere recognizable. In the western Dolomites, black fossiliferous limestones, the shallow-shelf Badiota facies of authors, are dominant in the upper part of the formation. To the east, in Cadore and southwest Carnia, this facies is commonly thicker and also occurs in the lower part of the Bellerophon Formation.

The topmost segment of the Bellerophon Formation consists of a sequence, 0.5 to 2 m thick, of dark marls and limestones with fusulinids and other foraminifers, calcareous algae, and brachiopods (*Comelicania*, *Ombonia*). This facies records the spread of shallow-marine conditions over a very wide area of the Southern Alps and also into the marginal basin previously dominated by restricted lagoonal and sabkha conditions (Fiammazza facies; Fig. 8.4).

The Bellerophon Formation is overlain, with no evidence of a stratigraphic gap, by oolitic units of the Tesero Horizon of the Werfen Formation (Fig. 8.2). This wedge-shaped unit is about 25 m thick in the Trento area, but thins gradually eastward and northward and in Cadore and western Carnia is replaced laterally by marly micritic sediments included in the Mazzin

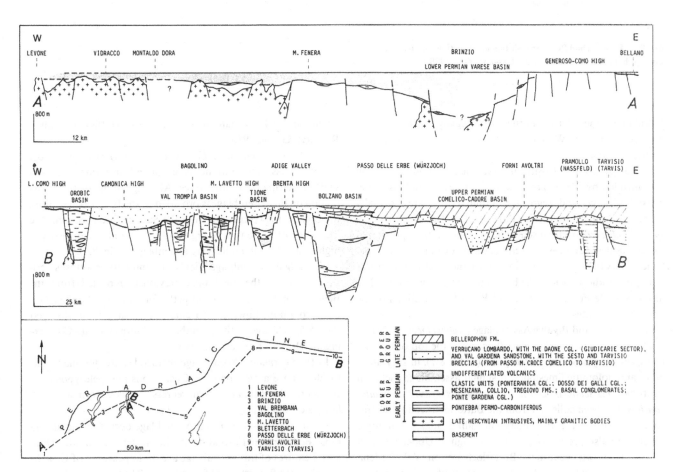

Fig. 8.1. Schematic nonpalinspastic section through the Permian of the Southern Alps, before deposition of overlying sedimentary units. (After Cassinis *et al.*, 1988.)

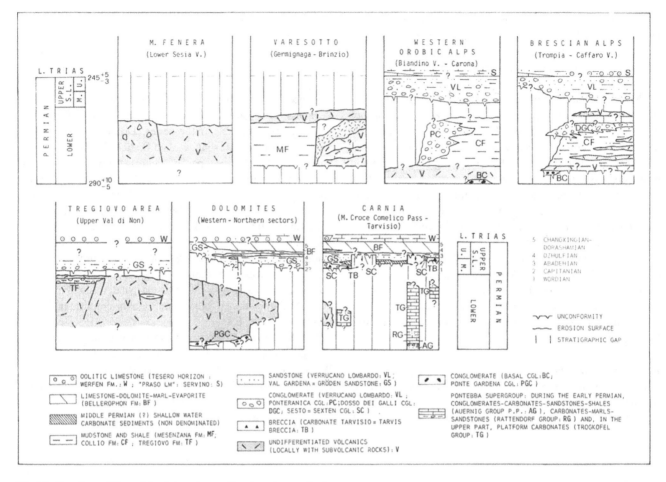

Fig. 8.2. Permian and Permotriassic boundary sections in various parts of the Southern Alps. (Modified from Cassinis et al., 1988.)

Member of the Werfen Formation. Regionally, these basal members of the Werfen interfinger and are completely intergradational.

The distribution of fossils within the Bellerophon Formation is controlled by facies. Those representing restricted lagoonal, coastal sabkha, and regressive terminal fans are normally barren, although rare fossils (Bakevellidae; *Earlandia, Glomospira, Cyclogyra*) occur at some places in the subtidal portion of sabkha cycles. Rich fossil assemblages occur in open-lagoonal facies below the middle regression, and in the black limestones of the upper badiota facies. The biostratigraphy is most complete in the Sass de Putia (Fig. 8.5) and Russisbach sections of the western Dolomites.

A Nautiloid and Bivalve Assemblage (*Neocycloceras, Tirolonautilus crux, Towapteria wahneri*) occurs at the top of the first transgresive cycle of the Bellerophon (Table 8.1), in silty dolomites and dolomitic limestones. A similar assemblage, richer in nautiloids (?*Pseudorthoceras, Tirolonautilus crux, ?Germanonautilus*) has been collected from the Butterloch section. Ostracodes, algae, and foraminifers, including some stenotopic genera, are also present (Fig. 8.3). The highest Upper Permian footprints known occur in the Butterloch section (Val Gardena sandstone tongue) (Fig. 8.3) and represent *Pachypes dolomiticus* and *Rhyncosauroides pallinii*. The age of this assemblage is Late

Dzhulfian or Dorashamian–Changxingian (Pittau, in Italian Research Group 1986).

An Algal Assemblage (Table 8.2), the richest and most diversified association in the Bellerophon Formation, occurs in shallow-marine dark limestones. Although some of the information on this assemblage is provided by the Butterloch section, the bulk of the data come from the Sass de Putia section. The frequency of algae is very high and the taxonomic diversity is higher than that in underlying assemblages (Fig. 8.5).

The Algal Assemblage yields foraminifers, mollusks, crinoids, and all the algal taxa previously reported from the Bellerophon Formation. Among the foraminifers, *Nankinella* is quite common; however, no colaniellids have been observed (Fig. 8.5). The age of the assemblage is Dorashamian–Changxingian (Broglio Loriga et al., 1988).

In the Ostracode Assemblage ostracodes are dominant and fusulinids are absent or rare. Gastropods and pelecypods are common in beds that represent regressive-phase storm accumulations on tidal flats.

A *Comelicania–Nankinella* Assemblage occurs in the uppermost Bellerophon Formation in the western Dolomites (Fig. 8.4). Features of this assemblage are the appearance of *Comelicania* and the flourishing of foraminifers. The brachiopods occur about 1.5 m below the top of the formation. Three main

Table 8.1. *Macrofossils of the Nautiloid-Bivalve Assemblage (Bellerophon Formation) from Russisbach, near Sass de Putia, and Butterloch–Bletterbach sections*

		R	B
gastropods	Bellerophon sp.	*	
	Trachynerita sp.	*	*
	? Natiria sp.		*
	Naticopsis pusiuncola (STACHE)	*	
	'Loxonema' tortile MERLA		*
bivalves	Towapteria wahneri (KITTL)	*	
	Permophorus jacobi (STACHE)	*	*
	Sanguinolites bellerophontianum KITTL	*	
	Edmondia dubia (WAAGEN)	*	*
	Aviculopecten sp.	*	*
nautiloids	Tirolonautilus crux (STACHE)	*	*
	Metacoceras sp.	*	*
	? Germanonautilus sp.or Stearoceras sp.		*
	? Neocycloceras sp.	*	
	? Pseudorthoceras sp.		*

Source: (From Broglio Loriga *et al.*, 1988.)

Table 8.2. *Macrofossils of the Algal Assemblage (Bellerophon Formation) from Sass de Putia*

gastropods	Bellerophon ulrici STACHE
	Naticopsis depressa (STACHE)
	Pseudozygopleura sp.
	Omphaloptycha sp.
	Trachynerita sp.
bivalves	Aviculopecten comelicanus (STACHE)
	Pernopecten tirolensis (STACHE)
	Towapteria wahneri (KITTL)
	Permophorus sp.
nautiloids	Tirolonautilus sp.
	Tainionautilus sp.
	? Peripetoceras wanneri (HANIEL)

Source: (From Broglio Loriga *et al.*, 1988.)

morphogroups have been recognized: *C. ladina* (Stache), *C. megalotis* (Stache), and *C. haueri* (Stache), which is represented by specimens 12–15 cm wide. Other brachiopods include *Janiceps*, ?*Araxathyris*, *Orthotetina* and *Ombonia*. In the top bed of the Bellerophon Formation *Comelicania* appears to be absent, whereas *Ombonia* and *Orthotetina* become more abundant. The same vertical range is reconigzable at different localities in the Dolomites (Sass de Putia, Agordo, Avoscan) where dark limestones of the Badiota facies are thicker. In parts of the western Dolomites farther southwest, the Badiota lithofacies is thinner and only *Ombonia* and *Orthotetina* are represented. Farther west, in the Adige Valley, aviculopectinid bivalves increase in number and become the dominant fossils and brachiopods are lacking (Posenato, 1988b).

The microfossil population of the *Comelicania* and *Nankinella* Assemblage is dominated by foraminifers, among which the fusulinid *Nankinella* is commonest (Fig. 8.5). Colaniellids are rare and small, and specimens of *Paraglobivalvulina* and *Paraglobivalvulinoides* exhibit almost aberrant form and are smaller in average size than in other Tethyan regions.

The lithologic unit that yields the *Comelicania* and *Nankinella* Assemblage is well recognizable in the southwest part of the western Dolomites (Butterloch, Tesero; Figs 8.3, 8.6) and in Carnia (Fig. 8.7), where *Comelicania* is generally lacking or rare. At the top of the unit in Carnia, the conodont *Hindeodus typicalis* appears (Perri & Andraghetti, 1987).

Remains of the fungus *Tympanocysta* flourish in the uppermost segment of the Bellerophon Formation, in an interval corresponding to that of the *Comelicania-Nankinella* Assemblage. This 'fungal event' is widespread geographically and *Tympanocysta* is of inter-regional importance. Its high fre-

quency in the Permo-Triassic boundary interval is a striking event and its dominance at the top of the Bellerophon Formation and in the basal Werfen Formation is unique (Visscher & Brugman, 1988, Fig. 1).

Age of the *Comelicania–Nankinella* Assemblage is Dorashamian–Changxingian. We will discuss this matter further in a later segment of our report.

Werfen Formation

The Lower Triassic Werfen Formation (= Servino Formation in Lombardy) is an interbedded sequence of carbonate and terrigenous deposits that ranges in thickness from about 150 m in Lombardy to 600–700 m in Cadore and, in the Dolomites, includes nine lithostratigraphic units. From west to east, the formation overlies the Verrucano Lombardo (or Val Gardena Sandstone), the lagoonal, sabkha, and finally the shallow-shelf facies of the Bellerophon Formation (Figs 8.1, 8.2). East of the Adige Valley, at least, the basal contact may be regarded as conformable.

In eastern Lombardy, the Adige Valley, and the central western Dolomites there is an oolitic unit, the Tesero Horizon, at the base of the Werfen Formation. The Tesero wedges out eastward and disappears in the Cadore-Comelico area (Italian Research Group, 1986; Broglio Loriga *et al.*, 1983, 1988).

Most of the Werfen Formation was deposited in a shallow-shelf setting that was generally below fair-weather wave-base but was frequently affected by storm waves. This resulted in a closely spaced alternation of 'muddy' lithotypes (e.g., marls, marly limestones, fine-grained siltstones) and 'sandy' lithotypes (e.g. oolitic-bioclastic-intraclastic calcarenites, calcirudites, arenaceous limestones, sandstones).

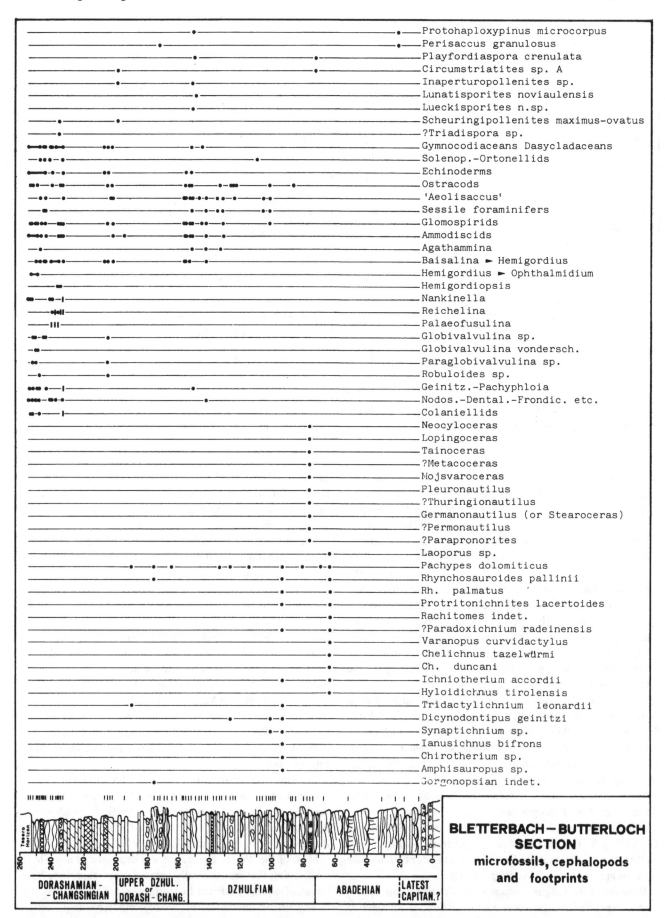

Fig. 8.3. Distribution of microfossils, cephalopods, and tetrapod footprints in Bletterbach–Butterloch section. Only zone-marking palynomorphs included. Bellerophon Formation cephalopods not identified specifically; lithostratigraphy and identifications of foraminifers and palynomorphs from Conti *et al.*, in Italian Research Group, 1986, and Broglio Loriga *et al.*, 1988; footprints from Ceoloni *et al.*, 1988.)

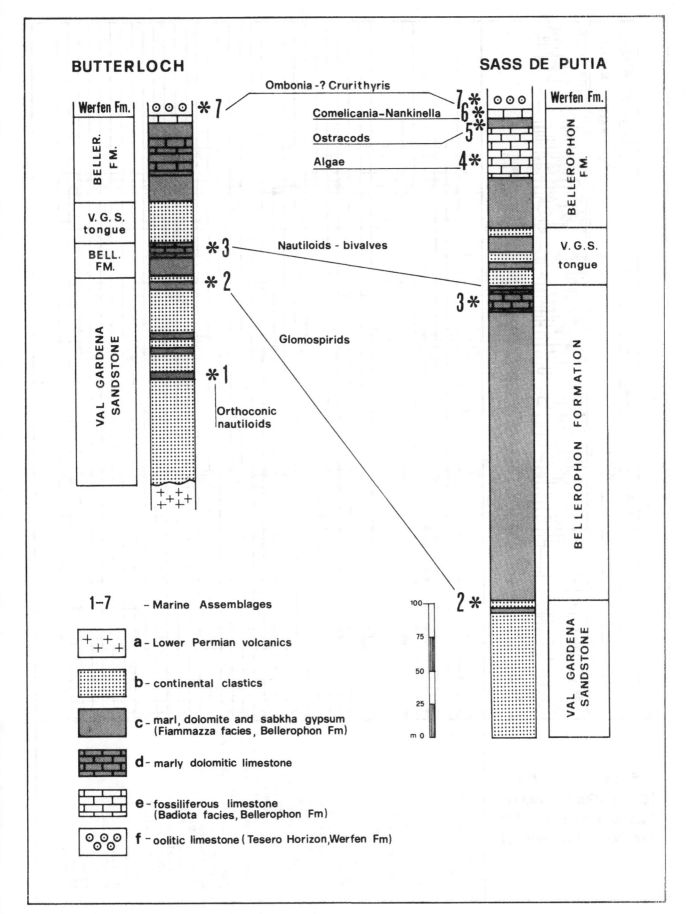

Fig. 8.4. Positions and correlation of marine fossil assemblages in two sections of the western Dolomites (Bolzano Province). (From Broglio Loriga *et al.*, 1988.)

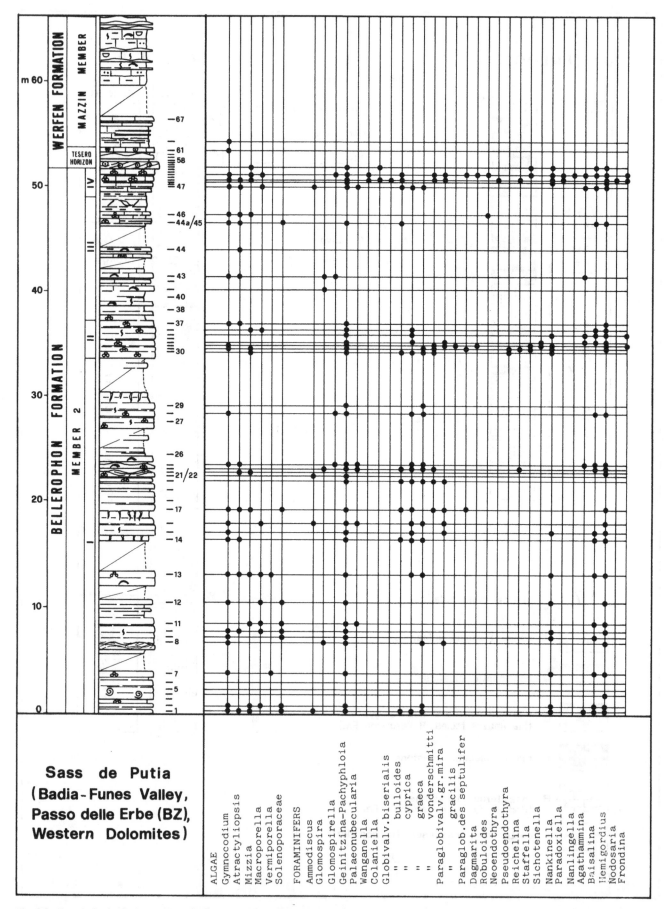

Fig. 8.5. Sass de Putia section, with vertical ranges of foraminifers and algae. (Data from Neri, Posenato, and Broglio Loriga in Italian Research Group, 1986, and Broglio Loriga *et al.*, 1988.)

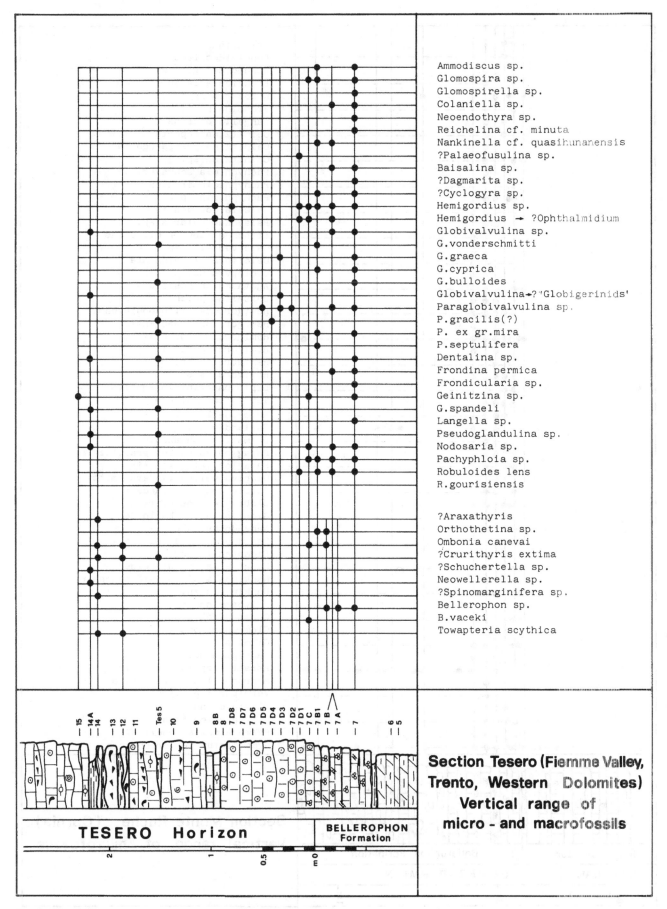

Fig. 8.6. Vertical ranges of microfossils, brachiopods and mollusks in Tesero section (Fiemme Valley, Trento). (Data from Italian Research Group, 1986; Broglio Loriga *et al.*, 1988; Posenato, 1988a.)

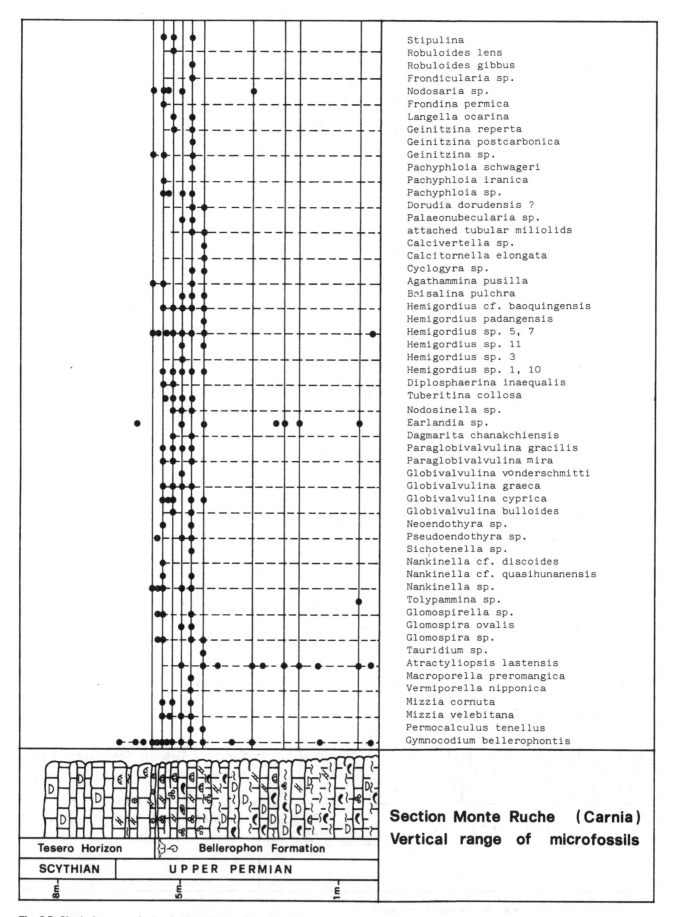

Stipulina
Robuloides lens
Robuloides gibbus
Frondicularia sp.
Nodosaria sp.
Frondina permica
Langella ocarina
Geinitzina reperta
Geinitzina postcarbonica
Geinitzina sp.
Pachyphloia schwageri
Pachyphloia iranica
Pachyphloia sp.
Dorudia dorudensis ?
Palaeonubecularia sp.
attached tubular miliolids
Calcivertella sp.
Calcitornella elongata
Cyclogyra sp.
Agathammina pusilla
Baisalina pulchra
Hemigordius cf. baoquingensis
Hemigordius padangensis
Hemigordius sp. 5, 7
Hemigordius sp. 11
Hemigordius sp. 3
Hemigordius sp. 1, 10
Diplosphaerina inaequalis
Tuberitina collosa
Nodosinella sp.
Earlandia sp.
Dagmarita chanakchiensis
Paraglobivalvulina gracilis
Paraglobivalvulina mira
Globivalvulina vonderschmitti
Globivalvulina graeca
Globivalvulina cyprica
Globivalvulina bulloides
Neoendothyra sp.
Pseudoendothyra sp.
Sichotenella sp.
Nankinella cf. discoides
Nankinella cf. quasihunanensis
Nankinella sp.
Tolypammina sp.
Glomospirella sp.
Glomospira ovalis
Glomospira sp.
Tauridium sp.
Atractyliopsis lastensis
Macroporella preromangica
Vermiporella nipponica
Mizzia cornuta
Mizzia velebitana
Permocalculus tenellus
Gymnocodium bellerophontis

Section Monte Ruche (Carnia)
Vertical range of microfossils

Tesero Horizon   Bellerophon Formation

SCYTHIAN   UPPER PERMIAN

8m   5m   1m

Fig. 8.7. Vertical ranges of microfossils in M. Ruche section, Carnia.
(From Noé, 1987).

Table 8.3. *Macrofossils of the* Ombonia–?Crurithyris *Assemblage (Tesero Horizon T14) and* Unionites–Lingula *Assemblage (Mazzin Member) Tesero section, Lower Triassic*

	TESERO Horizon		MAZZIN Member	
	? *Crurithyris* and *Ombonia* Ass. (faunal list from layer T-14)		*Unionites* and *Lingula* Ass. (faunal list from layer T-37)	
brachiopods	? ·*Crurithyris* *extima* GRANT	(ff)	*Lingula* sp.	(rr)
	Ombonia cf.*canevai* MERLA	(f)		
	? *Spinomarginifera* sp.	(rr)		
	? *Araxathyris* sp.	(r)		
bivalves	*Towapteria* cf.*scythica* (WIRTH)	(f)	*Unionites* *fassaensis* (WISSMANN)	(r)
	Aviculopectinds	(r)	*U.* cf. *canalensis* (CATULLO)	(ff)
			Aviculopectinids	(r)
gastropods	*Bellerophon* *vaceki* BITTNER	(r)	*Bellerophon* *vaceki* BITTNER	(ff)
	Poligyrina *gracilior* (SCHAUROTH)	(r)	*Coelostylina* *werfensis* WITTENBURG	(f)
			Coelostylina sp.	(f)
			Neritaria sp.	(r)
annelids	*Spirorbis* *valvata* (GOLDFUSS)	(f)	*Spirorbis* *valvata* (GOLDFUSS)	(r)
echinoids	? *Miocidaris* sp.	(f)		

Source: (From Broglio Loriga *et al.*, 1988.)

'Sandy' lithotypes occur as centimeter to decimeter thick storm layers, commonly with an erosional base, graded bedding and even and hummocky lamination. Wave ripples are common on the tops of beds. The Mazzin, Siusi, Gastropod Oolite, Campil, and Val Badia members are characterized by such sedimentary facies and differ from one another in mud:sand ratio, the degree of carbonate or terrigenous composition, and in degree of bioturbation.

The Cencenighe Member consists mainly of oolitic-bioclastic calcarenites, which represent littoral bars and shoals. The calcarenite bodies were controlled in thickness and distribution by both waves and tidal currents.

Shallow-marine Werfen deposits are here and there interrupted by strata that indicate deposition on supratidal mud flats of regional extent. Such rocks make up the Andraz Horizon, the uppermost Siusi Member, the uppermost Campil Member, and the S. Lucano Member. The occurrence of these almost isochronous units, as well as the distribution of terrigenous clastic bodies within the sequence, makes it possible to subdivide the Werfen Formation into major transgressive-regressive cycles, which were probably controlled by both eustatic oscillations and tectonics (Broglio Loriga *et al.*, 1983; Farabegoli & Viel, 1982).

Werfen fossil associations are primarily low-diversity assemblages of benthic taxa. Mega- and microfossils of Permian aspect have been collected from the Tesero Horizon. Ammonites occur only in the Val Badia and Cencenighe members (Broglio Loriga *et al.*, 1983); conodonts are rare (Staesche, 1964; Assereto

et al., 1973; Perri & Andraghetti, 1987), but range through much of the formation; and a few palynomorphs have been reported (Visscher & Brugman, 1988).

A biostratigraphy based on brachiopods and the bivalves *Claraia*, *Eumorphotis*, and *Costatoria* has been proposed for the Werfen (Broglio Loriga *et al.*, 1983, 1986; Italian Research Group, 1986, p. 33). In addition, Sweet (1988) has used the distribution of the conodonts reported by Staesche (1964) to effect a graphic correlation of the Werfen with Lower Triassic strata in Pakistan, Kashmir, Japan, Primorye, and the western United States. Here we are concerned primarily with fossil assemblages that bear on placement of the Permo-Triassic boundary.

The *Ombonia–?Crurithyris* Assemblage (Table 8.3) (in the *Lingula* Zone of Broglio Loriga *et al.*, 1983) characterizes the lowermost Tesero Horizon, the basal oolitic unit of the Werfen Formation. This layer, 10–20 cm thick, yields smooth pectinid bivalves, and brachiopods and foraminifers of Permian aspect (Fig. 8.6; Broglio Loriga *et al.*, 1988; Posenato, 1988a). In marly intercalations 0.5 to 2 m above the formational boundary there are sculptured aviculopectinid bivalves, '*Bellerophon*' *vaceki*, *Polygirina gracilior*, ?*Crurithyris extima*, *Ombonia* sp. cf. *O. canevai*, ?*Spinomarginifera* sp., and foraminifers of Permian type (i.e. *Nankinella*, *Globivalvulina*, *Hemigordius*, *Geinitzina*). *Gymnocodium* and *Mizzia* are rarely present. '*B.*' *vaceki* and *P. gracilior* are members of the Lower Triassic fauna. The conodont *Hindeodus latidentatus* is represented 2 m above the

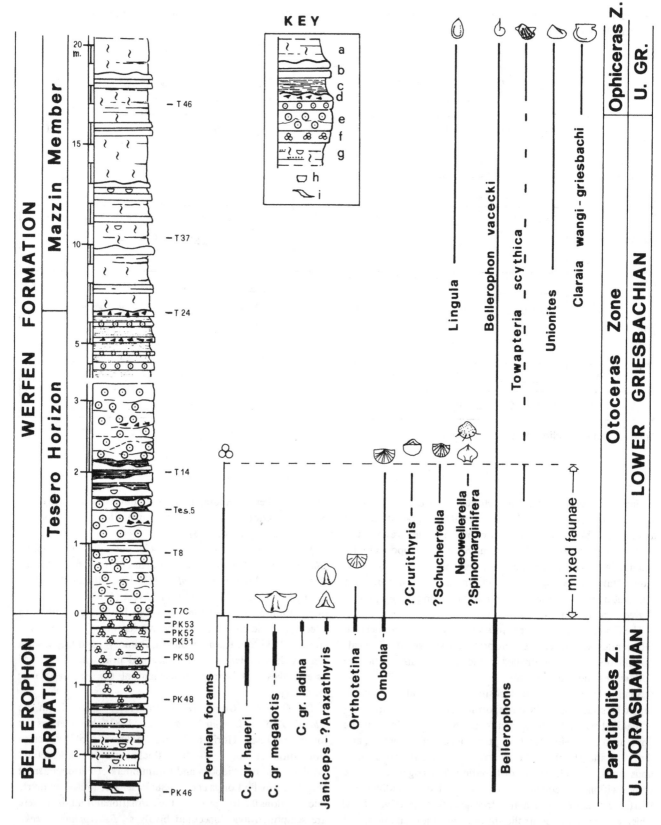

Fig. 8.8. Distribution of stratigraphically significant fossils and groups of fossils in uppermost Bellerophon Formation and lowermost Werfen Formation, Southern Alps. (From Broglio Loriga et al., 1986; Italian Research Group, 1986; Posenato, 1988a; modified.)

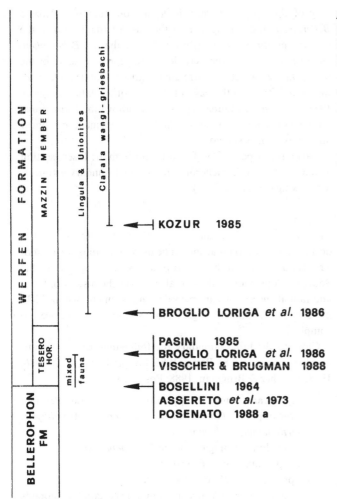

Fig. 8.9. Some proposals for placement of the Permo-Triassic boundary in the Southern Alps. (Modified from Posenato, 1988a.)

Bellerophon–Werfen contact. The mixed faunas have been reported previously (Neri & Pasini, 1985; Broglio Loriga *et al.*, 1988; Posenato, 1988a).

In the Tesero section the highest occurrence of a palynomorph assemblage (*Veryhachium/Micrhystridium*, Striatiti sp. div., *Klausipollenites schaubergeri*) with dominant fungal remains, coincides with the highest record of Late Permian elements. This event is also recognizable in Hungary and Israel (Visscher & Brugman, 1988; Góczán, Haas & Oravecz-Scheffer, 1988; Haas *et al.*, 1988; Hirsch & Weissbrod, 1988; Eshet, this volume).

The *Ombonia–?Crurithyris* Assemblage is thought to be Dorashamian–Changxingian, or earliest Griesbachian in age (Broglio Loriga *et al.*, 1986).

Next above the *Ombonia–?Crurithyris* Assemblage, in the lower Mazzin Member, is the *Unionites–Lingula* Assemblage (Table 8.3), which is of lower taxonomic diversity and quite different composition than assemblages in the Tesero Horizon and Bellerophon Formation. *Lingula* sp., *Unionites fassaensis*, *Unionites* sp. cf. *U. canalensis*, and '*Bellerophon*' *vaceki* are the most characteristic fossils; articulate brachiopods, algae, and foraminifers of Permian aspect appear to be absent; and the conodont *Isarcicella? parva* has been reported (as *Hindeodus*

parvus) from the lower Mazzin (Brandner *et al.*, 1986). The palynofacies appears 'to be dramatically changed' from the lower part of the Mazzin Member. The organic content consists largely of extreme abundances of marine acritarchs (*Scythiana, Veryhachium/Micrhystridium*; Visscher & Brugman, 1988). A Griesbachian age has been assigned by Assereto *et al.* (1973), Broglio Loriga *et al.* (1983), and Noé (1987).

The index bivalve *Claraia* appears 15 to 20 m above the base of the Werfen (Fig. 8.8), at a level roughly equivalent to that at which the conodont *Isarcicella isarcica* makes its debut (Assereto *et al.*, 1973).

Permo-Triassic boundary

The *Comelicania–Nankinella* Assemblage from the uppermost Bellerophon Formation and the *Ombonia–?Crurithyris* Assemblage from the lowermost Werfen Formation are involved in locating and defining the Permian/Triassic boundary in the Southern Alps (Fig. 8.8). However, before discussing this problem it should be pointed out that opinions differ as to the chronostratigraphic position of the boundary between the Bellerophon and Werfen formations (Fig. 8.9). The traditional view (Assereto *et al.*, 1973) holds that the formational contact is unconformable and that a gap representing early Dorashamian through earliest Griesbachian time occurs between the two formations. On the other hand, another group of authors (Pasini, 1985; Broglio Loriga *et al.*, 1986, 1988) believes that the two formations are transitional in lithology and paleontology and that no time gap is recorded.

As a result of detailed biostratigraphic and sedimentologic research interpretation of the Bellerophon–Werfen boundary has changed greatly in the last three years. A short history of the evolution of the boundary concept will help to explain the problem.

1964

The lithic unit known as the Tesero Horizon is recognized in the basal Werfen Formation (Bosellini, 1964). Permian foraminifers in the unit are considered to be reworked and the Permo-Triassic boundary is drawn just below it based on lithic features.

1973

Comelicania beds of the Bellerophon Formation are correlated with the *Comelicania–Phisonites* beds of Transcaucasia (Armenia and northwest Iran), regarded as uppermost Dzhulfian (*sensu* Ruzhentsev & Sarytcheva, 1965) or lower Dorashamian in age (Rostovtsev & Azaryan, 1973). Thus the black limestones with *Comelicania* (the *Comelicania–Nankinella* Assemblage of this report) are considered to be early Dorashamian in age. A gap representing the late Dorashamian and earliest Griesbachian is thus thought to separate the Bellerophon and Werfen formations (Assereto *et al.*, 1973).

1985

Tesero foraminifers are determined not to be reworked (Pasini, 1985) and it is shown that there is no difference in composition between microfaunal associations in the upper Bellerophon Formation and lowermost Werfen Formation. Microfossils in this interval indicate correlation with the *Palaeofusulina–Colaniella parva* Zone of authors; a Dorashamian–Changxingian age is inferred for the lithic interval including the formational boundary; the conodont *Hindeodus latidentatus* is discovered 2 m above the formational boundary; and research suggests there is no gap between the Bellerophon and Werfen formations.

1986–88

Research on macrofossils outlines the succession of Permian assemblages (Broglio Loriga *et al.*, 1986; 1988). The Scythian mollusk '*Bellerophon*' *vaceki* is associated with Permian brachiopods in the Tesero Horizon (Broglio Loriga *et al.*, 1986), hence the occurrence of a mixed fauna, first detected in 1985 by Neri & Pasini, is confirmed. Correlation with the mixed faunas of South China and Pakistan suggests that the Permo-Triassic boundary might lie within the Tesero Horizon (Broglio Loriga *et al.*, 1986; Noé, 1987) and the concept of a paleontologically defined boundary separate from a lithologic one arises.

Also at this time sedimentologic research emphasizes a gradual transition from the Bellerophon Formation to the Werfen Formation. Bioclastic packstones and grainstones with the *Comelicania–Nankinella* Assemblage coarsen and grade into oolitic grainstones of the overlying Tesero Horizon, the basal unit of the Werfen, in which the *Ombonia–?Crurithyris* Assemblage occurs. Evidence of erosion and clear indications of emergence supporting a gap occur only in the middle part of the Southern Alps. Several authors, from different countries, emphasize the gradual lithic transition (Italian Research Group, 1986; Noé, 1987; Brandner *et al.*, 1986; Broglio Loriga *et al.*, 1986, 1988). This conclusion raises a question as to the chronostratigraphic value of *Comelicania* at the generic level.

Extensive research emphasizes the peculiarity of the Bellerophon Formation's microfauna by comparison with those reported from other Tethyan areas, and the present-day Southern Alps is thought to represent a distinct bioprovince, which lacks the common index microfossils but has similarities with associations from Changxingian sequences in South China (Broglio Loriga *et al.*, 1986, 1988).

Palynomorphs are described from strata near the Bellerophon–Werfen boundary. A change in floral associations is recognized in the Tesero Horizon, 1.5 to 2 m above the formational boundary and mixed-fauna beds (Visscher & Brugman, 1988).

In 1987, at the final conference of IGCP Project 203, in Beijing, Broglio Loriga emphasized that, within the Bellerophon-Werfen sequence, the drop in taxonomic diversity coincides with the formational boundary and is easily recognizable in the field.

Representatives of *Comelicania* from the Southern Alps are revised at the species level and the stratigraphic ranges of mollusks in the boundary interval are re-examined (Posenato, 1988a, 1988b). Transcaucasian comelicaniids are referable to *Gruntallina*, whereas those from the Southern Alps belong in *Comelicania* and *Alatothyris*. A characteristic comelicaniid assemblage (*C. megalotis*, *C. ladina*) seems to support a latest Dorashamian age for the uppermost part of the Bellerophon Formation.

'*Bellerophon*' *vaceki* is a Triassic species whose first occurrence is in the basal oolitic layers of the Tesero Horizon. The morphology of this species is very different from that of Paleozoic *Bellerophon* and it ranges from the base of the Tesero to the *Claraia aurita* Subzone (Lower Nammalian). '*Bellerophon*' *vaceki* is considered the heralder of the Early Triassic in the mixed fauna, which is correlated with that of the *Otoceras woodwardi* Zone on the basis of this and other data (Posenato, 1988a). The mixed fauna occurs 1–2 m above the formational boundary (Figs 8.6, 8.8); above this Permian elements disappear and Triassic ones spread.

Carbon-isotope profiles (Fig. 8.10) through the Tesero section also indicate that the Bellerophon–Werfen boundary is transitional (Magaritz *et al.*, 1988).

Criteria

From the historical summary just outlined it is clear we need to decide on the criteria that should be used in defining the Permo-Triassic boundary in the Bellerophon–Werfen sequence of the Southern Alps. Obviously several events in the history of floral and faunal successions are recorded in a segment 3–4 m thick, but changes in sedimentologic history seem not to have been so complex.

Close to the lithic boundary the following occur, in stratigraphic order (Fig. 8.8) (Pasini, 1985; Broglio Loriga *et al.*, 1986, 1988; Visscher & Brugman, 1988; Posenato, 1988a):

1 Appearance and disappearance of several Permian brachiopods (*Comelicania*, *Janiceps*, *?Araxathyris*, *Orthotetina*, *Ombonia*);
2 flourishing of fungal remains (*Tympanocysta*);
3 drop in taxonomic diversity;
4 appearance of Werfen-like fossils;
5 disappearance of fungal remains and certain brachiopods (*Orthotetina*, *Ombonia*);
6 appearance and disappearance of other Permian-like brachiopods and the appearance and development of Triassic acritarchs 1–2 m above the base of the Werfen Formation; and disappearance in this interval of Permian foraminifers.

Thus, at present, the following proposals have been made with respect to placement of the systemic boundary:

1 The drop in taxonomic diversity is thought to represent the acme of the biologic crisis and to correspond to a clear sedimentologic event in which hydrodynamic energy increased and oolite sedimentation began. Both events are recognizable anywhere the marine sequence crops out, hence may easily be observed by the field geologist. In this case, the Bellerophon Formation–Werfen Formation contact becomes the systemic boundary.
2 The co-occurrence of fungal remains and of the *Veryhachium/Micrystridium* complex coincides with the highest Permian brachiopods and foraminifers from the *Ombonia–?Crurithyris* Assemblage (Figs. 8.4, 8.8). In this case, the Permo-Triassic boundary lies within the Tesero Horizon. Co-occurrence of these events has been identified at an analogous level in the transitional Permo-Triassic sequence of the Transdanubian Central Range, in Hungary, and in Greenland (Góczán *et al.*, 1988; Visscher & Brugman, 1988).

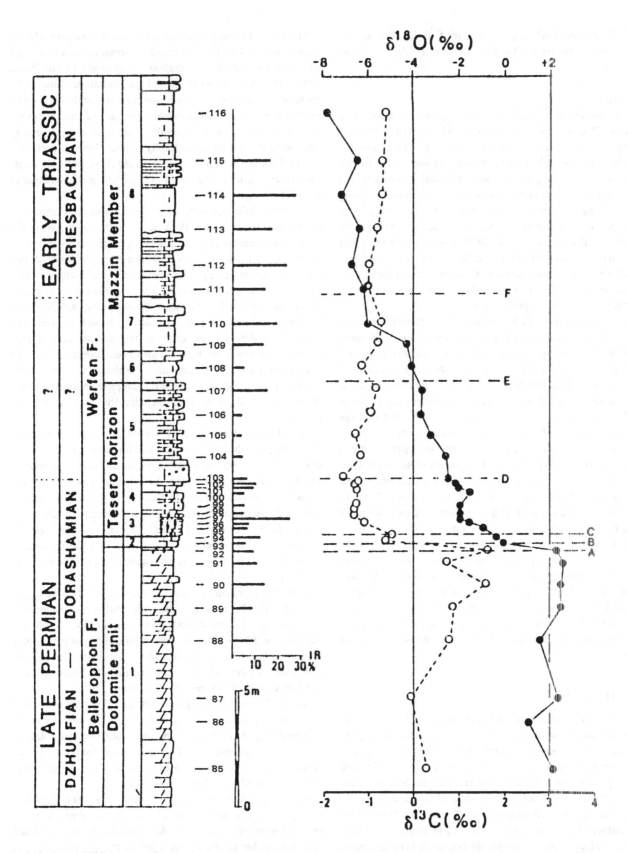

Fig. 8.10. Logs of $\delta^{13}C$ (—●—), $\delta^{18}O$ (--o--), and insoluble residues (IR) across the Permo-Triassic boundary in the Tesero section, western Dolomites. Stratigraphy based on Neri (in Italian Research Group, 1986). A = beginning, and B = end of sharp drop in $\delta^{13}C$ within Bellerophon Formation; C = contact of Bellerophon Formation and Tesero Horizon, with a change in sedimentation; D = highest Permian fossils; E = contact between Tesero and Mazzin members; and F = lowest beds of definite Triassic aspect. (From Magaritz *et al.*, 1988; see Fig. 8.6 for paleontologic details.)

3 If '*Bellerophon*' *vaceki* is regarded as heralder of the Triassic, the base of the Tesero Horizon is the base of the Triassic.

Conclusions

Clearly, choice of a boundary level has been affected by a number of factors, among which are knowledge of the geological region, personal specialization, and authority of the researcher or school of research. Paleontologists support index fossils, whereas sedimentologists emphasize physical parameters and paleogeographers rely on physiographic events. Further, among fossils, several different groups have been studied in the Southern Alps and a number of proposals have been made (Fig. 8.9).

After the Brescia meeting of IGCP Project 203, in 1986, and the final conference of the Project, in Beijing a year later, all data have been analyzed with the view of finding an event that is easy to recognize in the field and allows correlation between sequences on a wide geographic scale.

In considering possibilities, we note that because the appearance and disappearance of a single taxon was commonly controlled by regional tectonics, type of sedimentation, and other physical factors, the position in sedimentary sequences of the records of these events may be more or less variable from place to place. Thus, if there was no hiatus between deposition of the Bellerophon and Werfen formations, we suggest that the major event in the Permo-Triassic boundary interval was the essentially simultaneous drop in taxonomic diversity, the flourishing of fungi, and the alteration in pattern of sedimentation. These events are all recognizable in detail on a wide geographic scale and correlate with the base of the *Otoceras* Zone. In addition, it is also at this level that the first Triassic newcomers appear. This multiple event represents the first strong change in ecologic systems – in short, a biologic revolution.

In the view just expressed, the systemic (and erathemic) boundary would coincide with the contact between the Bellerophon and Werfen formations. Fossils of Permian aspect in the Tesero Horizon record the last attempt by survivors of the biologic revolution to adapt to a new ecosystem, which was very different from the one to which they were adapted in earlier times.

Yugoslavia

Sequences involving both Permian and Triassic rocks occur in Slovenia, southwest Croatia, Bosnia, Montenegro, and western Serbia. Uppermost Permian rocks are dark, fossiliferous limestones with a benthic biota, whereas lowermost Triassic strata are very poor in fossils. Clastic rocks occur here and there in the Upper Permian of Yugoslavia, and in Bosnia they are difficult to distinguish from lowermost Triassic lithofacies. As in the Southern Alps, ammonites are lacking in Yugoslavia. However, in the latter region *Palaeofusulina* occurs in Permian strata.

Slovenia

Upper Permian beds are well known in central Slovenia for their 'Caucasian (or South Tyrolian) and Indo-Armenian' brachiopods; gastropods; and pelecypods. Corals are rare, but foraminifers and algae are abundant (Buser *et al.*, 1988; Pešić *et al.*, 1988; Ramovš, in Italian Research Group, 1986).

The Upper Permian is represented in the Karawanke Alps by carbonate rocks of the Karawanke Formation and in central and western Slovenia by carbonates assigned to the Žažar Formation. Both units overlie and are gradational with clastic rocks of the Val Gardena Sandstone. Textures of the carbonates and thickness and sequence of limestone and dolomite beds vary from place to place as a result of changes in sedimentary environments and paleogeography. Upper Permian rocks are 300 m thick in the southern Karawanke Alps, but only a few dozen meters thick in the Sava Fold, east of Ljubljana (Buser *et al.*, 1988, fig. 2).

Dolomitic lithofacies prevail in the Karawanke Formation, but a limestone bed 1.5 m thick occurs in the lower part of the uppermost member in a section along a forest road north of Medvogje, in the southern Karawanke Alps. Fossils from this limestone bed indicate a Late Permian age and include *Velebitella triplicata*, *Mizzia cornuta*, *Gymnocodium bellerophontis*, *Permocalculus fragilis*, *Vermiporella nipponica*, *Agathammina* sp., *Glomospira* sp., and *Hemigordius* sp., as well as indeterminate bellerophontids and other gastropods. A recrystallized coral (?*Eugonophyllum* sp.) has also been collected from the uppermost part of the sequence. Fusulinids are apparently absent. However, studies are not yet complete on the part of the Karawanke Formation that includes the Permian-Triassic boundary.

In the Žažar Formation of central and western Slovenia limestone prevails over dolomite. The formation rests on sandstones, shales, and intercalated yellowish-brown dolomites, which are an equivalent of the Val Gardena Sandstone. Žažar limestone beds are dark grey to black, 10–30 cm thick (rarely 50 cm), and composed of recrystallized biomicrite, wackestone, and packstone (Buser *et al.*, 1988). Calcareous algae form the framework in most beds, which also contain the remains of echinoderms, foraminifers, ostracodes, gastropods, and brachiopods.

In the Žažar region of central Slovenia (Fig. 8.11), the Upper Permian Žažar Formation begins with a *Palaeofusulina* horizon that yields a brachiopod fauna of Caucasian type (i.e., *Spinomarginifera*, *Linoproductus*). This is followed upward by nine additional biohorizons, among which are limestones bearing *Edmondia permiana*; *Richtofenia* bioherms with *Notothyris*; and biostromes with *Waagenophyllum indicum*.

A *Comelicania–Paramarginifera* horizon, 1 m thick and 400 m below the top of the Žažar Formation, is the seventh in the sequence (Fig. 8.11). It yields specimens of *Comelicania haueri*, *C. vultur*, *Comelicania* sp. cf. *C. avis*, *Comelicania* sp. aff. *C. doryphora* and rare specimens of *Janiceps*, all of 'South-Tyrolian' type according to Pešić *et al.* (1988).

Horizon 8, which is about 400 m thick, includes beds rich in red algae (*Gymnocodium bellerophontis*, *Permocalculus fragilis*), smaller foraminifers and echinoderm remains. Horizon 9, a thin unit with sulfur clods, is succeeded at the top of the Žažar Formation by dolomitic Horizon 10, in which calcareous algae, smaller foraminifers, and crinoidal remains are common and specimens of the coral *Waagenophyllum* occur sporadically.

All levels in the Žažar Formation except Horizon 9 are rather rich in microfossils. Among the fusulinids, *Nankinella* and *Staffella* occur at nearly all horizons; others (*Codonofusiella*?, *Minojapanella*?, or *Boultonia*?) are more sporadic in occurrence

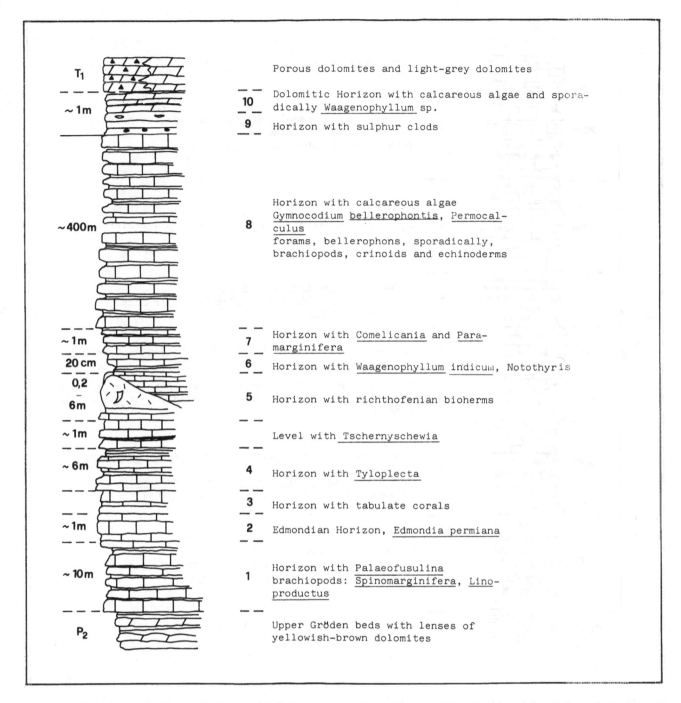

Porous dolomites and light-grey dolomites

10 Dolomitic Horizon with calcareous algae and sporadically Waagenophyllum sp.

9 Horizon with sulphur clods

8 Horizon with calcareous algae
Gymnocodium bellerophontis, Permocalculus
forams, bellerophons, sporadically,
brachiopods, crinoids and echinoderms

7 Horizon with Comelicania and Paramarginifera

6 Horizon with Waagenophyllum indicum, Notothyris

5 Horizon with richthofenian bioherms

Level with Tschernyschewia

4 Horizon with Tyloplecta

3 Horizon with tabulate corals

2 Edmondian Horizon, Edmondia permiana

1 Horizon with Palaeofusulina
brachiopods: Spinomarginifera, Linoproductus

Upper Gröden beds with lenses of
yellowish-brown dolomites

Fig. 8.11. Schematic section of Upper Permian rocks in Žažar region, central Slovenia, Yugoslavia. (From Pešić et al., 1988.)

and represent the subfamily Boultoniinae. Smaller foraminifers are referable to *Agathammina*, *Ammovertella*, *Climacammina*, *Globivalvulina graeca*, *Glomospira*, *Hemigordius*, *Nodosaria*, *Pachyphloia*, *Textularia*, *Nodosinella*, and *Ichtyolaria*. Among the algae, some species of *Vermiporella*, *Gymnocodium* and *Permocalculus* have been reported from Horizon 8.

According to Buser *et al.* (1988, p. 204):

> The *Permian/Triassic* boundary is placed at the end of the sedimentation of the Upper Permian type (black well bedded shallow marine platform limestone or dolomite with red algae, smaller foraminifers, rare fusulinids *Staffella* and *Nankinella*, and very rare highly developed

Boultoniinae), marked partly by a marly or clayey bed, 1 cm in thickness.

The lowermost Triassic is marked by beds that yield only tubelike fossil remains (Pešić *et al.*, 1988). These same fossils occur in the Karawanke Alps and are also useful as guide fossils for the basal Lower Triassic in Slovenia (Ramovš, 1982; Buser *et al.*, 1988).

Western Serbia
Upper Permian rocks are well known from the Jadar region of western Serbia (Fig. 8.12). There, a sequence of a little more than 100 m of bituminous and biohermal carbonates with a few thin

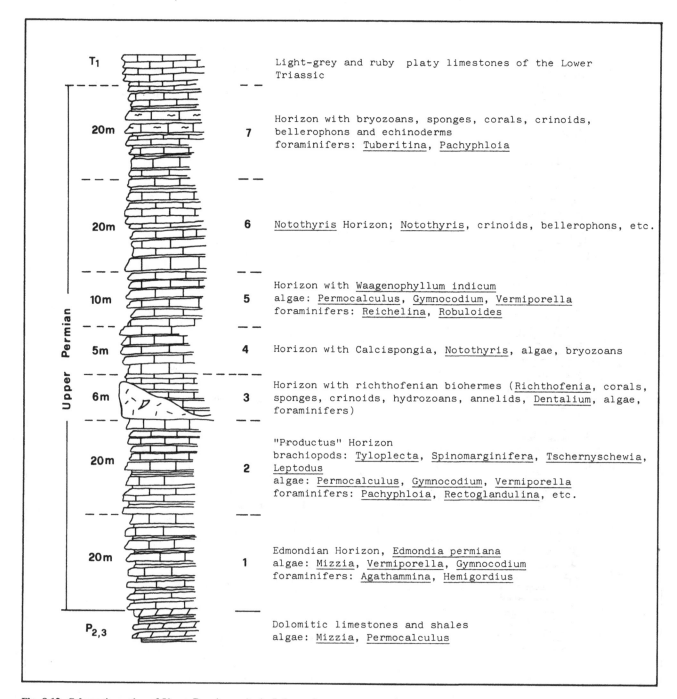

Fig. 8.12. Schematic section of Upper Permian rocks in Jadar region, western Serbia, Yugoslavia. (From Pešić _et al._, 1988.)

interbeds of shale and siltstone overlies a succession of clastic rocks corresponding to the Val Gardena Sandstone (Pešić _et al._, 1988; Ramovš, in Italian Research Group, 1986). In the most complete sequence of outcrops, along the middle course of the Jadar River, a succession of seven fossiliferous horizons has been recognized.

In the Upper Permian of the Jadar region (Fig. 8.12), the lowermost limestones are mostly bituminous and composed of finely crystalline calcite. These beds include a few algae and a sparse microfauna. The macrofauna consists of _Edmondia permiana._

Horizon 2 in the Jadar succession, the '_Productus_' Horizon,

includes flaggy light-grey dolomitic and marly limestones intercalated with shales and siltstones. These rocks are rich in macro- and microfossils. The abundance of brachiopods gives the name to this horizon.

Horizon 3 is a succession of biohermal limestones several meters thick, which yield abundant specimens of _Richthofenia_ as well as corals, crinoids, sponges, hydrozoans, foraminifers and an elongated form of the scaphopod _Dentalium_. The extent of this horizon is difficult to estimate because of the vegetative cover and complex tectonic fabric.

Horizon 4 has yielded calcareous algae, calcisponges, bellerophontids, and the brachiopod _Notothyris_. Above it is a sequence of flaggy limestones with algae, foraminifers, and banks of the coral *Waagenophyllum indicum*. These make up Horizon 5, which is succeeded upward by Horizon 6, composed of black, flaggy bituminous limestones with abundant specimens of brachiopods such as *Notothyris* and *Spiriferina* and the alga *Permocalculus*.

The seventh horizon consists of black, argillaceous and sandy limestones with sponges, bryozoans, corals, bellerophons, echinoderms and foraminifers. Above this are pink or light bluish, poorly fossiliferous, stratified limestones that mark the beginning of Triassic sedimentation (Pešić *et al.*, 1988).

Horizons 2–7 have yielded representatives of the following microfossils: *Agathammina, Hemigordius, Pachyphloia, Nodosaria, Reichelina, Codonofusiella, Permocalculus fragilis, Gymnocodium bellerophontis, Vermiporella nipponica*, and *Atractylioposis* sp.

The Upper Permian rocks of Yugoslavia have not been assigned to a stage and no markers have been identified among the macro- or microfossils known from the sequence. In Serbia and Slovenia the Permo-Triassic boundary is marked by a lithologic change, and also, according to the literature, by a strong drop in taxonomic diversity. However, there has been no detailed study within the lithologically defined boundary region to determine which or how many biologic events are recorded.

By way of comparison with the Southern Alps, it may be noted that there are some fossils in common:

1 Foraminifers and calcareous algae in the Karawanke Alps.
2 Foraminifers, calcareous algae, and *Comelicania* in central Slovenia.
3 Foraminifers and calcareous algae in western Serbia.

The sedimentary environments and fossil assemblages of Slovenia and Serbia are quite different from those of the Southern Alps. For example, no biohermal limestones or beds with abundant brachiopods and calcisponges are known in the Dolomites or Carnia. Foraminifers and calcareous algae were apparently the most widespread marine inhabitants, and the brachiopod *Comelicania* was widespread geographically in Tethys. Although the composition of fossil assemblages is very distinct in Slovenia and Serbia with respect to the Southern Alps, a similarity with fossil associations from the Bükk Mountains of Hungary has been demonstrated (Pešić *et al.*, 1988).

Conclusions

The biostratigraphy of the marine uppermost Permian and lowermost Triassic of the Southern Alps is documented in detail. Vertical ranges of foraminifers, brachiopods, mollusks, and palynomorphs are known bed by bed near the boundary between the Bellerophon and Werfen formations in the Butterloch-Bletterbach, Sass de Putia, and Tesero sections of the western Dolomites. Investigations in the eastern Dolomites (Cadore) and in Carnia support the events recognized at localities in the western Dolomites.

Studies of the Upper Permian and Lower Triassic rocks have not yet been carried out in the same detail in Yugoslavia. The common presence of certain foraminiferal and algal taxa is not a suitable basis for chronostratigraphic correlations because they have long vertical ranges. Further, since no index fossils have been identified, no stadial assignments have been suggested. *Palaeofusulina nana* has been reported from one horizon in the Žažar Formation in Slovenia, but conodonts and palynomorphs are unknown and *Comelicania* seems not to be useful at the generic level to effect stadial correlation. It is clear, however, that different ecosystems prevailed in the latest Permian in the Southern Alps and in Serbia-Slovenia.

All of the events recognized near the Bellerophon–Werfen boundary in the Southern Alps have not been recognized in Yugoslavia. However, in Yugoslavia, a strong drop in taxonomic diversity at the top of the Permian sequences may be inferred from the literature. This change, which coincides with the lithologic change that marks the transitional Permo-Triassic boundary, corresponds to one of the several events that marks the Permo-Triassic boundary in the Southern Alps.

The continental Permian of the Southern Alps and its boundary with the Triassic

Permian continental deposits crop out from the western to the eastern extremity of the South Alpine region. However, they are well developed only west of the Adige Valley, where no marine influence is known. They are readily divisible into two groups.

The lower group of rocks consists of intermediate and acid volcanics, which alternate in some areas with alluvial and lacustrine deposits. Intrusive bodies, concentrated in the area west of Lake Maggiore and along the Giudicarie, Valsugana and Pusteria tectonic lines, are also attributed to the same igneous activity.

The upper group consists of typical red beds, which are clearly present only in the central and eastern parts of the South Alpine region.

After the Hercynian orogeny, the region of the present-day Southern Alps was affected by a great tensional climax, interpreted either as a response to simple crustal extension (e.g., Cassinis *et al.*, 1980; Wopfner, 1984) or to transtensional movements in a generally transcurrent regime (see Massari, 1988 with bibliography). These active tectonics were responsible for development of a basin-and-swell topography and thus strongly controlled erosional and depositional processes during Late Carboniferous and Permian times (Fig. 8.1).

Some key sequences, thought to be important to a general knowledge of the Permian, are illustrated in the stratigraphic schemes of Fig. 8.2. These are drawn from the literature, but also include data provided by the author (G. C.) for the area west of the Brenta Group, by F. Massari and C. Neri for the Dolomites, and by C. Venturini for the Carnic Alps. The figure clearly shows the presence in the Permian of two tectono-sedimentary cycles, separated by a distinct unconformity.

At most places in Lombardy and the Dolomites, basin fills of the lower cycle consist of calcalkaline volcanic products and coarse- to fine-grained continental sediments (e.g., Collio Formation, Dosso dei Galli Gonglomerate), which may exceed 1000 m in thickness (Fig. 8.1). The volcanic products, made primarily of rhyolites and andesites in the form of ignimbrites, pyroclastites and lava flows, prevail west of Lake Como and between the Rendena and Sesto valleys. In the Carnic Alps, however, the lower cycle is formed of deltaic, paralic and marine deposits of the Auernig, Rattendorf and Trogkofel groups.

All the above-mentioned lower deposits, ranging in age from Westphalian to Early Permian, lie unconformably on the basement. Graben or semigraben geometry and tilting of the basin floor commonly caused asymmetry in facies development, and growth faulting is reflected locally in the character of sedimentary sequences laid down in marginal areas.

The end of the first Permian cycle is marked by a regional unconformity that represents a chronologic gap of variable extent. The mid-Permian tectonic activity responsible for the unconformity marked an important turning point in history, the results of which are easily recognized through the Southern Alps, from Carnia to part of Lombardy (Fig. 8.2).

Deposits of the upper Permian cycle begin with ruditic facies (Daone? and Sesto valleys, Tarvisio), which is covered in most places by continental red beds. Evolution of this younger Permian cycle was also tectonically controlled (Italian Research Group. 1986).

Compared with deposits of the first Permian cycle, those of the second are more widespread, thinner, and form an almost continuous blanket in the central-eastern part of the South Alpine region (Fig. 8.1). Syngenetic tectonic activity is marked by sharp changes in thickness (from a few meters to about 600 m) and by lateral facies changes.

The upper cycle, ranging from Middle (Wordian? to Capitanian) to Upper Permian (Dorashamian/Changxingian), records a global transgressive trend. East of the Adige Valley, this transgression superposed transitional and marine facies of the Bellerophon Formation on Val Gardena continental red clastics (Fig. 8.1), probably at different times from place to place, so that, in the Dolomites and Carnia the two formations interfinger (Fig. 8.2). The Bellerophon Formation is succeeded upward by the Lower Triassic Werfen Formation, the characters of which are given in the first part of this chapter.

In the Adige Valley and surrounding areas, a transition from 'Gardena' or 'Bellerophon' to 'Werfen' is not everywhere clear. Some authors (e.g. Neri, 1982 and pers. comm.; Farabegoli & Viel, 1982) emphasize structural discontinuity between the Permian and Triassic of this area.

West of the Adige Valley, in the Rendena and Giudicarie valleys, transgressive deposits similar to the Werfen Formation (also termed Servino; e.g., Peloso & Vercesi, 1982) occur above the Val Gardena Sandstone (also known as Verrucano Lombardo). These are generally composed of continental terrigenous sediments and include, at the base, oolitic limestones or dolostones of various thickness (the 'Praso Limestone' of Dozy, 1935). Correlation of these oolitic carbonates with the Tesero Horizon of the Dolomites can at present only be surmised (Fig. 8.2) because detailed stratigraphic and paleontologic studies are lacking.

Conditions more or less analogous to those just described also prevail in nearby Lombardy. Although studies of these rocks are numerous (see Assereto & Casati, 1965, 1966; Casati & Gnaccolini, 1967; Cassinis, 1968; Assereto et al., 1973; Casati, 1978; De Zanche & Farabegoli, 1983; Farabegoli & De Zanche, 1984; Gaetani, 1986; Gaetani et al., 1987), data useful for a wide and detailed interpretation of the Permo-Triassic transition zone are still lacking. As a rule, authors follow tradition and locate the Permo-Triassic boundary at the base of the Werfen (or Servino) Formation. Assereto et al. (1973) were the first to note the presence of an unconformity, and thus a gap that becomes wider west of the Adige Valley, between the Servino and underlying Permian clastics. Casati (1978), Casati & Bini (1982) and several others agree, although the gap has not yet been documented by paleontological or other types of evidence. The matter requires confirmation.

In upper Valsassina, and above all in the western sector of Lake Como, there were one or several uplifts before establishment of Middle Triassic carbonate platforms (e.g. Gianotti, 1968; Assereto et al., 1973; Casati, 1978; Farabegoli & De Zanche, 1984). Theses serve further to complicate reconstruction and interpretation of the Permo-Triassic boundary sequence. That is, local deposits, with a certain lithic affinity to the Verrucano and Servino, were built here at the expense of preceding units during Anisian time (e.g., Gaetani et al., 1987). Thus, while accepting the presence of a gap between the basement and its Triassic cover, we are unable to identify the original stratigraphy of this area and, consequently, to confirm the existence of a gap at the Permo-Triassic boundary.

Farther west, in the Varese area (Govi, 1960; Casati, 1978), in the Lake Orta area (Kälin & Trümpy, 1977), in the Sesia Valley area (Farabegoli & De Zanche, 1984), and in the Canavese area sensu stricto (Cassinis, personal observation), that is, as far as the extreme western limit of the Southern Alps, the geological situations were characterized by uplift from Permian to Jurassic. Thus sequences are interrupted by hiatuses and Lower Permian volcanics are commonly in contact with different Triassic or even less widespread Jurassic units. In these areas it is unfortunately impossible yet to formulate any general historical hypothesis on the transition from Permian to Triassic time.

References

Assereto, R., Bosellini, A., Fantini Sestini, N. & Sweet, W. C. (1973). The Permian–Triassic boundary in the Southern Alps (Italy). *Mem. Canad. Soc. Petrol. Geol.*, **2**: 176–99.

Assereto, R. & Casati, P. (1965). Revisione della stratigrafia permo-triassica della Val Camonica meridionale (Lombardia). *Riv. Ital. Paleont. Stratigr.*, **71**: 990–1097.

Assereto, R. & Casati, P. (1966). Il 'Verrucano' nelle Prealpi Lombarde. *Atti Symposium sul Verrucano, Soc. Toscana Sci. Nat.*, pp. 247–65, Pisa.

Bosellini, A. (1964). Stratigrafia, petrografia e sedimentologia delle facies carbonatiche al limite Permiano-Trias nelle Dolomiti Occidentali. *Mem. Mus. Storia Nat. Veneto-Tridentino*, **15**: 59–110.

Brandner, R., Donofrio, D. A., Krainer, K., Mostler, H., Nazarow, M. A., Resch, W., Stingl, V. & Weissert, H. (1986). Events at the Permian–Triassic boundary in the Southern and Northern Alps [Abstract]. In: *Field Conference on Permian and Permian-Triassic Boundary in the South-Alpine Segment of the Western Tethys, and Additional Regional Reports*, Abstracts vol., p. 15. Pavia: Soc. Geol. Ital. and IGCP Project 203.

Broglio Loriga, C., Masetti, D. & Neri, C. (1983). La Formazione di Werfen (Scitico) delle Dolomiti Occidentali: Sedimentologia e Biostrtragrafia. *Riv. Ital. Paleont. Stratigr.*, **88**: 501–98.

Broglio Loriga, C., Neri, C. & Posenato, R. (1986). The Early macrofaunas of the Werfen Formation and the Permian–Triassic boundary in the Dolomites (Southern Alps, Italy). *Studi Trentini Sci. Nat., Acta Geol.*, **62**: 3–18.

Broglio Loriga, C., Neri, C., Pasini, M. & Posenato, R. (1988). Marine fossil assemblages from Upper Permian to lowermost Triassic in the western Dolomites (Italy). *Mem. Soc. Geol. Ital.*, **34**(1986): 5–44.

Buser, S., Grad, H., Ogorelec, B., Ramovš, A., & Šribar, L. (1988). Stratigraphical, paleontological and sedimentological characteristics of Upper Permian beds in Slovenia, NW Yugoslavia. *Mem. Soc. Geol. Ital.*, **34**(1986): 195–219.

Casati, P. (1978). Tettonismo e sedimentazione nel settore occidentale delle Alpi Meridionali durante il tardo Paleozoico, il Triassico e il Giurassico. *Riv. Ital. Paleont. Stratigr.*, **84**: 313–26.

Casati, P. & Bini, A. (1982). Itinerari geologici nel Gruppo delle Grigne (Prealpi Lombarde). *CAI, Comit. Sci. Itiner. nat. e geogr.*, Publ. 18: 117.

Casati, P. & Gnaccolini, M. (1967). Geologia delle Alpi Orobie occidentali. *Riv. Ital. Paleont. Stratigr.*, **73**: 25–162.

Cassinis, G. (1968). Studio stratigrafico del 'Servino' di Passo Valdi (Trias inferiore dell'Alta Val Caffaro). *Atti Ist. Geol., Univ. Pavia*, **19**: 15–39.

Cassinis, G. (Ed.) (1988). Proceedings of the field conference on: Permian and Permian-Triassic boundary in the south-Alpine segment of the western Tethys, and additional regional reports. *Mem. Soc. Geol. Ital.*, **34**(1986): 1–366.

Cassinis, G., Elter, G., Rau, A. & Tongiorgi, M. (1980). Verrucano: A tectofacies of the Alpine-Mediterranean southern Europe. *Mem. Soc. Geol. Ital.*, **20**(1979): 135–49.

Cassinis, G., Massari, F., Neri, C., & Venturini, C. (1988). The continental Permian in the Southern Alps (Italy). A review. *Z. Geol. Wiss., Berlin*, **16**: 1117–26.

Ceoloni, P., Conti, M., Mariotti, N. & Nicosia, U. (1988). New Late Permian tetrapod footprints from the Southern Alps. *Mem. Soc. Geol. Ital.*, **34**(1986): 45–65.

De Zanche, V. & Farabegoli, E. (1983). Anisian stratigraphy in the northern Grigna area (Lake Como, Italy). *Mem. Sci. Geol., Univ. Padova*, **36**: 283–91.

Dozy, J. F. (1935). Über das Perm der Südalpen. *Leidse Geol. Meded.*, **7**: 41–62.

Farabegoli, E. & De Zanche, V. (1984). A revision of the Anisian stratigraphy in the western Southern Alps, west of Lake Como. *Mem. Sci. Geol., Univ. Padova*, **36**: 391–401.

Farabegoli, E. & Viel, G. (1982). Litostratigrafia della Formazione di Werfen (Trias inf.) delle Dolomiti occidentali. *Ind. Min.*, **6**: 3–14.

Gaetani, M. (1986). Il Triassico dell'Adamello meridionale. *Mem. Soc. Geol. Ital.*, **26**(1983): 105–18.

Gaetani, M., Gianotti, R., Jadoul, F., Ciarapica, G., Cirilli, S., Lualdi, A., Passeri, L., Pellegrini, M. & Tannoia, G. (1987). Carbonifero superiore, Permiano e Triassico nell'area Lariana. *Mem. Soc. Geol. Ital.*, **32**(1986): 5–48.

Gianotti, R. (1968). Considerazioni sul margine settentrionale del Gruppo delle Grigne (Lombardia centrale). *Atti Ist. Geol., Univ. Pavia*, **18**: 82–101.

Góczán, F., Haas, J. & Oravecz-Scheffer, A. (1988). Permian–Triassic boundary in the Transdanubian Central Range. *Acta Geol. Hung.*, **30**: 35–58.

Govi, M. (1960). Geologia del territorio compreso tra il Lago di Lugano e la Val Marchirolo. *Studi Ric. Div. Geomin. C. N. R.*, **3**: 160–217.

Haas, J., Góczán, F., Oravecz-Scheffer, A., Barabás-Stuhl, Á., Majoros, G. & Bériczi-Makk, A. (1988). Permian–Triassic boundary in Hungary. *Mem. Soc. Geol. Ital.*, **34**(1986): 221–41.

Hirsch, F. & Weissbrod, T. (1988). The Permian-Triassic boundary in Israel. *Mem. Soc. Geol. Ital.*, **34**(1986): 253–6.

Italian Research Group (Ed.) (1986). Permian and Permian-Triassic boundary in the South-Alpine segment of the western Tethys. Field guidebook. *Field Conf. SGI–IGCP Project 203*, July, 1986, Brescia (Italy), 180 pp., Pavia.

Kälin, V. O. & Trümpy, D. M. (1977). Sedimentation und Paläotektonik in den westlichen Südalpen: Zur triassich-jurassichen Geschichte des Monte Nudo-Beckens. *Eclogae Geol. Helvetiae*, **70**: 295–350.

Kozur, H. (1985). Biostratigraphic evaluation of the Upper Paleozoic conodonts, ostracods and holothurian sclerites of the Bükk Mts. Part II: Upper Paleozoic ostracods. *Acta Geol. Hung.*, **28**: 226–56.

Massari, F. (1988). Some thoughts on the Permo-Triassic evolution of the South-Alpine area (Italy). *Mem. Soc. Geol. Ital.*, **34**(1986): 179–88.

Magaritz, M., Bär, R., Baud, A. & Holser, W. T. (1988). The carbon-isotope shift at the Permian-Triassic boundary in the Southern Alps is gradual. *Nature*, **331**(6154): 337–9.

Neri, C. (1982). Il paleoalto di Lavis (Trento) e i suoi rapporti con la serie permo-scitica (Nota preliminare). *Ann. Univ. Ferrara*, n. ser., sez. IX-Sci. Geol. Paleont., **8**: 21–7.

Neri, C. & Pasini, M. (1985). A mixed fauna at the Permian-Triassic boundary, Tesero section, western Dolomites (Italy). *Boll. Soc. Paleont. Ital.*, **23**: 113–17.

Noé, S. U. (1987). Facies and paleogeography of the marine Upper Permian and of the Permian–Triassic boundary in the Southern Alps (Bellerophon Fm., Tesero Horizon). *Facies*, **16**: 89–142.

Pasini, M. (1985). Biostratigrafia con i Foraminiferi del limite Formazione a Bellerophon/Formazione di Werfen fra Recoaro e la Val Badia (Alpi Meridionali). *Riv. Ital. Paleont. Stratigr.*, **90**: 481–510.

Peloso, G. F. & Vercesi, P. L. (1982). Stratigrafia e tettonica della porzione di SW del Gruppo di Brenta, tra la Val Rendena e la Val d'Algone (Trentino occidentale). *Mem. Sci. Geol., Univ. Padova*, **35**: 377–95.

Perri, M. C. & Andraghetti, M. (1987). Permian/Triassic boundary and Early Triassic conodonts from the Southern Alps, Italy. *Riv. Ital. Paleont. Stratigr.*, **93**: 291–328.

Pešić, L., Ramovš, A., Sremac, J., Pantić-Prodanović, S., Filipović, J., Kovács, S., & Pelikán, P. (1988). Upper Permian deposits of the Jadar region and their positions within the western Paleotethys. *Mem. Soc. Geol. Ital.*, **34**(1986): 211–19.

Posenato, R. (1988a). The Permian/Triassic boundary in the western Dolomites, Italy. Review and proposal. *Ann. Univ. Ferrara*, sez. Sci. Terra, **1**: 31–45.

Posenato, R. (1988b). Chronological and geographic distribution of the Fam. Comelicaniidae Merla, 1930 (Brachiopods). *Riv. Ital. Paleont. Stratigr.*, **94**: 383–98.

Ramovš, A. (1982). The Permian-Triassic boundary in Jugoslavia. *Rud.-Metal. Zb.*, **29**: 29–31.

Rostovtsev, K. O. & Azaryan, N. R. (1973). The Permian-Triassic boundary in Transcaucasia. *Mem. Canadian Soc. Petrol. Geol.*, **2**: 89–99.

Ruzhentsev, V. E. & Sarytcheva, T. G. (Eds) (1965). Development and change of marine organisms at the Paleozoic–Mesozoic boundary. *Akad. Nauk USSR, Trudy Paleont. Inst.*, **108**: 431 pp.

Staesche, U. (1964). Conodonten aus dem Skyth von Südtirol. *N. Jb. Geol. Paläont., Abh.*, **119**: 247–306.

Sweet, W. C. (1988). A quantitative conodont biostratigraphy for the Lower Triassic. *Senckenb. lethaea*, **69**: 253–73.

Visscher, H. & Brugman, W. A. (1988). The Permian-Triassic boundary in the Southern Alps: A palynological approach. *Mem. Soc. Geol. Ital.*, **34**(1986): 121–8.

Wopfner, H. (1984). Permian deposits of the Southern Alps as products of initial alpidic taphrogenesis. *Geol. Rundschau*, **73**: 259–77.

9 Permo-Triassic brachiopod successions and events in South China

XU GUIRONG AND RICHARD E. GRANT

Introduction

In the last decade, several important reports have been published on the stratigraphy of sections that include the Permo-Triassic boundary in South China (e.g., Hou *et al.*, 1979; Sheng *et al.*, 1984; Yang, Wu & Yang, 1981; Yang *et al.*, 1987; Yin, 1983, 1985; Liao, 1980; Liao & Meng, 1986) and the authors have in press a study of the brachiopods collected from the boundary interval at 32 localities in South China. In the present contribution, we summarize our work with Permo-Triassic brachiopods, outline a zonal scheme based on them, and suggest correlations with sections in other parts of the Tethys region.

A number of schemes of brachiopod zonation have been suggested since 1979 when three Permian faunas were recognized by Hou *et al.* in the upper Wujiapingian through Changx-

ingian interval (Table 9.1) of the Lianxian area of Guangdong Province. A year later, on the basis of a study of brachiopods from the western part of Guizhou Province, Liao described three brachiopod assemblages, two in the Wujiapingian and one in the Changxingian (Table 9.1). In 1987, Xu (in Yang *et al.*, 1987) reported on a study of brachiopods from 31 sections in South China and in that study set up five assemblage zones in the Wujiapingian through lower Griesbachian interval. In this report, we suggest a new scheme of brachiopod zonation in the same interval (Table 9.1) and note that the nature of brachiopod faunas in the Changxingian is affected to a great extent by lithofacies.

In our discussion of correlations with ammonoid, conodont, and fusulinid biozones, we use information not only from the 32

Table 9.1. *Comparison of brachiopod zonations near the Permo-Triassic boundary in South China*

SEQUENCES		Hou *et al.*, 1979	Liao, 1980	Xu, 1987	This chapter	
SERIES	STAGE				clastic lithofacies	limestone lithofacies
LOWER TRIAS.	LOWER GRIES-BACHIAN			*Crurithyris speciosa*-- *Lingula subcircularis*	*Crurithyris pusilla*-- *Lingula subcircularis* Assemblage Zone	
UPPER PERMIAN	CHANGXINGIAN	*Cathaysia sinuata, Oldhamina minor, Crurithyris pusilla, & Hustedia indica*	*Enteletina zigzag**--	*Waagenites barusiensis*-- *Crurithyris pusilla* Assemblage Zone	*Cathaysia sinuata*-- *Waagenites barusiensis* Assemblage Zone	*Spirigerella discusella*-- *Acosarina minuta* Assemblage Zone
UPPER PERMIAN	CHANGXINGIAN	*Tschernyschewia geniculata, Enteletina** *sinensis, Meekella kueichowensis*	***Cathaysia sulcatifera* assemblage	*Enteletina zigzag**-- *Neowellerella**** *pseudoutah* Assemblage Zone	*Cathaysia chonetoides*-- *Chonetinella substrophomen-oides* Assemblage Zone	*Peltichia zigzag*-- *Prelissorhynchia triplicatioid* Assemblage Zone
UPPER PERMIAN	WUJIAPINGIAN	*Edriosteges, Permophricodothyris, Asioproductus, Tyloplecta yangtzeensis*	*Squamularia grandis*-- *Orthothetina ruber* assemblage	*Orthothetina ruber*-- *Squamularia grandis* Assemblage Zone	*Orthothetina ruber*-- *Squamularia grandis* Assemblage Zone	
UPPER PERMIAN	WUJIAPINGIAN		*Edriosteges poyangensis* assemblage	*Squamularia indica*-- *Haydenella wenganensis* Assemblage Zone	*Squamularia indica*-- *Haydenella wenganensis* Assemblage Zone	

* *Enteletina = Peltichia,* ** *Cathaysia sulcatifera = C. sinuata,* *** *Neowellerella = Prelissorhynchia*

sections we have studied ourselves, but also data from localities such as the Jiaozishan Section (Fig. 9.1, locality A; Yao *et al.*, 1980) and the Huatang Section (Fig. 9.1, locality C; Liao & Meng, 1986). Additional data for use in correlation of sections in South China have been derived from excellent studies of localities in north Tibet (Jing & Sun, 1981), Transcaucasia (Rostovtsev & Azaryan, 1973), northwest Iran (Teichert, Kummel & Sweet, 1973; Iranian–Japanese Research Group, 1981), the Salt Range of Pakistan (Grant, 1970; Pakistani–Japanese Research Group, 1981, 1985), Kashmir (Nakazawa *et al.*, 1975), the Southern Alps (Assereto *et al.*, 1973), East Greenland (Teichert & Kummel, 1976), and northwest Nepal (Waterhouse, 1978).

Lithofacies

There are two primary lithofacies in the Wujiapingian Stage in South China. One is a limestone lithofacies, termed the Wujiaping Formation and the other is a coal-measure lithofacies named the Longtan Formation. The composition of brachiopod faunas is somewhat variable in both formations because limestone or calcareous mudstone beds are commonly intercalated in the Longtan Formation and calcilutite or marly beds occur here and there in the Wujiaping Formation. However, a suitable brachiopod zonation is available that enables correlation of the two formations in South China.

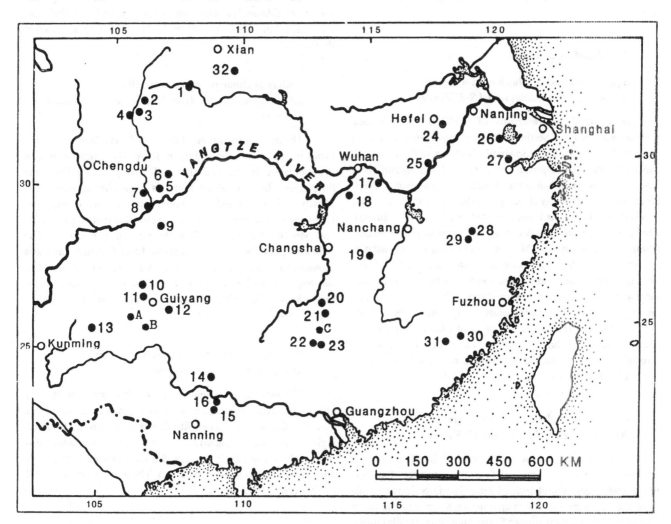

Fig. 9.1. Location of Permo-Triassic boundary sections in South China. 1, Kucaoping section, Xixiang County, Shaanxi Province; 2, Mingyuexia section, Guangyuan County, Sichuan Province; 3, Xindianzi section, Guangyuan County, Sichuan Province; 4, Shangsi section, Guangyuan County, Sichuan Province; 5, Xinjiacao section, Guangyuan County, Sichuan Province; 6, Huayingshan section, Linshui County, Sichuan Province; 7, Yanjingxi section, Hechuan County, Sichuan Province; 8, Liangfengya section, Zhongqing City, Sichuan Province; 9, Banzhuyuan section, Nantong County, Sichuan Province; 10, Liuchang section, Qingzhen County, Guizhou Province; 11, Yingpanpo section, Guiyang City, Guizhou Province, 12, Xiaochehe section, Guiyang City, Guizhou Province; 13, Huopu section, Pan County, Guizhou Province; 14, Longtoujiang section, Yishan County, Guangxi Province; 15, Paoshui section, Laibin County, Guangxi Province; 16, Penglaitan section, Laibin County, Guangxi Province; 17, Shatian section, Huangshi City, Hubei Province; 18, Guanyinshan section, Puqi County, Hubei Province; 19, Baimu section, Yichun County, Jiangxi Province; 20, Jueguangsi section, Laiyang County, Hunan Province; 21, Matian section, Yongxi County, Hunan Province; 22, Xiaoyuanchong section, Jiahe County, Hunan Province; 23, Meitian section, Yizhang County, Hunan Province; 24, Majiashan section, Chao County, Anhui Province; 25, Yueshan section, Huaining County, Anhui Province; 26, Meishan section, Changxing County, Zhejiang Province; 27, Huangzhishan section, Wuxing County, Zhejiang Province; 28, Minfa section, Guangfeng County, Jiangxi Province; 29, Xijia section, Shangrao County, Jiangxi Province; 30, Yading section, Zhangping County, Fujian Province; 31, Yanshi section, Longyan County, Fujian Province; 32, Xikou section, Zhenan County, Shanxi Province; A, Jiaozishan section, Anshun City, Guizhou Province; B, Shitouzhai section, Ziyun County, Guizhou Province; C, Huatang section, Chenxian County, Hunan Province.

The Changxingian Stage is represented by a variety of sedimentary rock types, of which three are the most typical. The Changxing Formation is dominated by limestone; the Dalong Formation by siliceous rocks; and the Xuanwei (or Yanshi) Formation by arenite. Because brachiopod occurrences are closely related to lithofacies, it is necessary to set up two parallel biozonations, one in the limestone lithofacies, the other in the clastic lithofacies.

Rocks included in the Lower Griesbachian Substage may be divided into two general types. The lower Daye Formation is dominantly carbonate strata, whereas the lower Feixianguan Formation is primarily clastic. Brachiopods are poorly preserved, but are principally of the same types in these two formations.

Brachiopod zonations

Squamularia indica–Haydenella wenganensis
Assemblage Zone (abbreviated S–H Zone)

Brachiopods characteristic of the S–H Zone include *Araxathyris kandevani* Sestini and Glaus, *Spiriferellina orientalis* (Frech), *S. triquetra* Liao, *Squamularia elegantus* (Waagen), *S. indica* (Waagen), *Costispinifera striata* Liao, *Dielasma zhijinense* Liao, *Edriosteges subplicatilis* (Frech), *E. acuminatus* Liao, *Haydenella wenganensis* (Huang), *Orthothetina speciosa* Liao, *Meekella pusilloplicata* Liao, *Liosotella magniplicata* (Huang), *Transennatia gratiosa* (Waagen), *Alatoproductus truncatus* Ching, *Cathaysia uralica* (Moeller), and *Chonetella nasuta putingensis* Liao. The S–H Zone is in the lower part of the Wujiapingian Stage, and *S. indica* and *H. wenganensis* are the most widely distributed brachiopod species in South China. The fauna of the S–H Zone corresponds in large part to the *Edriosteges poyangensis* (Kayser) assemblage of Liao (1980), but the nominal species of that assemblage is abundantly represented throughout the Wujiapingian and ranges into the Changxingian in some areas of South China.

Orthothetina ruber–Squamularia grandis
Assemblage Zone (abbreviated O–S Zone)

Liao named this zone on the basis of data from several sections in the western part of Guizhou Province, but it is recognizable throughout South China. The O–S Zone extends from the upper part of the Wujiapingian Stage into the lower part of the lower Changxingian Stage. Significant fossils limited to this zone include: *Araxathyris ogbinensis* Grunt, *A. shuizhutangensis* Chan, *Araxathyris* sp. cf. *A. lata* Grunt, *A. bisculcata* Liao, *A. filina* (Arthaber), *Tschernyschewia geniculata* Chan, *Waagenites deplanata* (Waagen), *Notothyris subnucleolus* Zhang & Ching, *N. minuta* Waagen, *Notothyris* sp. cf. *N. irregularis* Grabau, *Orthothetina deminuta* Chang, *O. provecta* Liao, *O. ruber* (Frech), *Orthothetina* sp. cf. *O. curvata* Ustriski, *Derbyia disalata* Liao, *Enteletina kwangtungensis* Chan, *Leptodus canceriniformis* Liao, *Oldhamina jiaozishanensis* Liao, *O. interrupta* Chan, *O. lianyangensis* Chan, *Poikilosakos dzhulfensis* Sarytcheva, *Haydenella kiangsiensis* (Kayser), *Costispinifera anshunensis* Liao, *Haydenoides orientalis* Chan, *Marginifera magniplicatus* Huang, *Squamularia grandis* Chao, *Tyloplecta*

costiferinoides Fong, *Martinia martini* Waagen, *Spiriferellina octoplica* (Sowerby), *Cathaysia speciosa* Chan, and *C. parvulia* Chang. One of the typical species, *Orthothetina ruber* is widespread in the clastic lithofacies and is particularly abundant in mudstone, siltstone, and calcareous mudstone. *Squamularia grandis*, on the other hand, is commonly found in limestone, cherty limestone, and calcareous mudstone.

In the previous two zones there are a number of common elements, such as *Edriosteges poyangensis*, *Streptorhynchus pelargonatus* (Schlotheim), *S. kayseri* Schlotheim, *Derbyia acutangula* (Huang), *Paraspiriferina alpheus* (Huang), *Punctospirifer oritata* (Schlotheim), *Licharewiella costata* (Waagen), and *Meekella abnormalis* Huang, as well as five species of *Leptodus* and six of *Oldhamina*. Thus the Wujiapingian fauna differs in aspect substantially from that of the Maokouian Stage (Lower Permian).

Cathaysia chonetoides–Chonetinella substrophomenoides *Assemblage Zone (abbreviated C–C Zone)*

The brachiopod fauna of the C–C Zone is found in clastic facies in the lower Changxingian Stage. By comparison with Wujiapingian faunas and those of Changxingian limestone lithofacies, the C–C Zone fauna is somewhat monotonous. Mostly it is composed of Chonetacea such as *Waagenites*, *Chonetinella* and *Fanichonetes*, Chonetellidae such as *Cathaysia*, and Meekellidae such as *Orthothetina* and *Perigeyrella*. The following play the important roles in the C–C Zone: *Cathaysia chonetoides* (Chao), *C. spiriferoides* Xu & Grant, *Chonetellina substrophomenoides* (Huang), *Waagenites wongiana* (Chao), *Orthothetina regularis* (Huang), and *Prelissorhynchia pseudoutah* (Huang).

Cathaysia sinuata–Waagenites barusiensis
Assemblage Zone (abbreviated C–W Zone)

The brachiopod fauna of the C–W Zone characterizes upper Changxingian clastic lithofacies and is more monotonous than the fauna of the same lithofacies in the lower Changxingian. It consists of *Cathaysia sinuata* Chan, *Waagenites barusiensis* (Davidson), *Crurithyris pusilla* Chan, *Orthothetina regularis* and *Leptodus nobilis* (Waagen). The zone is characterized by the rarity of brachiopods.

Peltichia zigzag–Prelissorhynchia triplicatioid
Assemblage Zone (abbreviated P–P Zone)

The brachiopod fauna in the limestone lithofacies of the lower Changxingian is composed of *Peltichia zigzag*, *P. sinensis* (Huang), *P. schizoloides* Xu & Grant, *Prelissorhynchia triplicatioid* Xu & Grant, *P. pseudoutah*, *Araxathyris beipeiensis* Xu & Grant, *Uncinunellina multicostifera* Xu & Grant, *Cyrolexis antearcus* Xu & Grant, *Squamularia formilla* Xu & Grant, and *Spinomarginifera kueichowensis* (Huang). The most important feature of the P–P Zone is that it is the one in which *Peltichia* of the superfamily Enteletacea reaches its greatest abundance. A second major feature of the zone is that members of the order Rhynchonellida are more abundant in it than in faunas of

Table 9.2. *Ammonoid, conodont, and fusulinid zonations near the Permo-Triassic boundary in South China*

SEQUENCES		AMMONOID ZONES	CONODONT ZONES	FUSULINID ZONES	
SERIES	STAGE				
LOWER TRIAS.	LOWER GRIES-BACHIAN	*?Otoceras, Hypophiceras*	*Isarcicella? parva*		
UPPER PERMIAN	CHANGXINGIAN	*Rotodiscoceras-Pseudotirolites*	*Neogondolella deflecta-N. changxingensis*	*Palaeofusulina* zone	*Palaeofusulina sinensis-Reichelina changxingensis*
		Tapashanites-Shevyrevites	*Neogondolella subcarinata-Neogondolella wangi*		*Palaeofusulina minima-Nanlingella simplex* subzone
	WUJIAPINGIAN	*Sanyangites, Araxoceras- Konglingites*	*Neogondolella orientalis*	*Codonofusiella* zone	*Codonofusiella* spp.
		Anderssonoceras-Prototoceras	*Neogondolella liangshanensis*		*Codonofusiella lui* subzone

*Refer to Yao *et al.*, 1980; Zhao *et al.*, 1981; and Yang *et al.*, 1987, emended

underlying or overlying zones. A third feature is the abundant occurrence of large-sized representatives of genera such as *Dictyoclostus*, *Spinomarginifera*, *Edriosteges*, and *Tyloplecta*.

Spirigerella discusella–Acosarina minuta
Assemblage Zone (abbreviated S–A Zone).

This zone, which characterizes upper Changxingian limestone lithofacies, is distinguished by small specimens of such species as *Acosarina minuta* (Abich), *A. indica* (Waagen), *Crurithyris pusilla* Chan, *Spirigerella discusella* Xu & Grant, *Uncinunellina theobaldi* Waagen, and *Rugosomarginifera chengyaoyensis* (Huang), and medium- and large-sized representatives of, for example, *Spinomarginifera alphus* (Huang), *Meekella langdaiensis* Liao, and *Perigeyerella altilosina* Xu & Grant.

Crurithyris pusilla Chan–Lingula subcircularis
Wirth Assemblage Zone (abbreviated C–L Zone)

Collections of brachiopods from lower Griesbachian rocks include an abundance of *Lingula* and *Crurithyris* as well as specimens of several Permian- or Changxingian-type survivors such as *Waagenites barusiensis*, *Cathaysia sinuata*, *C. orbicularis* Liao, and *Fusichonetes pigmaea* (Liao).

Relationship to other biozones

Typical elements of the S–H Zone, *Squamularia indica*, *Haydenella wenganensis*, and *Orthothetina speciosa* occur in the same beds with the fusulinid *Codonofusiella lui* Sheng in the Jiaozishan section (Fig. 9.1, locality A). The fusuline is an index

to the lower subzone of the *Codonofusiella* Zone, which is equivalent to the Wujiapingian Stage. The *Neogondolella liangshanensis* conodont zone (see Table 9.2), recognized in the Mingyuexia section (Fig. 9.1, locality 2) is usually regarded as lower Wujiapingian and thus corresponds to the S–H Zone.

Squamularia grandis is associated with species of *Codonofusiella* in several sections, such as those at Jiaozishan and Guanyinshan (Fig. 9.1, localities A and 18). *Orthothetina ruber* is not likely to coexist with *Squamularia grandis* because the two species are commonly represented in different lithofacies. However, specimens can be preserved in alternate beds in sections in which limestone and mudstone are interbedded. On the basis of stratigraphic position and the fact that its guide fossils mostly occur below those distinctive of the *Neogondolella subcarinata–N. wangi* conodont zone and the *Tapashanites–Shevyrevites* ammonoid zone, the O–S Zone generally correlates with the *Neogondolella orientalis*, and the *Sanyangites* ammonoid zone. The latter two zones are in the lower part of the Changxingian Stage (Table 9.2). For example, the ammonoids *Shevyrevites* and *Paratirolites* occur in beds above ones with the brachiopod *Orthothetina ruber* in the Paoshui section (Fig. 9.1, locality 15), and *Squamularia grandis* occurs in beds below ones yielding the conodont *Neogondolella wangi* in the Huayingshan and Jiaozishan sections (Fig. 9.1, localities A and 6).

No identifiable ammonoids or fusulinids have been found in the lower Yanshi Formation in the Yanshi section (Fig. 9.1, locality 31), which is the type section of the C–C Zone. However, a bed in the upper Yanshi Formation about 20 m above the C–C Zone has yielded a specimen of the ammonoid *Pseudotirolites*, which is an index to the upper Changxingian (Table 9.2). Several typical members of the C–C Zone fauna, *Cathaysia chonetoides*,

Prelissorhynchia pseudoutah, and *Waagenites wongiana* occur in the same beds with the ammonoids *Tapashanites* and *Pseudogastrioceras* and the fusulinid *Palaeofusulina* in the Paoshui Formation of the Meitian section (Fig. 9.1, locality 23). A rich ammonoid fauna of 17 species, including *Pseudogastrioceras gigantum* Zhao, *Pseudostephanites meishanensis* Yang & Huo, and *Tapashanites* spp., as well as conodonts of the *Neogondolella subcarinata* Zone occur in the same beds with brachiopods of the C–C Zone in the lower Changxingian of the Meishan section (Fig. 9.1, locality 26). In that section, however, the C–C Zone extends upward into beds with the conodont *Neogondolella changxingensis* Wang & Wang, which is regarded as an upper Changxingian species (Table 9.2). This implies that the upper boundary of the C–C Zone is above that of the *Neogondolella subcarinata* Zone. In the Paoshui section, some important members of the C–C Zone fauna, such as *Waagenites wongiana* and *Prelissorhynchia pseudoutah* are associated with the upper Changxingian ammonoids *Rotodiscoceras*, *Pseudotirolites*, and *Pleuronodocereas* in the lower part of the Paoshui Formation. This mixture suggests that the top of the C–C Zone is higher than that of the *Shevyrevites* Zone, which occurs only near the base of the Paoshui Formation in the Paoshui section.

The principal fossils of the P–P Zone, such as *Peltichia zigzag*, *Prelissorhynchia triplicatioid*, *P. pseudoutah*, and *Spinomarginifera kueichowensis* occur together with the fusulinid *Palaeofusulina* sp. and the conodonts *Hindeodus typicalis* and *Neogondolella wangi* in the lower part of the Changxing Formation in the Huayingshan section (Fig. 9.1, locality 6). In the Liangfengya section (Fig. 9.1, locality 8) *Prelissorhynchia triplicatioid*, *P. pseudoutah*, and *Araxathyris beipeiensis*, important elements of the P–P Zone fauna, are associated with the fusulinids *Nanglingella simplex* (Sheng & Chang) and *Reichelina changxingensis* (Sheng & Chang), which in most sections occur in the lower Changxingian (Table 9.2). *Peltichia zigzag*, however, is preserved together with the conodont *N. changxingensis* in the upper part of the Changxing Formation. This may imply that the P–P Zone extends into the upper Changxingian, as perhaps does the C–C Zone.

Cathaysia sinuata and *Waagenites barusiensis*, typical of the C–W Zone, occur together with the ammonoid *Pleuronodoceras* and the conodont *Neogondolella changxingensis* in the same bed of the upper Yanshi Formation in the Yanshi section (Fig. 9.1, locality 31). In the upper part of the Changxing Formation at Meishan (Fig. 9.1, locality 26), *Crurithyris pusilla* and *Orthothetina regularis*, members of the C–W Zone fauna, occur together in the same beds with the conodont *N. changxingensis*, and in a bed near the top of the Changxing Formation in the Majiashan section (Fig. 9.1, locality 24), the conodont *N. deflecta* Wang and Wang is associated with *Crurithyris pusilla* and *Waagenites barusiensis*. Moreover, *C. pusilla* and *Cathaysia sinuata* co-occur with the ammonoids *Pleuronodoceras* and *Pseudotirolites* and the conodont *N. deflecta* in the upper part of the Changxing Formation in the Shangsi section (Fig. 9.1, locality 4). Finally, in the Meitian section (Fig. 9.1, locality 23), beds that contain the C–W Zone overlie beds that yield the ammonoid *Tapashanites* sp. and the conodont *N. subcarinata*, which are elsewhere associated with brachiopods of the C–C Zone.

In the Huayingshan section (Fig. 9.1, locality 6) the S–A Zone fauna, including *Acosarina minuta*, *A. indica*, *Rugosomarginifera chengyaoyensis*, *R. sintanensis* (Huang), *Uncinunellina theobaldi*, and *Spirigerella discusella* is represented in the upper Changxing Formation, which also yields the conodonts *Neogondolella deflecta* and *N. changxingensis*. *Rugosomarginifera chengyaoyensis*, *Uncinunellina theobaldi*, and *Spinomarginifera alphus* are also associated with the fusulinid *Palaeofusulina* and the conodonts *N. deflecta* and *N. changxingensis* in the Liangfengya section (Fig. 9.1, locality 8). In a number of places, for example in the Longtoujiang section (Fig. 9.1, locality 14), principal members of the S–A Zone fauna, such as *Acosarina minuta* and *Uncinunellina theobaldi*, are mixed in the upper Dalong Formation with elements of the C–W Zone fauna, such as *Waagenites barusiensis* and *Crurithyris pusilla*, and commonly co-occur with the ammonoid *Pseudotirolites*.

As is well known, *Lingula* and *Crurithyris* have long ranges. However, the ranges of *Crurithyris pusilla* and *Lingula subcircularis* overlap only in the lower Griesbachian. Permian-type survivors are typical of the C–L Zone. In the Huayingshan section (Fig. 9.1, locality 6), *Crurithyris pusilla* is associated with *Lingula subcircularis* and *L. borealis* Bittner in the basal beds of the lower Daye Formation, which also contain *Acosarina minuta*. This brachiopod association co-occurs with the ammonoid *Ophiceras* spp. and the bivalves *Claraia griesbachi* (Bittner) and *Pseudoclaraia wangi* (Patte). Associations like this can be found in many sections, such as Xinjiacao, Yanjingxi, Liangfengya, Yading, Baimu, and Paoshui (Fig. 9.1, localities 5, 7, 8, 30, 19, 15, respectively). In several sections the Permian-type survivors are rather abundantly represented in the basal beds of the lower Griesbachian. For example, the basal beds of the Yinkeng Formation in the Meishan section (Fig. 9.1, locality 26) include many Permian-type brachiopods, such as *Waagenites barusiensis*, *Fusichonetes pigmaea*, *Cathaysia sinuata*, *C. orbicularis* Liao, *Acosarina* sp., and *Araxathyris* sp., which are associated with the ammonoids *Hypophiceras*, *Ophiceras*, and ?*Otoceras* (fide Sheng et al., 1984). At the base of the Yinkeng Formation in the Majiashan section (Fig. 9.1, locality 24), *Cathaysia orbicularis* and *Waagenites barusiensis* co-occur with *Ophiceras*, and a similar situation can be found in the Shatian, Guanyinshan, and Meitian sections (Fig. 9.1, localities 17, 18, 23).

In summary, the brachiopod assemblage zones recognized in this report may be correlated with ammonoid, conodont, and fusulinid zones in the upper Permian (Table 9.2). However, the upper boundaries of the C–C and P–P zones may be somewhat higher than those of the ammonoid and conodont zones in the lower Changxingian. The C–L Zone, in the lower Griesbachian, coincides grossly with the ammonoid zone of *Hypophiceras* and ?*Otoceras* (Table 9.2).

Correlations with other areas in the Tethyan region

Previously described localities at which upper Permian and lower Griesbachian strata form a continuous succession are virtually all within the Tethyan region. Thus, brachiopod faunas that may be correlated with those of South China are known from north Tibet, Transcaucasia, northwest Iran, the Salt Range of Pakistan, Kashmir, the Southern Alps, northwest Nepal, and central east Greenland.

North Tibet

A Changxingian brachiopod fauna has been discovered in the Shuanghu area of north Tibet (Long. 86.8 E, Lat. 33.6 N). Jing & Sun (1981) reported *Peltichia zigzag* and *Rugosomarginifera pseudosintanensis* (Huang) from the Reggyorcaka Formation and described additional brachiopods, including *Cathaysia chonetoides*, *Squamularia waageni* Loczy, and *Leptodus* sp. from the lower part of the same formation. In the Tanggula-Qamdo region, the *Peltichia sinensis–Squamularia superb* assemblage in the Toba Formation represents the Changxingian (Jing & Sun, 1981). The Tibetan faunas undoubtedly correlate with those of the P–P Zone of South China. Therefore, the lower part of the Reggyorcaka Formation and some part of the Toba Formation may belong in the lower Changxingian Stage.

Transcaucasia

In Transcaucasia, the Dzhulfa section was divided into 15 units by Stoyanow (1910; see Ruzhentsev & Sarycheva, 1965; Teichert *et al.*, 1973). Units 1 through 6 contain brachiopods. The fauna of Unit 5, with *Orthis indica* Waagen and *Marginifera spinocostata* (Abich), was named the 'zone of *Productus djulfensis* Stoyanow'. *Orthotichia indica* (Waagen) (= *Orthis indica*) has also been recorded from various levels in the lower and upper Permian of South China. Liao (1980) reported *Stepanoviella djulfensis* (Stoyanow) (= *Productus djulfensis*) from the upper Longtan and the Changxing Formation in Guizhou Province. When they synthesized information from five Transcaucasian sections (including the ones at Dorasham), Arakelyan, Grunt, & Shevyrev (1965, in Ruzhentsev & Sarycheva, 1965, p. 25) wrote that 'horizon 3 with *Bernhardites*' has *Araxathyris araxensis minor* Grunt, and that 'horizon 4 with *Paratirolites*' possesses *Enteletes dzhagrensis* Sokolov, *Orthotichia parva* Sokolov, *Orthothetina* sp., *Spinomarginifera pygmaea* Sarycheva, *Haydenella kiangsiensis* (Kayser), *H. minuta* Sarycheva, *Terebratuloidea* sp., *Araxathyris ogbinensis* Grunt, and *A. araxensis minor*. The brachiopod fauna shown by Sarycheva, Sokolov, & Grunt in table 9 of Ruzhentsev & Sarycheva (1965) has the same composition as the one mentioned from 'horizon 4', 454 specimens collected from the 'Induan Stage' at three localities (Dorasham, Ogbin, and Prochie). As Teichert *et al.* (1973) pointed out, 389 of the 454 specimens recorded from the 'Induan' were assigned to one species, *A. a. minor*. Representatives of this subspecies have allegedly been found in the Yinkeng Formation in the Meishan section of South China (Sheng *et al.*, 1984). In several sections in Guizhou Province, specimens of *Araxathyris araxensis* Grunt have been collected from the upper Longtan Formation and the lower Changxing Formation, and in sections in Sichuan Province representatives of the species have been found in the Changxing Formation. *Araxathyris ogbinensis* has been reported from the upper Longtan Formation of west Guizhou Province (Liao, 1980). *Haydenella kiangsiensis* (Kayser) occurs widely from lower to upper Permian in South China, but is best known from the Wujiapingian. Based on the foregoing, the brachiopod fauna of the Dzhulfa section (including the Dorashamian Stage) of Transcaucsia can be correlated with those of the O–S and C–C zones of South China.

Kuh-e-Ali Bashi, Iran

Teichert *et al.* (1973, p. 382) report that they collected 10 brachiopod specimens from the Ali Bashi Formation, at the Kuh-e-Ali Bashi locality in northwest Iran. Those specimens were identified by G. A. Cooper as *Araxathyris araxensis minor* Grunt and *Araxathyris* sp., which, as noted previously, are the main members of the Dorashamian brachiopod fauna.

Abadeh region, central Iran

In the Abadeh region of central Iran, the Surmaq, Abadeh, and Hambast formations contain numerous brachiopods. The Iranian–Japanese Research Group (1981) considers that the brachiopod fauna of Unit 1, in the Surmaq section, is intimately related to that of the Gnishik and Khachik beds of Transcaucasia. *Squamularia indica* (Waagen), which is represented from horizon R 18 to horizon R 21, and *Edriosteges poyangensis* (Kayser), which is recorded from horizon R 2 and from horizon R 18 to R 21 of Unit 1, are typical elements of the S–H Zone. Therefore, Unit 1 corresponds in part to the lower Wujiapingian of South China.

Twenty-one brachiopod species are represented in Unit 4 of the Abadeh Formation. Of these species, 6 range upward from Unit 1 of the Surmaq Formation and 6 range upward into Unit 6 of the Hambast Formation. The remainder are limited in their occurrence to Unit 4. The Iranian–Japanese Research Group (1981) pointed out that, based on common brachiopod species, Unit 4b can reasonably be compared with the Khachik Formation of Transcaucasia and part of the Nesen Formation of the Elikah Valley in the Alborz Mountains. Seven species from Unit 4 (*Phricodothyris asiatica* (Chao), *Tyloplecta yangtzeensis* (Chao), *Spinomarginifera lopingensis* (Kayser), *Orthothetina regularis* (Huang), and *Leptodus nobilis* (Waagen)) are also members of the Wujiapingian brachiopod fauna. If one notes that *P. asiatica* and *T. yangtzeensis* are abundantly represented in the O–S Zone, it is possible that Unit 4 of the Abadeh Formation correlates with the upper Wujiapingian of South China.

The Iranian–Japanese Research Group (1981) concluded on the basis of their fusulinids that Unit 5 of the Abadeh Formation and Unit 6 of the Hambast Formation correlate with the Wujiapingian, and that ammonoids from Unit 7 of the Hambast Formation indicate a comparison with the Dorashamian. Because Unit 6 has yielded specimens of *Tyloplecta yangtzeensis*, *Leptodus nobilis*, *Araxathyris araxensis*, and *A. a. minor*, the Hambast Formation appears to include equivalents of both the O–S and P–P zones.

Twenty-four brachiopod species were recorded from the Nesen Formation by the Iranian–Japanese Research Group. Many of these, such as *Phricodothyris asiatica* and *Tyloplecta yangtzeensis* are typical Wujiapingian forms. Hence the Nesen Formation can undoubtedly be correlated in part with the Wujiapingian of South China.

Salt Range, Pakistan

In the Chhidru Formation of the Salt Range and Trans-Indus ranges of Pakistan, brachiopods are poorly preserved and represent long-ranging species; hence the age of the Chhidru brachiopod fauna is difficult to determine. Identifiable specimens were found by Grant (1970) in the topmost bed of the white sandstone member at two localities in the Khisor Range, but most of the species represented have long ranges.

The Pakistani–Japanese Research Group (1981, 1985) recognized that five faunal assemblages, termed K and C1 to C4, can be distinguished in the Chhidru-I section. In their opinion, brachiopods from their zones C1 to C3, which are in Units 2 and 3 of the Chhidru Formation, indicate a correlation with the Wujiapingian and Changxingian of South China. In evaluating this opinion, however, we note that (1) most of the brachiopod species have long ranges (e.g., *Spirigerella derbyi* (Waagen) is recorded in South China from the Maokou to the Wujiaping formations; *Oldhamina decipiensis* (Koninck) and *Waagenoconcha abichi* Waagen range from the Wujiaping to the Changxing Formation; and most species of *Kiangsiella* and *Marginifera* have long pre-Changxingian ranges in South China); (2) several species, such as *Waagenites deplanata* (Waagen) and *Chonetella nasuta* (Waagen) are members of the brachiopod faunas of the S–H and O–S zones in South China, which we regard as largely or entirely Wujiapingian in age; and (3) no brachiopods typical of either the Changxingian or Dorashamian faunas have been reported from the Chhidru Formation. From these considerations, we conclude that brachiopods indicate that faunal assemblages C1 through to C3 of the Chhidru Formation are most likely Wujiapingian. Powerful collateral evidence is provided by the fact that, among fusulinids, only *Codonofusiella* occurs in the Chhidru; *Palaeofusulina* has not been reported. Faunal assemblage C4 includes only seven species, four with a long range, and three that range upward from faunal assemblage C3. It is thus difficult to say what is the age of faunal assemblage C4.

Several brachiopods were collected by Kummel & Teichert (1970) in the Salt Range and Surghar Range, Pakistan, from beds in the Kathwai Member of the Mianwali Formation that also include *Ophiceras connectens*. These specimens were identifed by G. A. Cooper, who pointed out that they are of Permian type and, because of their fragmentary preservation, might have been reworked from the underlying Chhidru Formation. Several reasonable arguments against the reworking hypothesis were enumerated by Grant (1970). Brachiopods identified from the basal beds of the Kathwai Member by Grant (1970) include *Crurithyris? extima* Grant, *Derbyia?* sp., dielasmatid undet., *Enteletes* sp. 2, *Lingula* sp., *Linoproductus* sp., *Lyttonia* sp., *Martinia* sp., *Ombonia* sp., *Orthothetina* sp. cf. *O. arakeljani* Sokotskaya, *Orthothetina* sp., *Spinomarginifera* sp., *Spirigerella* sp., and *Whitspakia* sp. 2. A lingulid, identifed by Rowell (1970) as *Lingula* sp. cf. *L. borealis* Bittner, and a specimen of *Orbiculoidea* were recorded by Kummel & Teichert (1970). Numerous specimens of *Crurithyris? extima* Grant occur 5 to 6 ft above the base of the dolomitic unit of the Kathwai Member at Khan Zaman Nala, well into the Triassic *Ophiceras* zone (Grant, 1970), and one specimen of *Spinomarginifera* was collected several inches above the lowest occurrence of the Triassic ammonoid *Ophiceras*. The position and mode of occurrence of the Kathwai brachiopod fauna thus resembles that of the lower Griesbachian C–L Zone of South China. That is, common to both faunas are *Crurithyris*, *Lingula*, and Permian-type survivors such as *Spinomarginifera*, *Enteletes*, *Orthothetina* and *Spirigerella* co-occurring with *Ophiceras*. However, typical Changxingian survivors, such as *Cathaysia*, *Waagenites*, and *Prelissorhynchia*, are not known from the Kathwai, and *Lyttonia* and *Linoproductus* have not been found in the C–L Zone of South China.

Kashmir

The brachiopod fauna of Division IV or E1 of the lowermost Khunamuh Formation, in the Guryul Ravine section, Kashmir, consists of *Athyris* sp. cf. *A. subexpansa* Waagen, *Dielasma?* sp., dictyoclostid?, *Derbyia* sp., *Linoproductus* sp. cf. *L. lineatus* (Waagen), *Lissochonetes morahensis* (Waagen), *Marginifera himalayensis* Diener, *Neospirifer* sp., *Pustula* sp., *Schellwienella* sp., and *Waagenoconcha purdoni* (Waagen), reported by Nakazawa *et al.* (1975), and *Chonetes lissarensis* Diener, *Chonetes?* sp. aff. *C. variolata* d'Orbigny, *Spinomarginifera* sp. cf. *S. helica* (Abich), *Waagenoconcha abichi* (Waagen), and *W. gangetica* (Diener), which were described by Diener (1915). This fauna is essentially similar to those of the underlying divisions of the Zewan Formation. *Waagenoconcha abichi*, *W. purdoni*, *Costiferina indica*, and *Linoproductus lineatus* are represented in K through to C3 assemblages of the Chhidru Formation, in the Salt Range of Pakistan, and in central Iran *Spinomarginifera helica* ranges upward from Unit 1 of the Surmaq Formation through Unit 7 of the Hambast Formation. As a whole, the brachiopod fauna of Division IV of the lowermost Khunamuh Formation may be correlated with those of the Chhidru Formation and thus with the Wujiapingian. In bed 52 of unit E2 of the Khunamuh Formation, specimens of *Marginifera himalayensis* Diener and *Pustula* sp. occur together with representatives of the ammonoids *Otoceras woodwardi* and *Glyptophiceras himalayanum* (Nakazawa *et al.*, 1975). The two Permian-type brachiopods are thus in the same stratigraphic position as the C–L Zone, but in the absence of common species the two intervals can not be compared with each other.

Southern Alps, Italy

In the Southern Alps of Italy, a brachiopod fauna that includes *Ombonia*, *Araxathyris*, *Janiceps*, *Comelicania* and rare spiriferaceans and dielasmataceans, occurs in the upper part of the Bellerophon Formation. In the Pusteria Valley (near Sesto), the brachiopod fauna is accompanied by the ammonoid *Paraceltites sextensis* (Diener) (Assereto *et al.*, 1973). Waterhouse (1976) considered this a Dorashamian fauna, but Assereto *et al.* (1973) thought it to be pre-Dorashamian in age. Species of *Araxathyris* dominate the brachiopod fauna in the Dorashamian of Transcaucasia; *Ombonia* sp. was described by Grant (1970) from the Kathwai Member of the Mianwali Formation in the Surghar Range, Pakistan; and *Janiceps janiceps* (Stache) has been reported from the Longtan Formation of Guizhou Province, South China, by Liao (1980). Thus, considering only the brachiopod fauna, the uppermost part of the Bellerophon Formation may be correlated with the Wujiapingian and part of the Changxingian. An analogous fauna, with *Crurithyris*, *Marti-*

nia, Comelicania, and licharewinids is known from the upper Permian of the Bukk Mountains of Hungary (Schreter, 1963). Assereto *et al.* (1973) considered the Hungarian fauna to be older than the one they reported from the Southern Alps.

East Greenland

The *Martinia* beds, which yield brachiopods and the ammonoid *Cyclolobus,* were considered to be the uppermost beds of the Paleozoic Erathem by Miller & Furnish (1940), and Trümpy (1961) believed that Permian-type brachiopods occur in beds with the ammonoids *Glyptophiceras* and *Otoceras* in the Kap Stosch area of East Greenland. Teichert & Kummel (1976) reported that the '*Martinia* Shale' on the north bank of River Zero (Ekstraelv) contains representatives of *Chonetina noenygaardi, Martinia greenlandica,* and *Liosotella hemispherica.* The latter is closely similar to *L. magniplicata,* which ranges from Unit 1 to Unit 4b in the section in the Abadeh region of central Iran. Otherwise, brachiopods from the Kap Stosch region of East Greenland are almost all endemic species and it is difficult to correlate the rocks in which they occur with those of other regions.

Northwest Nepal

In northwest Nepal, Waterhouse (1978) recognized and named a *Marginalosia kalikotei* Zone in the Nisal, Nambo, and Luri members of the Senja Formation. The brachiopod fauna of this zone is comprised of *Marginalosia kalikotei* (Waterhouse), *Megasteges nepalensis* Waterhouse, *Neospirifer ravaniformis* Waterhouse, *Platyconcha grandis* Waterhouse, *Spiriferella oblata* Waterhouse, and *S. rajah* (Salter). The latter two are dominant in the upper part of the zone, in the Luri Sandstone Member. Waterhouse suggests that the following pairs of species are similar: *Rugaria nisalensis* Waterhouse of Nepal and *R. soochowensis* (Chao) (= *Waagenites soochowensis*) of the Lopingian of China; *Marginalosia kalikotei* of Nepal and *M. planata* (Waterhouse) of the Stephens Formation (Vedian) of New Zealand; and *Martiniopsis* sp. aff. *M. inflata* Waagen and the same species from the Chhidru Formation of the Salt Range, Pakistan. Based on these comparisons, Waterhouse (1978) concluded that 'this could imply a Djulfian or early Dorashamian age'.

Waterhouse (1976) also named an *Aperispirifer nelsonensis* Zone, which he correlated with the Vedian of Transcaucasia and tentatively with the Changxing Formation of South China. However, he did not expound in detail about the zone. In his discussion of the age of the Stephens Limestone of South Island, New Zealand, Waterhouse concluded it should correlate with the Tatarian, based chiefly on a similarity between *Neospirifer nelsonensis* Waterhouse (= *Aperispirifer nelsonensis*) and Greenland shells described as '*Spirifer*' *striato-paradoxus* Toula by Dunbar (1955). Representatives of the Greenland species occur with *Cyclolobus,* which Dunbar (1955) considered Tatarian in age (Waterhouse, 1976). However, according to Stepanov (1973), the Tatarian Stage includes both the Dzhulfian and Dorashamian; and, from consideration of the brachiopod fauna, correlation of the *Aperispirifer nelsonensis* Zone with the Changxingian is questionable.

Brachiopod events

In order to reconstruct the developmental history of organisms and improve the accuracy of correlation in stratigraphy, it is particularly important to study organic evolutionary events. These may be considered at several levels, as indicated in Table 9.3. Because the present contribution deals only with the stratigraphic significance of events in the evolutionary history of Permo-Triassic brachiopod faunas, reasons for the classification of events shown in Table 9.3 will be discussed in another place.

Biospheric events (of rank IV) can be used to mark systemic or erathemic boundaries. For example, the mass extinction of reef-dwelling brachiopods is an important character of the Permo-Triassic boundary interval. Phylogenetic events (of rank III), such as extinction of the order Strophomenida or the proliferation of *Lingula,* are natural markers of stadial or biozonal boundaries. These events are described in subsequent paragraphs. Microevolutionary events (of ranks I and II) in the history of Permo-Triassic brachiopod faunas are beyond the scope of this report, however, and are not discussed.

Mass extinction of reef-dwelling brachiopods

Based on huge collections from west Texas, Grant (1971) recognized three ecologic categories of brachiopods: reef dwellers, antireef dwellers, and neutral or ubiquitous forms. Major contributors to the reef framework included the Prorichthofeniidae, Scacchinellidae, and Lyttoniidae, which adopted a coralliform shape or clustered together by means of direct cementation or entanglement of spines to form reefy frameworks of their own. Other reef dwellers, such as the Meekellidae, Orthotetidae, Aulostegidae, and Enteletidae, attached to reef frameworks formed by sponges, bryozoans, or algae and may be termed reef-attaching brachiopods.

In South China, reefs or bioherms extend to the uppermost part of the Changxingian and are widely distributed on the Upper Yangtze Platform. Localities with Changxingian reefs or bioherms include Longdongchuan, Zhenan County, Shaanxi Province; Lichuan, Hubei Province; Huayingshan (Fig. 9.1, locality 6), Huatang (Fig. 9.1, locality C), Shitouzhai (Fig. 9.1, locality B), and Xiangbo, Longlin County, Guangxi Province.

Changxingian reef-dwelling brachiopods of South China are closely similar, at least at the family level, to those of west Texas. Forms such as *Richthofenia* or *?Richthofenia,* with coralliform shells, *Meekella* and *Perigeyerella,* with a higher interarea, and *Araxathyris, Peltichia, Enteletes, Notothyris,* and *Rostranteris,* which attached by the pedicle, are found in Changxingian reefs. At localities such as the ones at Huatang (Fig. 9.1, locality C) and in the Longdongchuan region, they are rather numerous. *Oldhamina* and *Leptodus,* with oyster-shaped shells, lived in reefs or their shells piled up to form bioherms. Most Changxingian reefs, however, were built up by sponges and algae, hence brachiopods living in them are only reef-attaching organisms.

At the Permo-Triassic boundary, Permian-type reefs, and thus reef-dwelling brachiopods, were completely extinguished in South China – perhaps all over the world. Specimens of *Richthofenia,* the last of the superfamily Richthofeniacea, have been found in the upper part of the Huatang Formation (Liao &

Table 9.3. *Different classifications of organic evolutionary events*

EVENT RANKS	Goldschmidt 1940	Simpson, 1953	Stanley, 1979	Gould, 1985	Liu, 1985		This paper
IV	Macro-evolution: evolution above species	Macro-evolution: research area of paleontologists	Macro-evolution: phylogenetic drift; directed speciation; species selection	Third level: characterized by mass extinction	Biospheric level: replacement of ecosystem and organic group	Macroevolution	Biospheric events: mass extinction, adaptive radiation and replacement of ecosystem
III				Second level: geologic time; characterized by punctuated equilibria	Population level: punctuated equilibria		Phylogenetic events: change above species and speciation
II	Micro-evolution: change within species	Micro-evolution: research area of biologists	Micro-evolution: 3. natural selection	First level: ecologic niche; adaptation to environment		Microevolution	Phyletic events: change in population size adaptation migration pseudoextinction
I			1. genetic drift 2. mutation pressure		Molecular level: mutation, gene recombination		Molecular events: genetic mutation

Meng, 1986), and questionable representatives of the same genus have been collected from the upper Longdongchuan Formation. Considering that the Family Prorichthofeniidae died out at a time corresponding to the end of deposition of the Word Formation, these occurrences mark extinction of the Superfamily Richthofeniacea. The last occurrence of the Scacchinellidae may be below the highest occurrence of the Prorichthofeniidae, but two families, the Aulostegidae and Tschernyschewiidae, which are closely related to the former, made attachment by means of entangling spines and are represented in the Changxingian, for example in the Huatang reef.

Large, oyster-shaped shells of lyttoniacean brachiopods such as *Oldhamina* and *Leptodus*, existed as reef-attaching dwellers in Changxingian reefs, for example at Shitouzhai (Fig. 9.1, locality B) and in the Huatang region (Fig. 9.1, locality C). However, several good specimens of *Lyttonia* have also been reported from the dolomite unit of the Kathwai Member of the Mianwali Formation at Narmia Spring, Pakistan, which is considered to be equivalent to the C–L brachiopod zone and thus probably Triassic. Also, specimens of *Bactrynium*, a survivor of the Lyttoniacea, have been recorded from the Rhaetic Stage (latest Triassic) from Austria. These survivors of the superfamily were obviously not reef dwellers, for all specimens of *Lyttonia* are well below the median size of shells representing other species at lower stratigraphic levels (Grant, 1970), and the Rhaetic shells of *Bactrynium* were attached by the ventral apex, lacked an interarea, and were thus to a large extent different from the large, oyster-shaped reef dwellers.

Coincident with extinction of reef-dwelling brachiopods, the overwhelming majority of the suborders Strophomenidina and Productidina also died out. Only a few representatives of these groups have been discovered in the lower Griesbachian of the Tethyan region.

Mass extinction of reef-dwelling brachiopods marked a great change in the megaecosystem at the biospheric level. This implies that conditions in the marine environment fluctuated greatly and rapidly near the boundary between the Paleozoic and Mesozoic and that reef-building and reef-attaching organisms could not adapt to these variations and largely died out. As is well known, all types of reefs require certain stable conditions in the marine environment: normal, shallow water; stable, warm temperature; ample illumination; and sufficient sources of nourishment. Thus we may deduce that conditions greatly worsened for reef dwellers at the beginning of the Triassic and did not improve, at least through the Griesbachian. We still do not know what caused the fluctuating and worsening conditions, but it is nevertheless important to point out that virtual extinction of reef dwellers at the end of the Changxingian was a major evolutionary event that greatly altered the megaecosystem.

Extinction of the Strophomenida

The cohort of Permian-type brachiopods that survived into the early Triassic is composed principally of species representing the order Strophomenida. About 50% of the surviving genera in South China are attributed to the superfamilies Chonetacea (*Chonetinella*, *Fanichonetes*, *Fusichonetes*, *Waagenites*), Productacea (*Cathaysia*, *Rugosomarginifera*), and Enteletacea (*Acosarina*). Except for members of the Strophomenida and Orthida, brachiopod genera represented in the lower Griesbachian have a long range and are really not Permian-type survivors. *Crurithyris*, for example ranges from Devonian to upper Griesbachian, and *Prelissorhynchia* (of the Wellerellacea) and *Paraspiriferina* (of the Spiriferininacea) are Mesozoic pioneers.

On the basis of their study of the Meishan section (Fig. 9.1, locality 26) Sheng *et al.* (1984) divided the 'transitional beds' of South China, in which Permian-type and Griesbachian faunas are mixed together, into three beds. Permian-type brachiopods were collected, in effect, only from beds 1 and 2, and the top of the latter was placed about 20 cm above the Permo-Triassic boundary. Only *Prelissorhynchia* was recorded from mixed bed 3. According to Yin (1983), who was the first to define them, the transitional beds are restricted to the lower Griesbachian; therefore, Sheng's bed 3 is not included in the transitional beds. In the Majiashan section (Fig. 9.1, locality 24), *Waagenites barusiensis* and *Cathaysia triquetra* were discovered about 15 cm above the boundary, in association with the ammonoid *Ophiceras*. *Fusichonetes pigmata* occurs about 90 cm above the boundary in the Guanyinshan section (Fig. 9.1, locality 18), and is found together with *Pseudoclaraia wangi* and *Cathaysia* spp. about 1 m above the boundary in the Shatian section (Fig. 9.1, locality 17). Thus, Permian-type brachiopods are mostly confined to the lower Griesbachian, despite co-occurrence in the Guanyinshan section with *Pseudoclaraia wangi*, which is considered as an important upper Griesbachian bivalve but occurs in the lower Griesbachian at some places in South China.

In the Kathwai Member of the Salt Range, Pakistan, Permian-type brachiopods are represented by *Lyttonia*; *Derbyia*? *Ombonia*, and *Orthothetina* of the superfamily Davidsoniacea; and *Spinomarginifera* and *Linoproductus* of the Productacea. No representative of the Chonetacea has been described (Grant, 1970), so the aspect of the fauna differs from that of South China. Extinction of the fauna is limited to the dolomitic unit, however, so the event took place in the early Griesbachian.

Proliferation of Lingula

In the Griesbachian, lingulids spread widely throughout the world. In addition to South China and Pakistan, species of *Lingula* have been reported from the Southern Alps, Italy, Greenland, Hungary, Japan, Australia, and Iran.

In the Southern Alps, specimens of *Lingula tenuissima* Bronn have been collected from the Mazzin Member of the Werfen Formation (Assereto *et al.*, 1973). The fauna it characterizes, termed the *Lingula–Neoschizodus* assemblage, includes *Bellerophon vaceki* Bittner, which occurs with *Claraia griesbachi* above the bed containing *Otoceras woodwardi* at Shalshal Cliff in the Himalaya. *Lingula* sp. cf. *L. borealis* Bittner occurs at the same stratigraphic level as *Claraia clarai* in the Siusi Member of the Werfen Formation (Broglio Loriga, Neri, & Posenato, 1980).

In Hungary, *Lingula tenuissima* occurs with *Costatoria costata* in Campil strata at Mecsek Mountain (Haas *et al.*, 1988);

and *L. borealis* has been reported from the lower *Ophiceras* bed in East Greenland (Spath, 1935).

In Iran, *Lingula* sp. is recorded from the lower part of the Elikah Formation, beneath a dolomite yielding the bivalve *Claraia*, which is associated with the conodonts *Hindeodus typicalis* and *Isarcicella isarcica* (Hirsch & Sussli, 1973).

In Japan *Lingula* is associated with the upper Scythian ammonoids *Columbites* and *Subcolumbites* in the Kitakami Massif (Murata, 1973). In the Maizuro Zone, *Lingula* sp. cf. *L. borealis* occurs in strata probably of early Scythian age (Bando, 1964), and *Lingula* is also represented in the Griesbachian (Broglio Loriga *et al.*, 1980).

In Australia, *Lingula* is present in the Perth, Canning, and Bonaparte Gulf basins. In the Perth Basin, the genus is represented about 1100 ft above the Permo-Triassic boundary, where it occurs with *Claraia stachei* and *C. perthensis* of the lower Scythian. In the Canning and Bonaparte Gulf basins, *Lingula* is associated with the plant microfossil *Kraeuselisporites saeptatus*, which is considered to be Griesbachian-Dienerian in age (Gorter, 1978).

In Siberia, *Lingula borealis* has been collected from the Induan Stage; and in western North America, Newell & Kummel (1942) report that a *Lingula* Zone overlies the basal siltstone of the Dinwoody Formation in western Wyoming, or rests on the Phosphoria Formation in the western part of the Owl Creek and Wind River Mountains. In the southeastern part of the Wind River Mountains this zone is overlapped by a *Claraia* Zone, in which rare *L. borealis* occurs. *L. borealis* is accompanied by *Ophiceras* and the brachiopods *Spiriferina mansfieldi* Girty and *Mentzelia* sp.? in the *Lingula* Zone.

Near the top of the Changxingian of South China, species of *Lingula* are represented in places that were a short distance away from oldlands, for example in west Guizhou near the Kangdian Oldland; in the Changxing area near the Cathaysia Oldland; and in the Yichun area near the Jiangnan Oldland. In such places, the lingulids are commonly preserved in fine-clastic rocks. These facts suggest that, as with living *Lingula*, Permo-Triassic species lived in the littoral zone. Thus, to a certain extent, one may safely deduce that the wide distribution of *Lingula* in the lower Griesbachian indicates that shallow water dominated in many regions.

References

Assereto, R., Bosellini, A., Fantini Sestini, N. & Sweet, W. C. (1973). The Permian–Triassic boundary in the Southern Alps (Italy). *Canadian Soc. Petrol. Geol. Mem.*, **2**: 176–99.

Bando, Y. (1964). The Triassic stratigraphy and ammonite fauna of Japan. *Sci. Rep. Tohoku Univ., ser. 2 (Geol.)*, **36**(1): 1–137.

Broglio Loriga, C., Neri, C. & Posenato, R. (1980). La 'Lingula Zone' delle Scitico (Triassico Inferiore). Stratigrafia e Paleoecologia. *Annali dell'Università di Ferrara, sez. 9*, **6**(6): 91–130.

Diener, C. (1915). The Anthracolithic fauna of Kashmir, Kanaur and Spiti. *Geol. Surv. India Mem., Palaeont. Indica, n. s.*, **5**(2): 135 pp.

Dunbar, C. O. (1955). Permian brachiopod faunas of central east Greenland. *Medd. om Grønland*, **110**(3): 1–169.

Goldschmidt, R. (1940). *The Material Basis of Evolution*. New Haven: Yale Univ. Press.

Gorter, J. D. (1978). Triassic environments in the Canning Basin, Western Australia. *BMR Jour. Austral. Geol Geophys.*, **3**(1): 25–33.

Gould, S. J. (1985). The paradox of the first tier: an agenda for paleobiology. *Paleobiology*, **11**(1): 2–12.

Grant, R. E. (1970). Brachiopods from Permian–Triassic boundary beds and age of Chhidru Formation, West Pakistan. In *Stratigraphic Boundary Problems: Permian and Triassic of West Pakistan*, eds. B. Kummel & C. Teichert, pp. 117–51. (Dept. Geol., Univ. Kansas, Spec. Pub. 4.) Lawrence, Kansas: Univ. Kansas Press.

Grant, R. E. (1971). Brachiopods in the Permian reef environment of West Texas. *Proc. N. Am. Paleont. Convention, 1969*, **J**: 1444–81.

Haas, J., Góczán, F., Oravecz-Scheffer, A., Barabás-Stuhl, Á., Majoros, G. & Bériczi-Makk, A. (1988). Permian-Triassic boundary in Hungary. *Mem. Soc. Geol. Ital.*, **34** (1986): 221–41.

Hirsch, F. & Sussli, P. (1973). Lower Triassic conodonts from the lower Elikah Formation, central Alborz Mountains (northern Iran). *Eclog. Geol. Helv.*, **66**: 525–35.

Hou, H. F., Zhan, L. P., Chen, B. W., *et al.* (1979). *The Coal-Bearing Strata and Fossils of Late Permian from Guantung*. Beijing: Geol. Publ. House.

Iranian–Japanese Research Group, 1981. The Permian and the Lower Triassic systems in Abadeh region, central Iran. *Mem. Fac. Sci. Kyoto Univ., ser. Geol. Mineral.*, **47**(2): 62–133.

Jing, Y. G. & Sun, D. L. (1981). *Palaeozoic brachiopods from Tibet. Tibet Palaeontology, Part III*. pp. 127–71. Beijing: Science Press.

Kummel, B. & Teichert, C. (1970). Stratigraphy and Paleontology of the Permian–Triassic boundary beds, Salt Range and Trans-Indus ranges, West Pakistan. In *Stratigraphic Boundary Problems: Permian and Triassic of West Pakistan*, eds. B. Kummel & C. Teichert, pp. 1–110. (Dept. Geol., Univ. Kansas, Spec. Pub. 4.) Lawrence, Kansas: Univ. Kansas Press.

Liao, Z. T. (1980). Upper Permian Brachiopods from western Guizhou. In *Stratigraphy and Palaeontology of Upper Permian Coal-bearing Formation in Western Guizhou and Eastern Yunnan*, ed. Inst. Geol. Palaeont., Academica Sinica, pp. 241–77. Beijing: Science Press.

Liao, Z. T. & Meng, F. Y. (1986). Late Changxingian brachiopods from Huatang of Chenxian County, South Hunan. *Mem. Nanjing Inst. Geol. Palaeont., Acad. Sinica*, **22**: 71–94.

Liu, D. Y. (1985). From mass extinction to mass replacement – a concurrent discussion on time-space levels of evolution and systems geology. *Acta Palaeont. Sinica*, **26**(3): 354–66.

Miller, A. K. & Furnish, W. M. (1940). *Cyclolobus* from the Permian of eastern Greenland. *Medd. om Grønland*, **112**(5): 1–8.

Murata, M. (1973). Triassic fossils from Kitakami Massif, northeast Japan. Part 2. Pelecypods and brachiopods of the Osawa and Fukkosi formations. *Sci. Reports Tohoku Univ., ser. 2 (Geol)*., Spec. vol. **6**: 267–76.

Nakazawa, K., Kapoor, H. M., Ishii, K., Bando, Y., Okimura, Y. & Tokuoka, T. (1975). The Upper Permian and the Lower Triassic in Kashmir, India. *Mem. Fac. Sci. Kyoto Univ., Ser. Geol. Mineral.*, **47**(2): 106 pp.

Newell, N. D. & Kummel, B. (1942). Lower Eo-Triassic stratigraphy, western Wyoming and southeast Idaho. *Geol. Soc. Am. Bull.*, **53**: 937–94.

Pakistani–Japanese Research Group (1981). *Stratigraphy and correlation of the marine Permian and Lower Triassic in the Surghar Range and the Salt Range, Pakistan*. Kyoto: Kyoto Univ.

Pakistani–Japanese Research Group (1985). Permian and Triassic systems in the Salt Range and Surghar Range, Pakistan. In *The Tethys – Her Paleogeography and Paleobiogeography from Paleozoic to Mesozoic*, eds. K. Nakazawa & J. M. Dickins, pp. 221–94. Tokyo: Tokai Univ. Press.

Rostovtsev, K. O. & Azaryan, N. R. (1973). The Permian–Triassic boundary in Transcaucasia. *Canadian Soc. Petrol. Geol. Mem.*, **2**: 89–99.

Rowell, J. A. (1970). *Lingula* from the basal Triassic Kathwai Member, Mianwali Formation, Salt Range and Surghar Range, West Pakistan. In *Stratigraphic Boundary Problems: Permian and Triassic of West Pakistan*, eds. B. Kummel & C. Teichert, pp. 111–16. (Dept. Geol., Univ. Kansas, Spec. Pub. 4.) Lawrence, Kansas: Univ. Kansas Press.

Ruzhentsev, V. E. & Sarycheva, T. G., (Eds) (1965). Development and change of marine organisms at the Paleozoic-Mesozoic boundary. *Akad. Nauk USSR, Trudy Paleont. Inst.* **108**: 431 pp.

Schreter, Z. (1963). Die Brachiopoden aus dem oberen Perm des Bükkgebirges in Nordungarn. *Geologica Hungarica*, **28**: 79–160.

Sheng, J. Z., Chen, C. Z., Wang, Y. G., Rui, L., Liao, Z. T., Bando, Y., Ishii, K. I., Nakazawa, K. & Nakamura, K. (1984). Permian–Triassic boundary in middle and eastern Tethys. *J. Fac. Sci., Hokkaido Univ., Ser. IV*, **21**(1): 133–81.

Sheng, J. Z., Chen, C., Wang, Y., Rui, L., Liao, Z. & Jiang, N. (1982). On the '*Otoceras*' beds and the Permian–Triassic boundary in the suburbs of Nanjing. *J. Stratigr.*, **6**(1): 1–8.

Simpson, G. G. (1953). *The Major Features of Evolution*. New York: Columbia Univ. Press.

Spath, L. F. (1935). Additions to the Eo-Triassic fauna of East Greenland. *Medd. om Grønland*, **98**(2): 1–120.

Stanley, S. M. (1979). *Macroevolution – Pattern and Process*. San Francisco: W. H. Freeman and Co.

Stepanov, D. L. (1973). The Permian System in the USSR. *Canadian Soc. Petrol. Geol. Mem.* **2**: 120–36.

Teichert, C. & Kummel, B. (1976). Permian–Triassic boundary in the Kap Stosch area, East Greenland. *Medd. Om Grønland*, **197**(5): 1–54.

Teichert, C., Kummel, B. & Sweet, W. C. (1973). Permian–Triassic strata, Kuh-e-Ali Bashi, northwestern Iran. *Bull. Mus. Comp. Zool.*, **145**(8): 359–472.

Trümpy, R. (1961). Triassic of East Greenland. In *Geology of the Arctic*, ed. G. O. Raasch, pp. 245–54. Toronto: Univ. Toronto Press.

Waterhouse, J. B. (1976). World correlations for Permian marine faunas. *Univ. Queensland, Dept. Geol., Papers*, **7**(2): 232 pp.

Waterhouse, J. B. (1978). Permian Brachiopoda and Mollusca from North-west Nepal. *Palaeontogr., ser. A*, **160**: 176 pp.

Yang, Z. Y., Wu, S. B. & Yang, F. Q. (1981). Permian–Triassic boundary in the marine regimes of South China. *Selected Papers and Abstracts of Papers Presented at the Fifth International Symposium*, p. 71–7.

Yang, Z. Y., Yin, H. F., Wu, S. B., Yang, F. Q., Ding, M. H. & Xu, G. R. (1987). *Permian–Triassic Boundary Stratigraphy and Fauna of South China*. Beijing: Geol. Publ. House.

Yao, Z. J., Xu, J. T., Zheng, Z. G., Zhao, X. H. & Mo. Z. G. (1980). Late Permian biostratigraphy and Permian–Triassic boundary problems in west Guizhou and east Yunnan. In *Stratigraphy and Paleontology of Upper Permian Coal-bearing Formation in Western Guizhou and Eastern Yunnan*, ed., Inst. Geol. and Paleont., Acad. Sinica. pp. 1–69. Beijing: Sci. Press.

Yin, H. F. (1983). Bivalves near the Permian–Triassic boundary in South China. *Geol. Review*, **29**(4): 303–20.

Yin, H. F. (1985). On the transitional bed and the Permian-Triassic boundary in South China. *Newsl. Stratigr.*, **15**(1): 13–27.

Zhao, J. K., Sheng, J. C., Yao, Z. J., Liang, X. L., Chen, C. C., Rui, L. & Liao, Z. T. (1981). The Changhsingian and Permian–Triassic Boundary of South China. *Bull. Nanjing Inst. Geol. Palaeont., Acad. Sinica*, **2**: 1–95.

10 Conodont sequences in the Upper Permian and Lower Triassic of South China and the nature of conodont faunal changes at the systemic boundary

DING MEIHUA

Introduction

In the Late Permian, a large-scale regression occurred worldwide and the extent of the sea shrank enormously. As a result, there is a stratigraphic gap between Permian and Triassic rocks in many areas of the world. South China, however, possesses the most widely distributed, fairly continuous marine Permo-Triassic sequences known anywhere in the world. Thus studies of South Chinese Permian and Triassic conodonts are helpful to research on conodont biostratigraphy, biofacies, and Permo-Triassic events and a great deal of work has been done by Wang & Wang (1981, 1983), Wang & Cao (1981), Wang & Dai (1981), Ding (1983), and Yang *et al.* (1987).

Wang & Wang (1981) divided the Permian and Triassic of South China into 15 conodont zones, four Upper Permian and 11 Lower Triassic, and this scale provides a good basis for future research. Wang & Wang's contributions deal mostly with the determination of stratigraphic age, however, and little attention

has been focused on the relationships between conodonts and sediment types or on the nature of the temporal changes in conodont faunas. The purpose of the present study is thus to discuss Permian and Triassic conodont faunas and to demonstrate that the distribution of conodonts in this interval is related to the distribution of sedimentary facies.

Stratigraphy

In South China, marine Permian and Triassic facies occur south of a line joining Lianyun Harbor (Jiangsu Province), Xiang Fan (Hubei Province), and Tianshui (Gansu Province). The area north of this line was continental; hence the Permo-Triassic sections considered in this report are all south of the line, on the Yangtze Platform, which lies east of the ancient Kangdian Land and west of Cathaysia. Permo-Triassic conodonts treated herein were collected from 13 measured sections (selected from 36) in eight provinces (Fig. 10.1). Correlation of

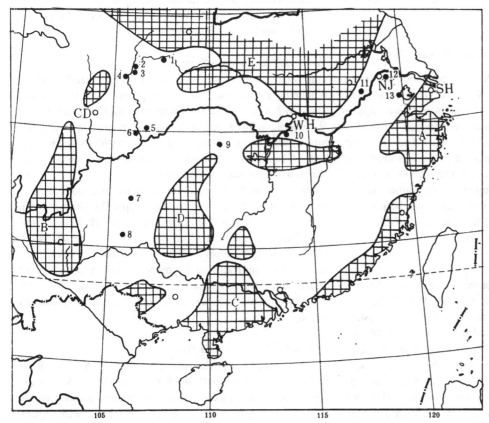

Fig. 10.1. Location of sections and paleogeographic sketch map of the Changxingian Stage in South China (from Yang *et al.*, 1987). 1, Kucaoping Co., Shaanxi Province; 2, Chaotian, Guangyuan Co., Sichuan Province; 3, Shahe, Guangyuan Co., Sichuan Province; 4, Shangsi, Guangyuan Co., Sichuan Province; 5, Huaying, Linshui Co., Sichuan Province; 6, Yanjing, Hechuan Co., Sichuan Province; 7, Liuchang, Qingzheng Town, Guizhou Province; 8, Ziyun Co., Guizhou Province; 9, Sangzhi Co., Hunan Province; 10, Puqi Co., Hubei Province; 11, Chaoxian Co., Anhui Province; 12, Hushan Mountain, Nanjing City, Jiangsu Province; 13, Changxing Co., Zhejiang Province. SH, Shanghai City; NJ, Nanjing City; WH, Wuhan City; CD, Chengdu City. A, Cathaysia; B, Kangdiania; C, Yunkainia; D, Jinnania, E, North China Old Land.

Table 10.1. *Conodont-based zonal schemes for the Upper Permian of South China*

Stage	This chapter			Wang Chengyuan & Wang Zhihao 1981, 1983	Zhao Jingke *et al.*, 1981	Yang Zunyi *et al.*, 1987
Changxingian	*N. changxingensis*	Upper subzone		*N. changxingensis–N. deflecta*	*N. changxingensis–N. deflecta*	*N. changxingensis–N. deflecta*
		Lower subzone		*N. subcarinata–N. wangi*	*N. subcarinata–N. wangi*	*N. subcarinata–N. wangi*
		N. subcarinata				
Wujiapingian		*N. orientalis*		*N. orientalis*	*N. orientalis*	*N. orientalis*
		N. liangshanensis		*N. liangshanensis*	*N. liangshanensis*	*N. liangshanensis*

these sections and the distribution of conodonts in them are shown in Figs 10.2 and 10.3.

Upper Permian conodont faunas and zones

The sequence of Upper Permian conodont faunas in South China was established by Wang & Wang (1981) and Wang (1978) from studies of the Changxing section (Zhejiang Province) and the Liangshan section (Shaanxi Province). Detailed work on Permian conodont faunas had only just begun, so information on the conodont fauna was limited to these two regions. Since 1978, however, there has been a great increase in our knowledge of conodont distribution in the Permian and the relationship between faunas has been recognized. Upper Permian conodonts define four zones (Table 10.1), which, in ascending order are as follows:

Neogondolella liangshanensis Zone
This zone, which is roughly coextensive with the lower Wujiaping Formation, is characterized by abundant *N. liangshanensis* Wang. Best-known occurrences of the zone are in the Liangshan Mountains (Hanzhong County, Shaanxi Province) (Wang, 1978) and in the town of Chaotian (Guangyuan County, Sichuan Province). In Chaotian, the beds with *N. liangshanensis* lie directly below the *N. orientalis* Zone, which is equivalent to the upper *N. bitteri* Zone and the *N. leveni* Zone of the Hambast Formation in the Abadeh area, Iran (Iranian–Japanese

Research Group, 1981). Thus, the *N. liangshanensis* Zone should be approximately equivalent to the upper Abadeh Formation and the lower Hambast Formation.

Neogondolella orientalis Zone
This zone begins at the base of the middle member of the Wujiaping Formation. The first appearance of *N. orientalis* (Barskov & Koroleva) defines its base and the first appearance of *N. subcarinata* Sweet defines its top. Lithostratigraphically, this zone coincides with the upper member of the Wujiaping Formation in Chaotian and Shangsi, and with the upper member of the Longtan Formation in Changxing.

The *N. orientalis* Zone can be correlated by conodonts with the *Vedioceras–Haydenella* bed of the upper Dzhulfian Stage in Transcaucasia (Kozur *et al.*, 1978), the upper Dzhulfian of Kuh-e-Ali Bashi in northwest Iran (Teichert, Kummel & Sweet, 1973), and with the upper *N. bitteri* Zone and the *N. leveni* Zone of Unit 6 of the Hambast Formation in the Abadeh area of central Iran (Iranian–Japanese Research Group, 1981). All these units contain *N. orientalis* and equivalent ammonoid faunas.

Neogondolella subcarinata Zone
This zone begins with the first appearance of *N. subcarinata* Sweet and terminates with the first appearance of *N. changxingensis* Wang & Wang. The nominate species ranges from the upper Wujiaping Formation almost to the top of the Changxing

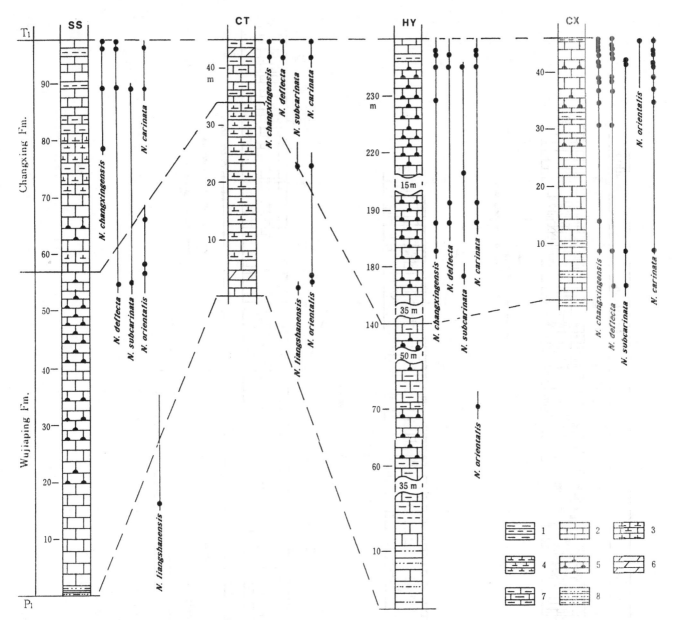

Fig. 10.2. Stratigraphic distribution of significant conodonts and correlation of four Upper Permian sections in South China. 1, mudstone; 2, limestone; 3, siliceous limestone; 4, chert; 5, cherty limestone; 6, dolomite; 7, marl; 8, siltstone. SS, Shangsi section; CT, Chaotian section; HY, Huaying section; CX, Changxing section.

Formation. The basal part of the zone has been identified in only two sections in Guangyuan County, one at Chaotian (Yang *et al.*, 1987) and another at Shangsi (Zhang, Dai & Tian, 1984). In these sections, the zone begins in the uppermost few meters of the Wujiaping Formation. Elsewhere, *N. subcarinata* is confined to the Changxing Formation. *N. subcarinata* is very rare in the Wujiaping Formation, in which *N. orientalis* is common. In the Changxing Formation, however, *N. subcarinata* becomes common and is represented in many sections. Thus although it appears that a lower subzone with rare *N. subcarinata* might be distinguished from an upper subzone characterized by the acme of occurrence of the nominate species, it is practical to use the acme primarily in South China. The acme subzone is well

represented in sections at Chaotian, Shangsi, Huaying, Puqi, Qingzhen, and Changxing.

The lower of the two *N. subcarinata* subzones is dominated in frequency by representatives of *N. orientalis*. The upper, or acme subzone of *N. subcarinata* is characterized by abundant *N. wangi* Zhang (= *N. subcarinata elongata* sensu Wang & Wang, 1981, non *N. elongata* Sweet, 1970). A few specimens of *N. deflecta* have been found toward the top of the acme subzone in the Changxing section.

The acme subzone of the *N. subcarinata* Zone is equivalent to the lower part of the Dorasham Formation of Transcaucasia (Kozur *et al.*, 1978); to the lower part of the Ali Bashi Formation at Kuh-e-Ali Bashi, northwest Iran (Teichert *et al.*, 1973); and to

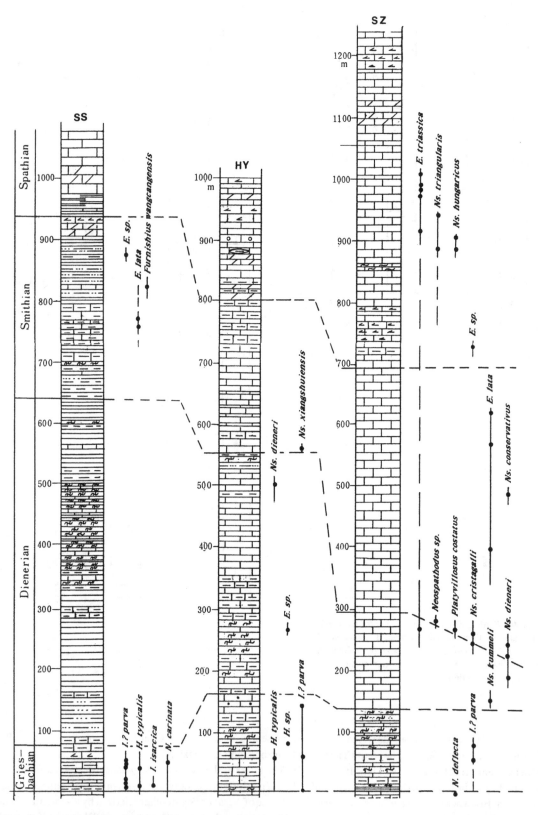

Fig. 10.3. Stratigraphic distribution of significant conodonts and correlation of seven Lower Triassic sections in South China. 1, shale; 2, mudstone; 3, siltstone; 4, marl; 5, oolitic limestone; 6, cumularspharolithic limestone; 7, nodular limestone; 8, limestone; 9, brecciolas; 10, breccias; 11, calcareous marl; 12, limestone; 13, dolomitic limestone; 14, dolomite; 15, myrmekitic limestone. SS, Shangsi section; HY, Huaying section; SZ, Shangzhi section; PQ, Puqi section; CX, Chaoxian section; HS, Hushan section; ZY, Ziyun section.

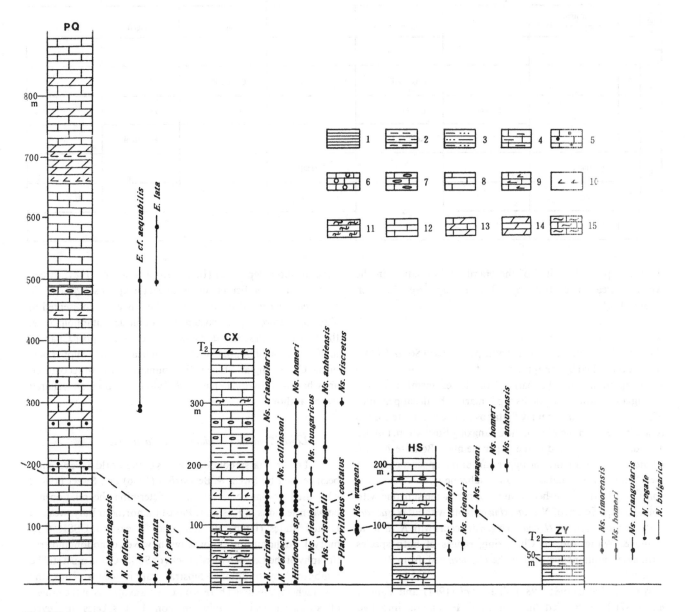

Fig. 10.3. (*cont*)

Table 10.2. *Sections in which various Lower Triassic conodont zones have been recognized in South China. Note that in no section has the complete sequence of 12 zones been recognized*

		GUANGYUAN SICHUAN PROV.	HUAYING SICHUAN PROV.	SANGZHI HUNAN PROV.	PUQI HUBEI PROV.	CHAOXIAN ANHUI PROV.	HUSHAN JIANGSU PROV.	ZIYUN GUIZHOU PROV.
Spathian	12							Timorensis
	11					Homeri	Homeri	
	10					Collinsoni		
	9			Triangularis		Triangularis		
Smithian	8	Furnishius wangcangensis		Conservativus		Waageni	Waageni	
	7							
Dienerian	6			Cristagalli		Cristagalli		
	5		Dieneri	Dieneri		Dieneri	Dieneri	
	4			Kummeli			Kummeli	
Griesbachian	3	Carinata?			Carinata?			
	2	Isarcica		Parvus				
	1	Parvus	Parvus					

the lower part of Unit 7 of the Hambast Formation in the Abadeh area of central Iran (Iranian–Japanese Research Group, 1981).

Neogondolella changxingensis Zone

This zone has been recognized in nine provinces in South China. Its base is marked by the appearance of *N. changxingensis* Wang & Wang in the middle part of the lower member of the Changxing Formation, and its top is marked by the appearance of *Isarcicella? parva* (Kozur & Pjatakova). The nominate species is rare in the lower member of the Changxing Formation, but it is abundantly represented in the upper member. Because *N. subcarinata* extends up into the upper part of the upper member of the Changxing Formation, a lower subzone (with *N. subcarinata*) may be distinguished from an upper, acme subzone with abundant specimens of *N. changxingensis* but without *N. subcarinata*. In most sections *N. changxingensis* occurs with *N. deflecta* Wang & Wang and, in Changxing, ranges of both species terminate simultaneously below the *I.? parva* bed (from Zhao *et al.*, 1981).

Wang & Wang (1981, 1983), Zhao *et al.* (1981), and Yang *et al.* (1987) have divided the Changxing Formation into two conodont zones, the *N. subcarinata–N. subcarinata elongata* [= *N. wangi*] Zone and the *N. changxingensis–N. deflecta* Zone. However, the boundary between these zones is not clear and they apparently overlap. Therefore, definitions of these zones by previous authors are revised in this paper (Table 10.1).

According to Wang & Wang (1981), *N. changxingensis* was derived from *N. subcarinata* and in India and Kashmir the two

have not been separated (Bhatt, Joshi & Arora, 1981; Sweet, 1979). Thus it is difficult to determine if the upper subzone of the *N. changxingensis* Zone of South China is also represented in Kashmir. However, the existence of a unique ammonoid zone, the *Pseudotirolites–Rotodiscoceras* Zone, in the upper Changxing Formation of South China, may indicate that the upper part of the Changxingian Stage is the youngest Permian recognized anywhere in the world (Zhao *et al.*, 1981). This opinion is shared by the author.

Lower Triassic conodont faunas and zones

Lower Triassic conodont zones of the world have recently been revised by Sweet & Bergström (1986) and Sweet (1988), who used graphic correlation to determine the relationship between sections in Kashmir, Pakistan, northern Italy, western United States (Utah, Nevada) and far-eastern USSR (Primorye). To the author's knowledge, however, conodont biostratigraphy of the Lower Triassic rocks of South China must still be established by traditional biostratigraphic means. Thus, the zonal scheme discussed here is based on a synthesis of the ranges of stratigraphically significant conodont species in seven sections in South China (Table 10.2). These faunas, which represent different sedimentary facies, are characterized, in ascending order, as follows (Table 10.3):

Isarcicella? parva Zone

This zone, the lowest conodont zone that can be recognized in the Lower Triassic, is characterized by abundant specimens of

Table 10.3. *Ranges of stratigraphically significant conodont species in the synthetic sequence of Upper Permian and Lower Triassic zones in South China*

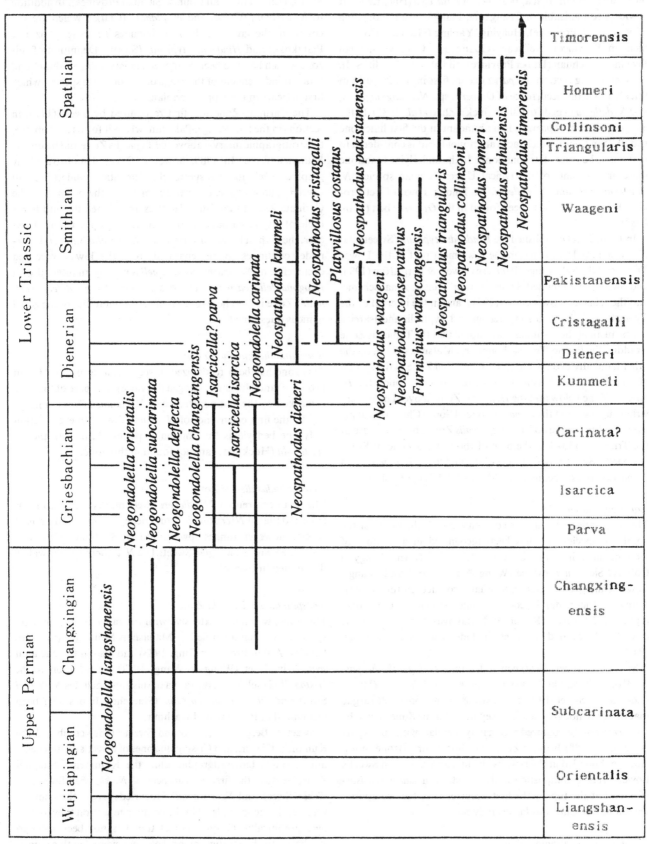

I.? parva (Kozur & Pjatakova) and by *Hindeodus typicalis* (Sweet). The zone begins with the appearance of *I.? parva* and terminates with the first appearance of *I. isarcica* (Huckriede). It is widely recognizable in Changxing, Liangfengya (Chongqing City, Sichuan Province), Huaying, Yanjing (Hechuan County, Sichuan Province), Xejiacao (Guangyuan County, Sichuan Province), Lichuan (Hubei Province), and at other localities. In the Changxing section, *I.? parva* occurs first in bed 27 (0.15 m thick), above the occurrence of *Otoceras* sp., *N. changxingensis*, and *N. deflecta* in bed 26 (0.07 m thick) (Zhao *et al.*, 1981). In the Yanjing section *I.? parva* is first abundant in the first limestone bed (0.25 m thick), which lacks ammonoids but is considered to be Triassic because of its conodonts. In the underlying limestone there are elements of the typical Upper Permian species, *N. changxingensis* and *N. deflecta*. At Tulong Town (Tibet Province), *I.? parva* has also been found in the *Otoceras* bed (Tian, 1982).

In 1970, Sweet established the *Hindeodus typicalis* (Sweet) and *N. carinata* (Clark) zones in the lowermost part of the Lower Triassic of the Salt Range (Pakistan). Sweet & Bergström (1986) and Sweet (1988) revised Sweet's Lower Triassic zonal scheme, and the scheme of Sweet *et al.* (1971), and in the interval of the Typicalis and Carinata zones recognized Typicalis and Isarcica zones. In all of these studies it was noted that the Typicalis Zone straddles the Permo-Triassic boundary. Matsuda (1981), on the other hand, divided the *Otoceras woodwardi* Zone in the section at Guryul Ravine, Kashmir, into two conodont zones, the *H. minutus* Zone below and the *H. parvus* Zone above. The present writer suggests that the *I.? parva* Zone of South China is roughly equivalent to the part of the *H. typicalis* Zone above the base of the Triassic and below the base of the Isarcica Zone of Sweet (1988) and Sweet & Bergström (1986), and to the *H. minutus* and *H. parvus* zones recognized by Matsuda (1981) in Kashmir.

Isarcicella isarcica Zone

This biostratigraphic unit is coextensive with the range of the nominate species and has been recognized in a number of sections including the ones in Beichuan Town, Jiangyou County, Sichuan Province (Wang & Dai, 1981) and Shangsi Town, Guangyuan County, Sichuan Province. In this zone, *I. isarcica* (Huckriede) is common and associated conodonts represent *I.? parva* (Kozur & Pjatakova) and *H. typicalis* (Sweet). Associated bivalves include *Pseudoclaraia wangi* (Patte).

Sweet (in Ziegler, 1977) concluded that *Anchignathodus parvus* Kozur & Pjatakova is synonymous with *I. isarcica* (Huckriede), hence Sweet (1988) and Sweet & Bergström (1986) argue that there are not grounds for a separate Parvus Zone. Thus the *I. isarcica* Zone recognized herein may correlate with only a part of Sweet's (1988) Isarcica Zone, which is a chronozone (not a biozone) bounded at the base by the first appearance in Sweet's Lower Triassic composite standard section of *Isarcicella isarcica*, and at the top by the first occurrence in that composite section of *Neospathodus kummeli* Sweet.

'Neogondolella carinata'? Zone

In South China, this zone begins with the disappearance of *Isarcicella isarcica* and ends with the first occurrence of *Neos-*

pathodus kummeli Sweet. The zone has been identified in sections in Lichuan County (Hubei Province; Wang & Cao, 1981) and at Shangsi (in Guangyuan County, Sichuan Province). In addition to *N. carinata* (Clark) and *N. planata* (Clark), which are not common, the fauna of the zone includes *I.? parva* (Kozur & Pjatakova) and *Hindeodus typicalis* (Sweet). Definition of this zone, which has also been recognized in Pakistan, Kashmir, and India, is independent of the range of its nominate species, which first appears in the Upper Permian.

The Carinata Zone was first recognized by Sweet (1970) in sections in the Salt Range, Pakistan, where it forms a distinctive biostratigraphic interval above the Typicalis Zone and below the Kummeli Zone. In 1988, however, Sweet demonstrated by graphic correlation that species characteristic of underlying and overlying zones appear at various places in the position of the Carinata Zone of Pakistan. He thus abandoned the unit in his table of conodont-based chronozones. In South China, however, the highest known occurrence of *Isarcicella isarcica* (Huckriede) continues to be separated from the lowest recorded occurrence of *Neospathodus kummeli* Sweet by an interval with *Neogondolella carinata*. The designation '*Neogondolella carinata*'? Zone is used here for that interval, but utility of the zonal name is questioned.

Neospathodus kummeli Zone

This zone has been recognized in sections at Songzhi (Hunan province) and Hushang (Jiangsu Province). The base of the zone is marked by the first occurrence of *N. kummeli* Sweet, and the top by the first occurrence of *N. dieneri* Sweet. In each section only one bed yields the nominate species. *N. dieneri* and *N. cristagalli* (Huckriede) are not found in this zone.

Neospathodus dieneri Zone

The first occurrence of *N. dieneri* Sweet marks the base of this zone and that of *N. cristagalli* (Huckriede) its top. The nominate species, however, ranges upward into the Smithian, where it is associated with *N. waageni* Sweet. The *N. dieneri* Zone is widely distributed in Eurasia.

Neospathodus cristagalli Zone

This zone, which is coextensive with the range of its nominate species, has been identified at Majiashan Mountain (Caoxian County, Anhui Province; Ding, 1983) and also in the Nielamu section in Tibet (Wang & Wang, 1977). In addition to *N. cristagalli* (Huckriede), the fauna of this zone includes *N. dieneri* Sweet and *Platyvillosus costatus* (Staesche). It has also been recognized in Pakistan and Kashmir.

Sweet & Bergström (1986) and Sweet (1988) combined the Kummeli, Dieneri, and Cristagalli zones of Sweet's 1970 scheme into a single biostratigraphic unit, the Kummeli–Cristagalli Zone, because the first appearances of *N. kummeli* Sweet, *N. dieneri* Sweet, and *N. cristagalli* (Huckriede) are at the same, or very nearly the same, level in the composite section they assembled graphically. Graphic correlation has not been used to synthesize the biostratigraphic scheme considered in this report, however, and in South China, as in the Salt Range of Pakistan, the three species of *Neospathodus* appear first at different stratigraphic levels. Thus, the zonal scheme proposed by Sweet in 1970 is the one used here.

Neospathodus pakistanensis Zone

In South China this zone has been recognized only in Lichuan County, Hubei Province (Wang & Cao, 1981). Because the *N. cristagalli* and *N. pakistanensis* zones have not been identified in sequence in South China, the relationship between them can not be described. From information assembled graphically by Sweet (1988), however, we know that the Pakistanensis Zone is above the Cristagalli Zone and that it probably represents Late Dienerian time.

Neospathodus waageni Zone

This conodont zone has been recognized in numerous sections in South China, including ones in the Majiashan Mountains of Chaoxian County, Anhui Province (Ding, 1983), in Lichuan County, Hubei Province (Wang & Cao, 1981), and in the Hushan section in Nanjing City, Jiangsu Province. As in Pakistan (Sweet, 1970) and elsewhere, the base of the zone is drawn at the first appearance of *N. waageni* Sweet and the top at the first appearance of *N. triangularis* (Bender). In the Lichuan section, the *N. waageni* Zone is above the *N. pakistanensis* Zone and ranges of the nominate species of these zones are not known to overlap in South China. Regionally, however, *N. pakistanensis* ranges well up into the *N. waageni* Zone and *N. waageni* itself probably ranges to the top of the Lower Triassic (Sweet, 1988).

In addition to its occurrences in South China, *N. waageni* has been identified in collections from Kashmir, Pakistan, western United States, and far-eastern USSR. *N. conservativus* (Müller), from a section in Shangzhi County (Hunan Province) and *Furnishius wangcangensis* Dai & Tian, from the Shangsi section of Guangyuan County (Sichuan Province) may also represent the *N. waageni* Zone, although they do not occur with *N. waageni* and their stratigraphic position can not be determined directly. Elsewhere in the world (Sweet, 1988), *N. conservativus* and species of *Furnishius* appear first in the interval here identifed as the *N. waageni* Zone. However, both are also known to range well above the Waageni Zone regionally.

Neospathodus triangularis Zone

In South China this zone has been recognized only in the Majiashan section in Chaoxian County, Anhui Province, where it is in sequence with the *N. waageni* Zone. The base of the zone is drawn at the first appearance of *N. triangularis* (Bender); the top is placed at the first appearance of *N. collinsoni* Solien. Although the nominate species has been identified widely in South China, this zone is thin and *N. triangularis* itself ranges upward into younger rocks, where it is associated with *N. collinsoni*, *N. homeri* (Bender), and *N. timorensis* (Nogami).

Neospathodus collinsoni Zone

In South China this zone has been identified only in the Majiashan section in Chaoxian County, Anhui Province. In South China, the base of this zone is drawn at the first appearance of *N. collinsoni* Solien and its top is placed at the level of first occurrence of *N. homeri* (Bender). Elsewhere in the world, however, conodonts identified as *N. homeri* first appear well

below the level of first occurrence of *N. collinsoni*, and the Collinsoni Zone is defined by the range of its nominate species (Sweet, 1988). *N. collinsoni* has been reported previously from sections in western United States, Japan, and Primorye.

Neospathodus homeri Zone

This zone begins with the first appearance of *N. homeri* (Bender) and has been recognized quite widely in South China. In the Majiashan section in Chaoxian County (Anhui Province), the *N. homeri* Zone occurs in sequence above the *N. collinsoni* Zone and, in Ziyun County (Guizhou Province) the top of the zone may be represented by an occurrence of specimens of *N. homeri* in a sample that also yields elements of *N. timorensis* (Nogami), the index to the next higher *N. timorensis* Zone. Of course, *N. homeri* is known to range completely through the *N. timorensis* Zone regionally (Sweet, 1988), so, although it is certain that the Ziyun occurrence represents the *N. timorensis* Zone, it is not certain that it represents the base of that zone and thus the top of the *N. homeri* Zone.

Neospathodus timorensis Zone

The nominate species of this zone has been identified only in a section in Ziyun County, Guizhou Province (Wang, 1982). The *N. timorensis* Zone is the youngest Lower Triassic zone in South China. Its base is drawn at the first appearance of *N. timorensis* (Nogami) and the first occurrence of *Neogondolella regale* Mosher marks its top. The fauna of the zone also includes *Neospathodus homeri* and *N. triangularis*, which make their first appearances in South China, and regionally, in rocks well below the base of the *N. timorensis* Zone.

The age of the *N. timorensis* Zone is debatable. Sweet (1970), Sweet *et al.* (1971), and Kozur & Mostler (1972, 1973) regarded the *N. timorensis* Zone as highest Scythian. However, Sweet & Bergström (1986) and Sweet (1988) followed Nicora (1977) and, with no particular conviction, drew the boundary between the Lower and Middle Triassic at the base of the Timorensis Zone. In the Ziyun section, *N. timorensis* occurs below a bed with ammonoids that Wang (1978) regarded as highest Scythian. Furthermore, at Ziyun, *N. timorensis* does not occur with representatives of species restricted to the Middle Triassic. It is associated at Ziyun only with *N. homeri* and *N. triangularis*, which range into the Middle Triassic but are also common in the upper Scythian. In the following level, *N. timorensis* occurs with elements of *Neogondolella regale*, which is restricted to the Middle Triassic.

Conodont faunal changes

In South China, changes in conodont faunas during the Late Permian and Early Triassic were prominent at the species level but not at the generic level (Table 10.4). *Neogondolella*, *Hindeodus*, and *Ellisonia*, three of the most important genera in the Early Triassic, extend upward from the Permian. *Neospathodus* originated in the Early Triassic and was represented in that epoch by numerous species. *Isarcicella*, *Platyvillosus*, and *Furnishius* also originated in the Early Triassic, but are represented by only a few species. Changes at the specific level occurred

Table 10.4. *Ranges of main species of Permo-Triassic conodonts in South China*

STAGES / SPECIES	WU	CH	GR	DI	SM	SP
N. liangshanensis	+					
N. orientalis	+	+				
N. bitteri	+					
N. leveni	+					
N. subcarinata	+	+				
N. wangi	+	+				
N. deflecta	+	+				
N. changxingensis		+				
N. carinata		+	+			
H. typicalis	+	+	+			
I. ?parva			+			
I. isarcica			+			
Ns. kummeli				+		
Ns. dieneri				+	+	
Ns. cristagalli				+		
Ns. pakistanensis				+	+	
Ns. waageni					+	
Ns. conservativus					+	
Ns. triangularis						+
Ns. collinsoni						+
Ns. homeri						+
Ns. anhuiensis						+
Ns. discretus						+
Ns. hungaricus						+
Ns. timorensis						+
Platyvillosus costatus				+		
Furnishius wangcangensis					+	
E. lata					+	
E. cf. aequabilis					−	

Note:

* N. – Neogondolella; H. – Hindeodus; I. – Isarcicella;
Ns. – Neospathodus; E. – Ellisonia

mainly at the end of the Permian, a fact emphasized in Clark's (1987) summary of species-level and generic-level diversity.

Explanations for the change in conodont faunas at the Permo-Triassic boundary are as varied as are interpretations of physical events in the boundary interval. Nevertheless, the most important reasons for changes in conodont faunas were surely environmental, whether caused by asteroid-impact or volcanic events. That is, the widespread occurrence of carbonate rocks indicates that normal marine environments were widespread in South China during the latest Permian. Most of the conodonts recovered from these rocks represent species of *Neogondolella*, which suggest shallow marine environments of normal salinity. At the end of the Permian, however, a worldwide regression caused shoaling of marine environments and extinction of Permian species of *Neogondolella*. Species of *Hindeodus* and *Ellisonia*, however, were evidently able to tolerate those shallower-water conditions and thus escaped the catastrophic extinction of the latest Permian to survive into the Early Triassic.

Conclusion

Permian and Triassic conodont features of South China are reconstructed by synthesis of information from several sections. Four Late Permian faunas are recognized. Those of the *Neogondolella liangshanensis* and *N. orientalis* zones are of early Late Permian age; the *N. subcarinata* fauna ranges in age from the late part of the early Late Permian to the early part of the late Late Permian; and the *N. changxingensis* fauna is of latest Late Permian age.

The Lower Triassic of South China is divided biostratigraphically into 12 conodont zones. As a general rule, the Parva, Isarcica, and 'Carinata'? zones are Griesbachian; the Kummeli, Dieneri, Cristagalli, and Pakistanensis zones Dienerian (although the latter may include Smithian strata in its uppermost part); the Waageni Zone Smithian; and the Triangularis, Collinsoni, Homeri, and Timorensis zones Spathian.

The author also suggests that environmental changes in the latest Permian were the most important of all factors in causing changes in conodont faunas, primarily at the species level.

References

Bhatt, D. K., Joshi, V. K. & Arora, R. K. (1981). Conodonts of the *Otoceras* beds of Spiti. *J. Pal. Soc. India*, **25**: 130–4.

Clark, D. L. (1987). Conodonts: the final fifty million years. In *Palaeobiology of Conodonts*, ed. Aldridge, R. J., pp. 165–74. Chichester: Ellis Horwood, Ltd.

Ding Meihua (1983). Lower Triassic conodonts from the Mountain Majiashan in Anhui Province and their stratigraphic significance. *J. Wuhan College Geol.*, **20**: 37–48. (In Chinese.)

Iranian–Japanese Research Group (1981). The Permian and the Lower Triassic systems in Abadeh region, central Iran. *Mem. Fac. Sci. Kyoto Univ. Ser. Geol. Min.*, **47**(2): 1–133.

Kozur, H., Leven, E. Y., Lozovskiy, V. R. & Pjatakova, M. V. (1978). Subdivision of Permian-Triassic boundary beds in Transcaucasia on the basis of conodonts. *Obshch. Ispytateley Prirody Byull., otdel geol.*, **53**(5): 15–24. (In Russian.)

Kozur, H. & Mostler, H. (1972). Die Bedeutung der Conodonten für stratigraphische und paläogeographische Untersuchungen in der Trias. *Mitt. Ges. Geol. Bergbaustud.*, **21**: 777–810.

Kozur, H. & Mostler, H. (1973). Beiträge zur Mikrofauna permotriadischer Schichtfolgen. Teil I: Conodonten aus der Tibetzone des Niederen Himalaya (Dolpogebiet, Westnepal). *Geol. paläont. Mitt. Innsbruck*, **3**(9): 1–23.

Matsuda, T. (1981). Early Triassic conodonts from Kashmir, India. *J. Geosci., Osaka City Univ.*, **24**(3): 75–108.

Nicora, A. (1977). Lower Anisian platform-conodonts from the Tethys and Nevada: Taxonomic and stratigraphic revision. *Palaeontographica*, **A157**: 88–107.

Sweet, W. C. (1970). Uppermost Permian and Lower Triassic conodonts of the Salt Range and Trans-Indus ranges, West Pakistan. In: *Stratigraphic Boundary Problems; Permian and Triassic of West Pakistan*, eds. Kummel, B. & Teichert, C., pp. 207–75. (Dept. Geology Spec. Pub. 4.) Lawrence, Kansas: Univ. Kansas Press.

Sweet, W. C. (1979). Graphic correlation of Permo–Triassic rocks in Kashmir, Pakistan and Iran. *Geologica et Palaeontologica*, **13**: 239–48.

Sweet, W. C. (1988). A quantitative conodont biostratigraphy for the Lower Triassic. *Senckenbergiana lethaea*, **69**(3/4): 253–73.

Sweet, W. C. & Bergström, S. M. (1986). Conodonts and biostratigraphic correlation. *Ann. Rev. Earth Planet, Sci. 1986*, **14**: 85–112.

Sweet, W. C., Mosher, L. C., Clark, D. L., Collinson, J. W. & Hasenmueller, W. A. (1971). Conodont biostratigraphy of the Triassic. *Geol. Soc. Am. Mem.*, **127**: 441–65.

Teichert, C., Kummel, B. & Sweet, W. C. (1973). Permian–Triassic strata, Kuh-e-Ali Bashi, northwestern Iran. *Bull. Mus. Comp. Zool.*, **145**(8): 359–472.

Tian, C. R. (1982). Triassic conodonts in the Tulong section from Nyalam County, Xizang (Tibet), China. *Contrib. to Geol. Quinghai-Xizong (Tibet) Plateau*, 7: 153–66. Beijing: Geol. Publ. House. (In Chinese.)

Wang Chengyuan & Wang Zhihao (1977). Triassic conodonts from the Mount Jolmo Lungma region. In *A Report of Scientific Expedition in the Mount Jolmo Lungma Region, 1966–1968*, Palaeontology, fasc. 2, 387–416. Beijing: Science Press.

Wang Chengyuan & Wang Zhihao (1981). Permian conodonts from the Longtan Formation and Changxing Formation of Changxing, Zhejiang and their stratigraphical and palaeoecological significance. *Selected Papers 1st Convention Micropaleont. Soc. China, 1979*, 114–20. (In Chinese.)

Wang Chengyuan & Wang Zhihao (1983). Review of conodont biostratigraphy in China. *Fossils and Strata*, **15**: 29–33.

Wang Yikang (1978). Latest Early Triassic ammonoids of Ziyun, Guizhou – with notes on the relationship between Early and Middle Triassic ammonoids. *Acta Palaeont. Sinica*, **17**(2): 151–79. (In Chinese.)

Wang Zhihao (1978). Permian–Triassic conodonts of the Liangshan area, southeast Shaanxi. *Acta Palaeont. Sinica*, **17**(2): 213–27. (In Chinese.)

Wang Zhihao (1982). Discovery of Early Triassic *Neospathodus timorensis* fauna in Ziyun of Guizhou. *Acta Palaeont. Sinica*, **21**(5): 584–7. (In Chinese.)

Wang Zhihao & Cao Yanyu (1981). Early Triassic conodonts from Lichuan, western Hubei. *Acta Palaeont. Sinica*, **20**(4): 363–75. (In Chinese.)

Wang Zhihao & Dai Jinye (1981). Triassic conodonts from the Jiangyou-Beichuan area, Sichuan Province. *Acta Palaeont. Sinica*, **20**(2): 138–50. (In Chinese.)

Yang Zunyi, Yin Hongfu, Wu Shunbao, Yang Fengqing, Ding Meihua & Xu Guirong (1987). Permian–Triassic boundary stratigraphy and fauna of South China. *Geol. Memoirs*, ser. 2, **6**: 379 pp. Beijing: Geol. Publ. House. (Chinese, with English summary.)

Zhang Jinghua, Dai Jingye & Tian Shugang (1984). Biostratigraphy of Late Permian and Early Triassic conodonts in Shangsi, Guangyuan County, Sichuan, China. *Scientific Papers on Geology for International Exchange, Prepared for 27th International Geological Congress*, 163–76. Beijing: Geol. Publ. House. (In Chinese.)

Zhao Jinke, Sheng Jinzhang, Yao Zhaoqi, Liang Xiluo, Chen Chuzhen, Rui Lin & Liao Zhuoting (1981). The Changxingian and Permian-Triassic boundary of South China. *Bull. Nanjing Institute Geol. Palaeont., Academia Sinica*, **2**: 1–85. (In Chinese.)

Ziegler, W. (Ed.) (1977). *Catalogue of Conodonts*. vol. III Stuttgart: Schweizerbart.

11 A conodont-based high-resolution biostratigraphy for the Permo-Triassic boundary interval

WALTER C. SWEET

Introduction

In 1979 I made a preliminary attempt to correlate three important sections of Permian and Triassic rock graphically (Sweet, 1979). In the sections considered in that study, data on the ranges of conodonts in strata thought to be Lower Triassic were adequate to suggest that further work might result in a high-resolution biostratigraphic scheme of worldwide significance. Consequently, considerable time was invested in assembling data on the ranges of conodonts in widely distributed sections of Lower Triassic strata and, in 1988 (Sweet, 1988b) a high-resolution biostratigraphic framework was established for the Lower Triassic. That scheme resulted from graphic correlation of sections in Kashmir, Pakistan, Japan, western United States, northern Italy, and far-eastern USSR. Correlation based on the measured ranges in those sections of 29 conodont species resulted in a framework that is divisible at the 95% confidence level into 21 chronozones (or Standard Time Units), each equivalent to a unit 5 m thick in the standard reference section at Guryul Ravine, Kashmir, and each representing approximately the same interval of time.

In 1979 data on the scaled ranges of stratigraphically significant invertebrate fossils were sparse for Permian rocks in the three sections I considered. Consequently, I used crude information on the ranges of scaphopods, bellerophontids, productacean brachiopods, and a few species of cephalopods and conodonts. Since 1979, however, there has been a flood of information on the measured ranges of conodonts and other invertebrates in rocks of the Permo-Triassic boundary interval, which it is now possible to use to expand the high-resolution framework begun in my 1988 report.

In the report at hand, I use the ranges of 11 additional conodont species and several cephalopods to extend the framework described in 1988 downward into the uppermost Paleozoic. This extension incorporates data from sections of marine Permian and Triassic strata in Soviet Dzhulfa, northwest and central Iran, Kashmir, the Salt Range of Pakistan, and the provinces of Sichuan and Zhejiang, in South China. All these sections have been studied intensively and recently by paleontologists and stratigraphers, who have reported their results in ways that make them amenable to synthesis by semiquantitative stratigraphic means. Further, of course, parts of sections in Soviet Dzhulfa and South China are stratotypes of stages (Dorashamian, Changxingian) that are widely used in international classification of strata in the Permo-Triassic boundary interval. Correlations developed here make it possible to evaluate stratigraphies that involve those stages and to assess the relationships between them.

It should be noted that conclusions reached in this study were partially predicted in my 1979 report, which was based on very few reliable data. Also, Yin *et al.* (1988) used graphic methods to support conclusions with respect to the correlation of two Chinese sections that are similar to mine. The graphic methodology employed here differs in certain respects from that used by Yin *et al.* (1988) however, and we use data of somewhat different quality. Thus the principal significance of our two efforts is that the results are similar.

The data

As detailed below, the biostratigraphy developed in this report is based on information from my own collections and from the recent literature. Dr Bruce R. Wardlaw, of the US Geological Survey, has also generously contributed an important file of unpublished data on the distribution of conodonts in the Permian of the Salt Range, Pakistan, which, in combination with published information on the ranges of Triassic conodonts in the same sections, form a significant part of the Permo-Triassic framework. Dr Wardlaw is currently preparing the Permian data for publication, so I have used data only from the Chhidru Formation, which forms the uppermost part of the Salt Range Permian sequence. Wardlaw's publication of the remaining Salt Range data will materially improve our knowledge of Permian conodont biostratigraphy.

In assembling the ranges in various sections of the species listed in Table 11.1., published information has been taken largely at face value. I have not tried to check and compare the materials involved. Specimens from most localities have been illustrated, but materials from a few others have not. It is difficult to confirm the identities of unillustrated forms; however section-to-section consistency in ranges of the taxa to which the conodonts have been assigned is taken as evidence of consistency in identification. On the other hand, the correlation procedures used here suggest that certain unillustrated forms may have been misidentified because the reported positions of these forms in some sections are strongly at variance with their positions in others.

Finally, I have emended the names by which several conodont

Table 11.1. *Conodont and other species used in graphic correlation*

Species no.	Species name
1	*Ellisonia clarki* Sweet 1970
2	*E. delicatula* Sweet 1970
7	*H. julfensis* (Sweet 1973)
8	*H. turgidus* (Kozur *et al.*, 1975)
9	*H. typicalis* (Sweet, 1970)
11	*Iranognathus tarazi* Kozur *et al.*, 1975
12	*Isarcicella isarcica* (Huckriede 1958)
13	*I.? parva* (Kozur & Pjatakova 1975)
16	*Merrillina* n. sp. Wardlaw
17	*N. bitteri* (Kozur 1975)
18	*N. carinata* (Clark 1959)
19	*N. changxingensis* Wang & Wang 1981
20	*N. deflecta* Wang & Wang 1981
21	*N. elongata* Sweet 1970
22	*N. jubata* Sweet 1970
23	*N. leveni* (Kozur *et al.* 1975)
25	*N. orientalis* (Barskov & Koroleva 1970)
27	*N. subcarinata* Sweet 1973
31	*Neospathodus cristagalli* (Huckriede 1958)
32	*N. dieneri* Sweet 1970
33	*N. homeri* (Bender 1968)
34	*N. kummeli* Sweet 1970
35	*N. pakistanensis* Sweet 1970
36	*N. peculiaris* Sweet 1970
37	*N. timorensis* (Nogami 1968)
38	*N. triangularis* (Bender 1968)
39	*N. waageni* Sweet 1970
45	*Stepanovites dobruskinae* Kozur & Pjat. 1975
49	*Xaniognathus curvatus* Sweet 1970
50	*X. deflectens* Sweet 1970
51	*X. elongatus* Sweet 1970
53	*X.* sp. cf. *X. elongatus* Sweet 1970
54	*Claraia* spp.
55	*Cyclolobus walkeri* Diener 1903
56	*Gyronites* spp.
57	*Meekoceras* spp.
58	*Ophiceras* spp.
59	*Otoceras woodwardi* Griesbach 1880
61	*Tapashanites chaotianensis* Zhao *et al.* 1978
62	*Changhsingoceras* sp.

Note:

Numbers in lefthand column are used in other tables and figures to identify these species.

species have been identified in several sections. These emendations merit brief mention here:

Anchignathodus and species of *Hindeodus*. In 1970 (Sweet, 1970b) I established *Anchignathodus* for one of the two types of P elements in the skeletal apparatus of several species of multielement *Hindeodus*. Ramiform elements of the apparatus of the typical species, *A. typicalis* Sweet (1970b), were assigned to *Ellisonia teicherti*, as were the nearly identical ramiform elements of *Hindeodus julfensis* (Sweet, 1973). Consequently, I here combine as *Hindeodus typicalis* conodonts reported by several authors as *Anchignathodus typicalis* and *Ellisonia teicherti*. I also note that many authors (e.g., Matsuda, 1981) follow Merill (1973) in regarding *Anchignathodus typicalis* as a subjective synonym of Pennsylvanian (Desmoinesian) *Hindeodus minutus* (Ellison, 1941). It is my view, however, that *H. minutus* and *H. typicalis* are different species of the same long-ranging genus, hence their skeletal elements have many features in common. However, I have compared topotype material of the complete apparatus of *H. minutus*, which has never been described, with the type material of *H. typicalis*, and have noted minor, but distinctive differences between homologous elements in the two apparatuses. Most of the students of Permian and Triassic conodonts who studied the same type materials at a meeting in Brescia, Italy, in 1986, agreed that apparatuses of the two species are different and that the differences justify recognition of two species.

Gondolella vs. *Neogondolella*. Platformed P elements thought by some authors to represent species of *Gondolella* are here taken to represent species of *Neogondolella*. I have listed the morphologic criteria by which I distinguish the P elements typical of these two genera in another place (Sweet, in Ziegler 1973) and note here only that those of *Gondolella* species lack the platform brim posterior of the cusp that distinguishes P elements of *Neogondolella*. Elements of all species included here in *Neogondolella* have a distinct posterior platform brim.

Species of *Isarcicella*. Closely related conodonts assigned to *Isarcicella isarcica* (Huckriede) and to a species variously described as *Anchignathodus parvus* Kozur & Pjatakova or *Hindeodus parvus* (Kozuir & Pjatakova) are important stratigraphically in the Permo-Triassic boundary interval. However, opinions differ as to how the two species are related and this affects how they are named. Elsewhere (Sweet in Ziegler, 1977; Sweet, 1988b) I treat *I. isarcica* and *A. parvus* as morphologically different elements of a single species, *I. isarcica*. However, most other authors (e.g., Matsuda, 1981) have assigned the two species to different genera but have not documented their assignment. That is, most authors treat '*parvus*' as a species of *Hindeodus*, but ignore the fact that species of that genus have a seximembrate skeletal apparatus like that of Permo-Triassic *H. typicalis* (Sweet). Because no ramiform elements of *Hindeodus* type have been described from any of the collections in which its characteristic P elements are common (e.g., collections from the Werfen of the Southern Alps described by Staesche, 1964), I regard it as highly unlikely that '*parvus*' is a species of *Hindeodus*. Elements typical of '*parvus*' and *I. isarcica* are morphologically similar, however, have closely similar stratigraphic distribution and biofacies preference, and represent species that I regard as late members of a long-ranging *Diplognathodus* stock. In my view (Sweet, 1988a) that stock was separate from the *Hindeodus* stock of the Ozarkodinida through most of the Late Paleozoic. Thus, pending further study of the two groups of conodonts involved, I assign '*parvus*' tentatively in this report to *Isarcicella* and therefore recognize two species.

Kashmir data

In building a high-resolution conodont-based biostratigraphy for the Lower Triassic (Sweet, 1988b) I used the section

Table 11.2. *Ranges of conodont and other species (in m) in Permian and Triassic strata at seven localities*

SP	GRK	ZCO	SHA	HAM	KAB	DOR	MEI
1	407–408	355–358	—	—	—	—	—
2	388–389	367–377	—	—	—	—	—
7	—	116–249	—	1017–1021	501–520	39	—
8	—	—	—	1028–1041	—	59–61	—
9	288–325	247–257	68–130	989–1027	468–521	0–55	1–41
11	—	—	—	1009–1017	489–499	—	—
12	305–315	253–256	126	1026–1040	529–551	59–65	—
13	303–307	—	120–164	1024–1039	522–551	57–72	43–44
16	—	196–249	—	*935–997	—	—	—
17	—	—	72–82	989–999	—	0	—
18	290–323	247–263	108–198	1021	499–552	—	11–42
19	—	—	96–116	—	504–515	—	11.5–41.6
20	—	—	68–115	—	—	—	7–41.5
21	379–415	269–313	—	—	—	—	—
22	361–415	314–356	—	—	—	—	—
23	—	—	72–80	992–1007	468–484	3–15	—
25	—	205	76–112	999–1016	484–521	5–56	0
27	307–308	—	68–116	1015–1027	499–520	39–55	1–42
31	316–355	258–265	—	—	—	—	—
32	314–369	255–323	—	1047–1052	—	—	—
33	370–437	316–360	—	—	—	—	—
34	314–346	254–257	—	—	—	—	—
35	330–355	265–299	—	—	—	—	—
36	307–356	259–264	—	—	—	—	—
37	405–532	363–364	—	—	—	—	—
38	361–422	317–357	—	—	—	—	—
39	338–406	296–343	—	—	—	—	—
45	—	—	—	993–1020	—	11	—
49	314–354	252–257	—	—	—	—	—
50	316–330	254–287	—	—	—	—	—
51	388–408	313–330	—	—	—	—	—
53	—	—	70–115	991–1020	499–520	—	1–42
54	298–384	—	—	1029–1053	—	58 + up	43–46
55	277	235	—	—	—	—	—
56	317	255–259	—	—	—	—	—
57	352–372	292	—	—	—	—	—
58	300–317	251–256	116	1037–1039	—	—	43–47
59	300–305	—	115?	—	—	—	42?
61	—	—	102	—	—	—	11
62	—	—	105	—	—	—	14.5

Notes:

Conodonts named and identified by number in Table 11.1. GRK = Guryul Ravine, Kashmir; ZCO = Zaluch Composite, Salt Range; SHA = Shangsi, Guangyuan Co., Sichuan; HAM = Kuh-e-Hambast, Iran; KAB = Kuh-e-Ali Bashi, Iran; DOR = Dorasham II, Nakhichevan A.S.S.R.; MEI = Meishan, Changxing Co., Zhejiang.
* Reported as *Merrillina divergens* by Iranian-Japanese Research Group, but probably *Merrillina* n. sp. of Wardlaw.

at Guryul Ravine, Kashmir, as Standard Reference Section (or SRS), primarily because it was then the leading candidate for stratotype of the base of the Triassic. Data on the distribution of conodonts in the Guryul Ravine section are derived from my own collections (Sweet, 1970a), which I have recently re-examined, and from reports by Matsuda (1981, 1982, 1983, 1984), which are keyed to the measured section reported by Nakazawa *et al.* (1975). Unfortunately, few conodonts have been collected from the Permian part of the Guryul Ravine section. Despite this it is the only long, unbroken section for which there is a detailed record of both Permian and Lower Triassic conodonts and the

key ammonoids, *Cyclolobus walkeri* and *Otoceras woodwardi*. For this reason the Guryul Ravine section is retained as SRS in this report.

The ranges of conodonts and other taxa named in Table 11.1 and listed in the column headed GRK in Table 11.2 are composite standard (CS) ranges from my 1988 report on Lower Triassic conodont biostratigraphy (Sweet, 1988b), which have been modified by adding 200 m to each value and by substitution of 'next-best' information for all ranges controlled in the CS by data from the Nammal Nala or Chhidru Nala sections in the Salt Range. The latter two sections, which contain a good record of

Permian conodonts, are dealt with separately here, hence have been removed from the Lower Triassic standard. The 200 m addition was made to remove negative range values from the final CS. Note also that *Neogondolella deflecta* Wang & Wang (1981), is also reported from the Guryul Ravine section. Two incomplete specimens of this species occur in sample K40 among elements I apparently identified originally (Sweet, 1970a) as *N. carinata*. Sample K40 is from about 105 m above the base of the Guryul Ravine section, although it is difficult to combine with any precision the measurements made by Kummel and Teichert (who collected sample K40) with those of Nakazawa *et al.* (1975).

The Guryul Ravine section is regarded in this report as a continuous, unbroken record of Permian and Triassic strata. Some who have studied this section and/or the fossils collected from it, infer a discontinuity of some sort between the top of Zewan unit D and the base of Khunamuh unit E. I have not seen a description of any features that suggest physical discontinuity between the top of the Zewan and the base of the Khunamuh, and no such features were indicated at this level by Nakazawa *et al.* (1975). Further, Kapoor *et al.* (1989), after a thorough review, recently concluded that in the Guryul Ravine section there is no discontinuity between the Zewan and Khunamuh formations. Thus, pending description and discussion of evidence in support of a discontinuity at the Zewan/Khunamuh contact, I defer to the authority of reports by Nakazawa *et al.* (1975) and Kapoor *et al.* (1989) in regarding the sequence as an essentially uninterrupted one.

The Guryul Ravine section is the standard reference section (SRS) for both the uppermost Permian biostratigraphy developed here and the Lower Triassic one outlined previously (Sweet, 1988b). It is also the section favored as stratotype for the Permo-Triassic boundary by a majority of the members of the Permian-Triassic Boundary Working Group polled informally at a meeting in Moscow in July, 1984. Consequently, it is important to point out that, with respect to many of the sections correlated with it graphically (Sweet, 1988b), the Guryul Ravine section is greatly condensed. That is, an interval of early Triassic time recorded in the Southern Alps of Italy, Primorye, or the western United States by 100 m of rock is represented in the Guryul Ravine section by as few as 9 to a maximum of only 29 m. Furthermore, conodonts from the Guryul Ravine section have a color-alteration index (CAI) of more than 4, which suggests that rocks in the section may have been heated through deep and prolonged burial to as much as 300°C. Thus palynomorphs have probably been destroyed, and the original distribution of various isotopes and magnetic properties may have been disrupted. These matters must be considered before the Guryul Ravine section is selected as stratotype of the basal Triassic boundary, but they do not affect its use as SRS in the study described here.

South China data

In this report I use data on the distribution of conodonts in just two sections in South China, although additional information, published recently in chart form by Yang *et al.* (1987), may ultimately make it possible to include many more sections.

The section at Shangsi, in Guangyuan County, Sichuan, is a continuous exposure, about 200 m thick, of primarily carbonate strata assigned to the Maokou, Wuchiaping, Dalong, and Feixianguan formations. The base of the section, the 'O' level for measurements listed in the column headed 'SHA' in Table 11.2, is a point 16 m below the top of the Maokou Formation; the highest sample in the section is from a point 198 m above the base, or from a level about 85 m above the Permo-Triassic boundary, which is drawn in this section just above the highest of three thin bentonitic beds. Conodonts from this section have been described in detail by Zhang, Dai, & Tian (1984), and I have abstracted distributional information from charts in their report that were drawn to scale.

Although a diverse conodont fauna was reported by Zhang *et al.* (1984) from the uppermost Maokou Fm. in the Shangsi section, no specimens were apparently collected from the lower half of the superjacent Wuchiaping Fm. However, from a point about 68 m above the base of the Shangsi section, the record of conodonts is nearly continuous through the Permo-Triassic boundary interval to a level 198 m above the base of the section. Conodonts are well illustrated in two plates and their identities are readily confirmed.

Because this report focuses on condont biostratigraphy in the Permo-Triassic boundary interval, the Shangsi data I list in the column headed 'SHA' in Table 11.2 are limited to those derived from samples 68 to 198 m above the base of the section. Data from the upper Maokou Fm. will be very useful, of course, when it becomes possible to extend the graphic network even farther down into the Permian.

Data listed in the column headed 'MEI' in Table 11.2 are from reports by Sheng *et al.* (1984) and Yin *et al.* (1988) and pertain to the distribution of condonts in the Changxing Formation as it is developed in six small, closely spaced artificial exposures at Meishan, in Changxing County, Zhejiang. In the Meishan district, the Changxing Formation is about 42 m thick and is composed primarily of limestone, from which a variety of brachiopods, snails, foraminifers and cephalopods has been collected. Distributional charts in the report by Sheng *et al.* (1984) suggest that the Changxing Limestone at Meishan is also distinguished by a relatively large and varied conodont fauna. However, names used on those charts indicate that different skeletal elements of the same multielement species have been listed as separate species. When appropriate combinations are made, the Changxing conodont fauna turns out to be more limited in variety, but clearly similar to that of the Dorasham beds and Ali Bashi Formation of Transcaucasia and the Dalong Formation of the Shangsi section in Sichuan. I have excluded from Table 11.2 the Changxing Limestone occurrence of *Isarcicella? parva* reported (as *Hindeodus parvus*) by Wang & Wang (1979). Sheng *et al.* (1984, p. 163) and Kozur (1989) concluded, and I concur, that the specimen illustrated by Wang & Wang is a fragmentary P element of *H. typicalis*. Yin (1985) and Yin *et al.* (1988) report *I.? parva* from the lowermost part of the Chinglung Group, which overlies the type Changxing Formation at Meishan.

Data from the Salt Range, Pakistan

Range data in the column of Table 11.2 headed 'ZCO' are from the Chhidru and superjacent Mianwali formations at four localities in the Salt Range, Pakistan. Range values for the

Chhidru Formation were generously made available by Dr Bruce Wardlaw, of the US Geological Survey. Those for the overlying Mianwali Formation are from my own studies (Sweet, 1970b) and from more recent reports by the Pakistani–Japanese Research Group (1981, 1985). Wardlaw's data were derived from his studies of the distribution of conodonts in the Zaluch Group (Amb, Wargal, and Chhidru formations) in sections at Kotla Lodhian, Zaluch Nala, Nammal Nala, and Chhidru Nala. Sweet (1970b) and the Pakistani–Japanese Research Group (1981, 1985) provided information on the ranges of conodonts in the Mianwali Formation at a number of Salt Range localities, including the ones at Zaluch Nala, Nammal Nala, and Chhidru Nala studied by Wardlaw.

Because conodonts are not abundant in Zaluch Group strata, I have combined information from four of Wardlaw's sections graphically to form a composite section named 'ZCO', for which the Zaluch Nala section is local SRS. In developing ZCO, data on conodont ranges in the overlying Mianwali Formation were also used. Hence, ZCO includes ranges of both Permian and Triassic conodonts. ZCO agrees closely with an unpublished composite section that was developed graphically and independently by Wardlaw. The similarity of our results suggests that our solutions are an accurate summary of the distribution of conodonts in Zaluch Group and Mianwali strata.

In Table 11.2, I list only that part of ZCO between the base of the Chhidru Formation (the uppermost member of the Zaluch Group) and the top of the Mianwali Formation, which is of Early Triassic age. Data on the ranges of conodonts in pre-Chhidru strata will be useful in developing a conodont biostratigraphy for older Permian strata. However, there is little with which to compare them in the other sections considered here.

Data from Iran and Soviet Dzhulfa

In this report I use data on the ranges of conodonts in two sections (Kuh-e-Hambast and Kuh-e-Ali Bashi) in Iran, and in one (Dorasham II) in the Nakhichevan ASSR.

Data from the Kuh-e-Hambast section, in the Abadeh region of central Iran, were taken from charts included in a report by the Iranian–Japanese Research Group (1981), which also include information on the ranges of other invertebrate fossils in the same section. Conodonts from the Hambast Valley section were identified by Murata, and in column 'HAM' of Table 11.2, I list the ranges of those represented between the base of Unit 5 of the upper Abadeh Formation, and a level in Lower Triassic strata about 28 m above the top of the Hambast Formation.

Ranges of conodonts in the uppermost Khachik, Julfa, Ali Bashi, and Elikah formations, in the section at Kuh-e-Ali Bashi, northwest Iran, are listed in column 'KAB' of Table 11.2. These ranges are from a report by Teichert, Kummel & Sweet (1973), who studied only the Ali Bashi and Elikah formations, and from one by Kozur et al. (1978), who considered conodonts from the uppermost Khachik beds, the Julfa beds, the Ali Bashi Formation, and the overlying Claraia beds (= Elikah Formation). Stepanov, Golshani, & Stöcklin (1969) and Teichert, et al. (1973) provide data on the ranges of other invertebrates in the Kuh-e-Ali Bashi section.

Kozur et al. (1978) chart ranges of conodonts in the Dorasham II section of Soviet Dzhulfa, in the Nakhichevan ASSR.

These ranges are listed in the column headed 'DOR' in Table 11.2 and pertain to conodont species represented at Dorasham between the base of the Codonofusiella–Reichelina beds, at the top of the Khachik Suite, and a level in the Lower Triassic Claraia beds about 16 m above the top of the Dorasham beds. The Dorasham beds of Kozur et al. (1978) are here considered to be the stratotype of the Dorashamian Stage of Rostovtsev & Azaryan (1973). However, Rostovtsev & Azaryan appear to have included at the top of the Dorashamian 1 to 2 m of clay and claystone, which Kozur and his colleagues assign to the lowermost part of the superjacent Karabaglar Suite. There is thus some question as to how the top of the Dorashamian Stage is to be fixed.

The Iranian–Japanese Research Group (1981) suggested that the marine environment in which the upper Hambast Formation was deposited may have 'freshened', or even have become supratidal, late in the Dorashamian, but that this was followed by a rapid post-Dorashamian transgression. In the Hambast Valley, the contact between the Dorashamian part of the Hambast Formation and overlying Lower Triassic rocks of 'Unit a' is said by the Iranian–Japanese Research Group (1981, p. 125) to be 'paraconformable', but it is shown in various diagrams in their report (e.g., p. 66) to be conformable. In a more recent report, Partowazar (1989) states unequivocally that the boundary is '. . . gradational and conformable', and this is also Kozur's (1989) opinion.

Stepanov et al. (1969) and Teichert et al. (1973) mention an abrupt lithic change between the Ali Bashi and Elikah formations in the Ali Bashi Valley area of northwest Iran, but do not characterize the change as an unconformity. Consequently, it is reasonable to treat the Julfa–Ali Bashi–Elikah sequence of northwest Iran and its equivalents in Soviet Dzhulfa and central Iran, as conformable because no compelling physical evidence of unconformity has been described and a conformable relationship is indicated by most who have studied the sections firsthand.

Graphic development of composite section

The technique of graphic correlation was fully described 25 years ago (Shaw, 1964) and has been used in essentially its original form to develop the network of correlated sections documented in this report. Because the arguments and procedures of the graphic-correlation method have been extensively discussed at numerous times and in various places during the last quarter-century, it should not be necessary to deal with them again here. However, many biostratigraphers are reluctant to explore or understand, let alone adopt, schemes of stratigraphic synthesis that depart from the qualitative methodology of traditional biozonal biostratigraphy and this has resulted in misunderstanding or misinterpretation of the results of graphic correlation. Consequently, I include a full account of my procedures in assembling the network of correlated sections depicted schematically in Fig. 11.7 and the Composite Standard Section included as the right-hand column of Table 11.3.

Standard reference section (SRS)

In a network assembled graphically, one section is designated standard reference section (or SRS) and all others are

Table 11.3. *GRK-equivalent values of ranges of species listed in Table 11.2 and identified in Table 11.1*

SP	GRK	ZCO	SHA	HAM	KAB	DOR	MEI	CS
1	407–408	406–408	—	—	—	—	—	406–408
2	388–389	367–376	—	—	—	—	—	367–389
7	—	148–291	—	301–302	301–305	300	—	148–305
8	—	—	—	306–311	—	306	—	306–311
9	288–325	290–316	294–321	289–305	294–305	289–305	298–305	288–325
11	—	—	—	298–301	298–300	—	—	298–301
12	305–315	313–315	307	305–311	307–312	306–308	—	305–315
13	303–307	—	306–315	304–310	305–312	305–310	305	303–315
16	—	235–292	—	265–292	—	—	—	235–292
17	—	—	295–297	289–293	—	289	—	289–297
18	290–323	290–322	303–323	302	300–312	—	300–305	290–323
19	—	—	300–305	—	301–304	—	300–305	300–305
20	305	—	294–305	—	—	—	299–305	294–305
21	379–415	327–367	—	—	—	—	—	327–415
22	361–415	368–406	—	—	—	—	—	361–415
23	—	177–238	295–297	290–297	294–297	290–293	—	177–297
25	—	244	296–304	293–301	297–305	291–305	299	244–305
27	307–308	—	294–305	300–305	300–305	300–305	298–305	294–308
31	316–355	317–324	—	—	—	—	—	316–355
32	314–369	314–376	—	314–316	—	—	—	314–376
33	370–437	370–410	—	—	—	—	—	370–437
34	314–346	314–316	—	—	—	—	—	314–346
35	330–355	324–355	—	—	—	—	—	324–355
36	307–356	318–323	—	—	—	—	—	307–356
37	405–432	413–414	—	—	—	—	—	405–432
38	361–422	371–407	—	—	—	—	—	361–422
39	338–406	352–395	—	—	—	—	—	338–406
45	—	—	—	290–302	—	292	—	290–302
49	314–354	312–316	—	—	—	—	—	312–354
50	316–330	313–344	—	—	—	—	—	313–344
51	388–408	367–383	—	—	—	—	—	367–408
53	—	—	295–305	289–302	300–305	—	298–305	289–305
54	298–384	—	—	306–316	—	306 + up	305–306	298–384
55	277	277	—	—	—	—	—	277
56	317	314–318	—	—	—	—	—	314–318
57	352–372	348	—	—	—	—	—	348–372
58	300–317	311–315	305	309–310	—	—	305–306	300–317
59	300–305	—	305	—	—	—	305	300–305
61	—	—	302	—	—	—	300	300–302
62	—	—	302	—	—	—	301	301–302

Note:
Composite Standard (CS) ranges given in right hand column. Localities identified in Table 11.2 and text.

related to it by means described in subsequent paragraphs. Ideally, the SRS is a long, unbroken section that has been sampled at closely spaced intervals, is uniformly fossiliferous, and has yielded, relative to the other sections compared with it, the maximum amount of information on the largest number of fossil species. In practice, of course, other factors must be considered.

For this exercise, I have chosen the section at Guryul Ravine, Kashmir (Nakazawa *et al.*, 1975) as SRS. My decision was founded on two primary considerations. First, because this section was favored as stratotype of the basal Triassic boundary in an informal poll of the Permian-Triassic Boundary Working Group taken in Moscow, in July, 1984, I chose it as SRS for the

Lower Triassic network I described in 1988 (Sweet, 1988b). Further, if the informal opinion of the Working Group ultimately becomes its formal recommendation, other sections will have to be compared with the Guryul Ravine section whether or not it has all the desirable characters of a SRS for graphic purposes.

Although the Guryul Ravine section is the most suitable choice for SRS of the 13 involved in assembling a high-resolution network for the Lower Triassic, it is inadequate as SRS for the Permian part of this network. Several sections considered here would serve to anchor a network of uppermost Permian sections, but most of them are poor in Triassic information. As noted in the section on 'Correlation strategy', however, the Guryul Ravine SRS can be extended into the Permian conceptually by overlapping correlation with sections in China, the Salt Range, and Transcaucasia, and this is the primary reason I retain it as SRS for the exercise described in this report. It is also convenient to have a Composite Standard Section for the uppermost Permian and Lower Triassic that can be related to the same SRS and thus can be expressed in the same units.

Correlation strategy

Following selection of the Guryul Ravine section as SRS, sections were compared with it graphically in the following order: Zaluch composite (ZCO), Shangsi (SHA), Kuh-e-Hambast (HAM), Kuh-e-Ali Bashi (KAB), Dorasham (DOR), and Meishan (MEI). As a general rule, the order of compiling a series of sections graphically is determined by decreasing numbers of species recognized in sections, as well as by decreasing adequacy of control on their ranges. That is, sections with the largest numbers of species in common and the best-controlled ranges are compiled first; less well-controlled sections are compiled later. The order of compilation adopted here conforms closely to this rule, and was modified only by compiling SHA before HAM and leaving MEI to last. These modifications were determined by a need for early compilation of sections with the best Triassic record, so that those with poor Triassic data but extensive Permian records could be added without requiring interminable later rounds of recorrelation.

Initial round of correlation

In the initial round of correlation, sections were compared graphically with the SRS in the order indicated in the preceding paragraph. Each step began with construction of a two-axis graph with the SRS arrayed along the X-axis and the section compared with it along the Y-axis. First occurrences of species common to the two sections were plotted as dots; last occurrences were plotted as crosses. The resulting array of dots and crosses is a useful visual means of sorting out the biologic events (i.e., first and last occurrences) that occur in the same order and with comparable spacing in the two sections compared.

If the distribution of fossils representing the same species is equally well controlled in the two sections compared graphically; if collections are adequate in size; and if depositional conditions were comparable in the two places, the array of dots and crosses should ideally define a straight line, with a slope representing the relative rate of rock accumulation between the two sections and an X intercept indicating the position of the base of the section on the Y-axis relative to the SRS. However, vagaries in preservation, frequency of occurrence, sample spacing, depositional conditions, and collection adequacy commonly conspire to make the determined distribution of fossils less than ideal in any section and the resulting array of dots and crosses more or less diffuse. Nevertheless, if the best-controlled section is kept on the X-axis, dots representing first occurrences and crosses representing last occurrences will tend to group in different segments of the graphic space. Those that represent the best-controlled events in the two sections will define the interface

between the dot and cross sectors, and a line fit to this subset of the total array will best express the relationship between the two sections.

An object of the initial round of graphic correlation is thus to determine, in the manner just described, the dots and crosses that represent the best-controlled and most uniformly spaced succession of biologic events in each section. If these subsets of the total array are visually linear, a line of correlation (or LOC) may be fit to them through use of any of several statistical procedures. Here, as in Shaw's (1964) more elaborate exposition of the graphic method, LOCs are fit by the method of least squares.

LOCs are expressed visually by straight lines, but symbolically by equations. For example, the final relationship between the Shangsi section in South China (SHA) and the SRS-based composite standard section (Fig. 11.2) is expressed by the equation, CS = 0.222 SHA + 279, in which the coefficient 0.222 indicates that for every meter of rock deposited at SHA only 0.222 m accumulated at GRK, the greatly condensed SRS. The intercept coefficient, 279, indicates that the base of the section at Shangsi (a point some 16 m below the top of the Maokou Fm.) is equivalent to a point 79 m above the base of the section at Guryul Ravine, Kashmir (recall that 200 m have been added to the base of GRK to avoid negative range values).

Following derivation of an equation for the LOC range data from the section on the Y-axis may easily be converted into terms of the SRS. In the equation of Fig. 11.2, for example, SRS values of the section at Shangsi (SHA) are determined by multiplying original SHA values by 0.222 and adding 279. SRS-equivalent values of SHA calculated in this way are listed in the column headed SHA in Table 11.3.

SRS-equivalent values for the section plotted on the Y-axis are next compared with range values in GRK, the SRS, and the first of several composite sections is compiled by choosing from either GRK or SRS-equivalent values of the Y section the lowest first-occurrence values and the highest last-occurrence values. Range values from this composite section (or CS) are the ones against which range values from the next section in the compilation schedule are plotted.

Subsequent sections were compared in the order indicated above with a CS composed of the maximized range values of all previously compiled sections. Thus, during the compilation process each new section is compared with best information on range from all previously compiled sections. Ultimately, all sections for which range data have been summarized have been compared with a composite section, based on the SRS, and range values from all sections are now stated in terms of the SRS.

Recorrelation rounds

In succeeding rounds of recorrelation, sections were compared in the same order as previously with a CS stated in terms of the SRS and composed of maximized range values from all sections that were then part of the network. Prior to each step in every recorrelation round, however, the CS was adjusted to remove all values controlled by the section being compared with it. These adjustments involved substitution in the CS of 'next-best' values from another section in the network.

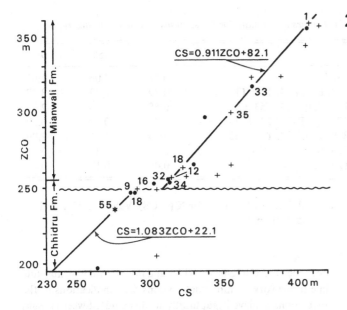

Fig. 11.1. Graphic correlation of Zaluch Composite Section (ZCO) of Salt Range, Pakistan, with seventh-round CS. Numbered dots (first occurrences) and crosses (last occurrences) make up the array subsets to which LOCs are fit.

During recorrelation rounds the CS changed substantially in the first recorrelation round, much less in the second through sixth, and reached stability in the seventh round. That is, during the seventh round there was no change in the first decimal place in any range value and the exercise was terminated.

Results

Figures 11.1 through 11.6 are seventh-round graphs for sections considered in this exercise, and the right-hand column of Table 11.3 lists the maximized range values of all species in the system in terms of the SRS at Guryul Ravine, Kashmir. Because the latter set of range values is a composite of information from all the sections now in the network, it is termed a CS.

Table 11.4 lists seventh-round LOC equations, the array-subsets to which LOCs were fitted, and other statistics that will be discussed later. Note that the rate of rock accumulation was nearly the same in Salt Range sections and the SRS, but that during given intervals of time rock accumulated to much greater thicknesses in all other sections than in the SRS. This observation is consistent with the results of my graphic assembly of Lower Triassic range data, with GRK as SRS, which indicate that GRK was also greatly condensed during the Early Triassic with respect to all sections except those in the Salt Range.

Resolution

In the column of Table 11.4 headed 'S' I list the standard errors of estimate for X (the CS) calculated for each section that is now part of the graphic network. None of these values has been computed for an array-subset of very great size, hence each has been adjusted for small sample size and values significant at the 95% and 99% confidence levels are given in the columns of

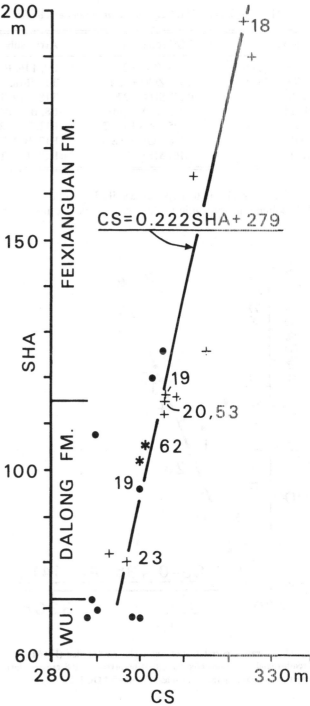

Fig. 11.2. Graphic correlation of Shangsi section (SHA) with seventh-round CS. Numbered dots (first occurrences) and crosses (last occurrences) make up the array subsets to which LOC is fit. Asterisks indicate single occurrences in SHA and CS.

Table 11.4 headed '95%' and '99%'. Values in the latter two columns are the basis for establishing basic correlation units (= standard time units, or STUs) in the graphic network that now includes six Permo-Triassic sections.

At the 95% confidence level, which is probably rigorous enough for the arguments made here, the maximum value of S in Table 11.4 is 3.06. This means that if the Guryul Ravine SRS is divided into units uniformly 3.06 m thick, equivalents of those

Table 11.4. *Summary of LOC equations, array subsets, and standard errors of estimate for six sections considered in this report*

Section	LOC equation	Array subset	S	95%	99%
ZCO (< 249 m)	1.083 ZCO + 22.4	B9, T16, B18, 55	0.71	3.06	7.02
ZCO (> 249 m)	0.911 ZCO + 82.1	T12, T18, B32, B33, B34, T35, T1	0.79	2.02	3.17
SHA	0.222 SHA + 279	T18, B19, T19, T20, T23, T53, 62	0.58	1.50	2.35
HAM	0.435 HAM − 141.7	B8, B11, B12, T16, B17, B23, T23, B32	0.24	0.59	0.89
KAB	0.225 KAB + 188.2	B11, T11, B19, T19, T23, T25, T53	0.74	1.91	2.99
DOR	0.286 DOR + 289.1	B17, B23, T25, B8	0.08	0.34	0.78
MEI	0.164 MEI + 298.1	B19, T19, T20, T53	0.05	0.18	0.41

Note:
Localities identified in text; species identified in Table 11.1. S is Standard Error of Estimate for X (= CS). Values in 95% and 99% columns are of S adjusted for small sample size at 95% and 99% fiducial limits.

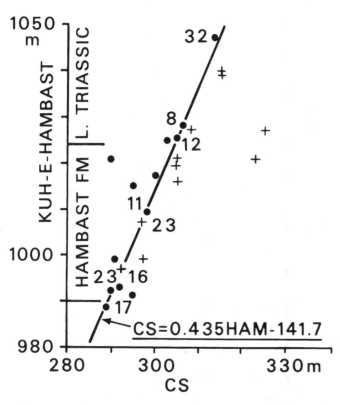

Fig. 11.3. Graphic correlation of Kuh-e-Hambast section (HAM) with seventh-round CS. Numbered dots (first occurrences) and crosses (last occurrences) make up the array subsets to which LOC is fit.

the range of 12 GRK-based STUs and thus provides an absolute time framework within which to consider ramifications of the correlations achieved graphically in this study. Sweet (1988b) chose to divide the Early Triassic portion of the framework into STUs based on segments of GRK 5 m thick; however, data were also presented to show that STUs based on SRS segments 3 m thick could also be recognized at the 95% confidence level in 11 of the 13 sections considered in that report. Thus it is reasonable to base STUs in the Permo-Triassic boundary interval on SRS units 3 m thick, which may readily be resolved in all the sections added in this exercise.

In terms of the International Stratigraphic Guide, STUs established in this report and shown schematically along the left-hand margin of Fig. 11.7, are objective reference standards for a succession of chronozones of conceptually equal temporal value. It is not possible at present to assign a value in years to these STUs, but a crude estimate can be made. That is, if each of the 27 STUs in the Early Triassic scale represents something like 185,000 years of Triassic time (Sweet, 1988b), then each STU in the Permo-Triassic boundary interval, anchored in a unit just 3 m thick at GRK, represents about 111,000 years of Permo-Triassic time.

Biostratigraphic interpretation

Fig. 11.7 shows schematically the correlations achieved graphically in the exercise just described. Fig. 11.8 gives the CS ranges of biostratigraphically significant conodonts, brachiopods, and cephalopods and indicates along the right-hand margin the extent of several conodont zones and stages. The vertical scale of Figs 11.7 and 11.8 is that of the CS, and the ladder scale along the left-hand margin of Fig. 11.7 marks off the extent of 12 of the STUs into which the CS may now be divided at the 95% confidence level.

Most of the correlations indicated in Fig. 11.7 merely confirm those achieved by traditional biostratigraphic procedures. The principal novelty is demonstration that the Dorasham Beds of Soviet Dzhulfa and the coeval Ali Bashi Formation of northwest Iran represent the same interval of time as the *Otoceras wood-wardi* Zone (= Unit E2 of the Khunamuh Formation) in the SRS. In most biostratigraphies published in the last 20 years the *Otoceras woodwardi* Zone overlies the Dorasham Beds and Ali

units might be recognized at the 95% confidence level in all sections now included in the graphic network. Each of these units is a STU and each represents, at least conceptually, the same amount of time. Below 249 m in the Salt Range Composite section (ZCO) 2.77 m of rock accumulated during each STU; above 249 m in ZCO 3.29 m of rock accumulated during the same interval of time. At SHA, however, the rate of rock accumulation was much greater than at GRK so a single GRK-based STU is represented there by 13.5 m of rock. At HAM a GRK-based STU is represented by 6.9 m of rock; at KAB by 13.3 m; at DOR by 10.49 m; and at MEI by 18.3 m.

The ladder scale along the left-hand margin of Fig. 11.7 shows

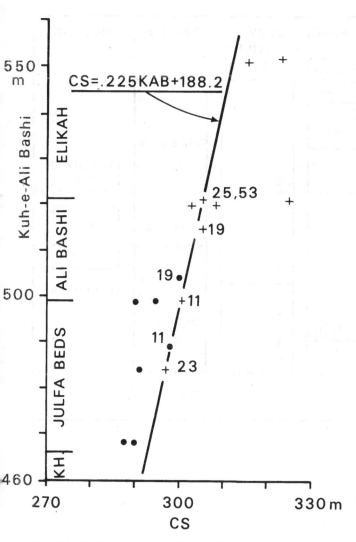

Fig. 11.4. Graphic correlation of Kuh-e-Ali Bashi section (KAB) with seventh-round CS. Numbered dots (first occurrences) and crosses (last occurrences) make up the array subsets to which LOC is fit.

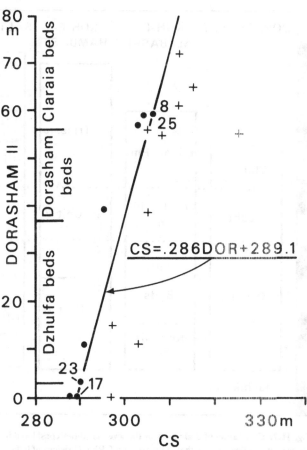

Fig. 11.5. Graphic correlation of Dorasham II (DOR) section with seventh-round CS. Numbered dots (first occurrences) and crosses (last occurrences) make up the array subsets to which LOC is fit.

Bashi Formation even though the former is nowhere known to be superimposed on either of the latter.

The conclusion that the Dorasham-Ali Bashi unit and Guryul Ravine *Otoceras woodwardi* Zone are of the same age and stratigraphic extent is not novel. It is, in fact, closely similar to the conclusion reached in a completely different way by Ruzhentsev & Sarycheva (1965), Li & Yao (1984), and independently by many other authors (summarized in Kozur, 1989). Equivalence of the Dorasham-Ali Bashi beds and the Guryul *O. woodwardi* Zone is vigorously opposed, however, by Tozer (1988a, 1988b, 1989), who has considerable firsthand experience with sections that might be considered as stratotypes for the Permo-Triassic boundary. Consequently, the correlations documented here merit extended discussion.

The data on conodont distribution used in graphic assembly of the biostratigraphic framework presented here are not new. However, they have been ignored, or their significance has been minimized, in most recent discussions of correlation within the Permo-Triassic boundary interval, which deal primarily with relationships suggested by the distribution of a few cephalopod

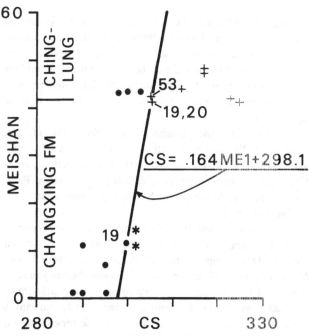

Fig. 11.6. Graphic correlation of Meishan (MEI) section with seventh-round CS. Numbered dots (first occurrences) and crosses (last occurrences) make up the array subsets to which LOC is fit. Asterisks indicate single occurrences in MEI and CS.

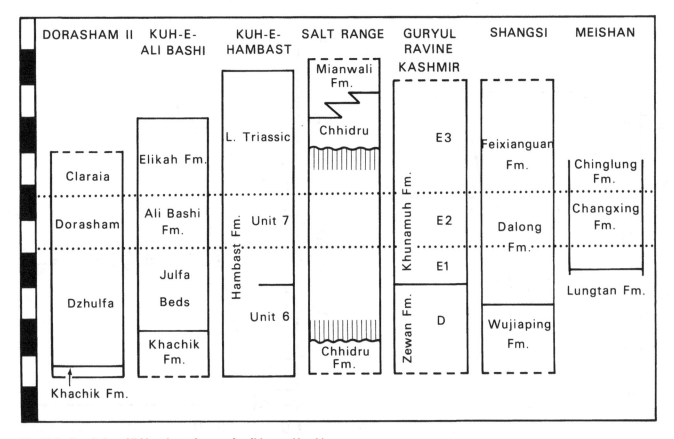

Fig. 11.7. Correlation of lithic units at the seven localities considered in this report. Vertical scale is that of the SRS (= GRK). Divisions of ladder scale on left are Standard Time Units (STUs) based at 95% confidence level on units 3 m thick at GRK. Horizontal dotted lines are actual or projected position of base and top of *Otoceras woodwardi* Zone as it is developed at GRK.

and/or pelecypod species. For this reason, it must be emphasized that the fauna of the Dorasham Beds, the Ali Bashi Formation, and the *Otoceras woodwardi* Zone also includes brachiopods, corals, foraminifers, and a number of distinctive conodont species. Distribution of the latter is evidently far more cosmopolitan, uniform, and predictable than that of the bivalves or cephalopods.

Conodonts from the Dorasham Beds of Soviet Dzhulfa and the Ali Bashi Formation of northwest Iran represent *Hindeodus julfensis* (Sweet), *H. typicalis* (Sweet), *Neogondolella carinata* (Clark), *N. changxingensis* Wang & Wang, *N. orientalis* (Barskov & Koroleva), *N. subcarinata* Sweet, and *Xaniognathus* sp. cf. *X. elongatus* Sweet.

In Transcaucasia and Iran, *Hindeodus julfensis* is confined to the Ali Bashi and Dorasham beds, but the species is also known from pre-Dorashamian strata in Pakistan and from Upper Permian strata on Hydra, in Greece. *H. typicalis*, the youngest member of the dominantly late Paleozoic Anchignathodontidae, makes its debut in Kashmir in Zewan Unit D and ranges through the Dorashamian into Lower Triassic rocks assigned loosely to the Dienerian (Sweet, 1988b).

Neogondolella carinata, like *H. typicalis*, has a long range in uppermost Permian and Lower Triassic strata. *N. subcarinata* has a somewhat shorter range; and *N. changxingensis* is res-

tricted in its known occurrence to the Ali Bashi Formation in Iran, the upper Changxing Limestone in South China, and presumably equivalent strata in Tibet. *N. orientalis* appears first in Dzhulfian strata below the Dorasham beds and Ali Bashi Formation, but is not known to range higher than the tops of those units.

Xaniognathus sp. cf. *X. elongatus* is a distinctive, unnamed species that has now been recognized in northwest and central Iran and at several localities in South China. The species is known to range from near the base to the top of the Hambast Formation of central Iran; from the base to the top of the Ali Bashi Formation in northwest Iran; and through rocks of comparable age at several localities in South China.

From the range summaries just outlined, it appears that rocks of Dorashamian age may be recognized regionally by a combination of distinctive conodont species, of which *Neogondolella changxingensis* is perhaps the most diagnostic. In Soviet Dzhulfa, northwest Iran, and South China the upper range limit of this aggregation of distinctive Dorasham-Ali Bashi conodont species is in rocks immediately and conformably below strata distinguished by *Isarcicella? parva* and only a few meters below rocks with *Isarcicella isarcica*.

In the greatly condensed Selong section of Nyalam County, Tibet (Yao & Li, 1987) a Dorashamian conodont assemblage

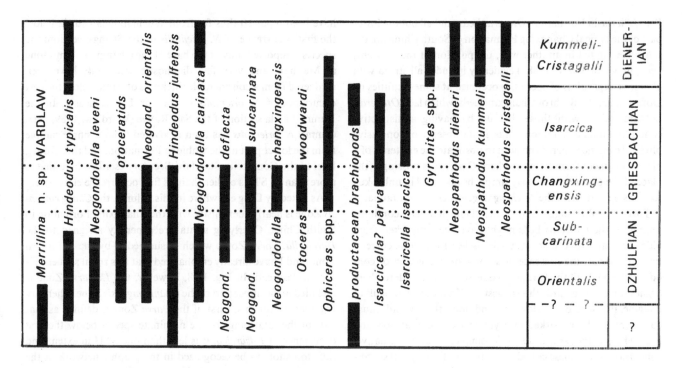

Fig. 11.8. Ranges of conodonts and other stratigraphically significant invertebrate species in Permo-Triassic boundary CS created by graphic correlation of sections in South China, Kashmir, Pakistan, Iran, and Soviet Dzhulfa. Conodont zones recognized in this interval indicated at right, as is extent of stages in which these zones are provisionally included.

with *Neogondolella changxingensis* ranges to the top of the *Otoceras latilobatum* Zone, which Tozer (1988b) and others equate with the *Otoceras woodwardi* Zone of the Guryul Ravine section, Kashmir. *Isarcicella? parva* appears about midway through the *Otoceras* beds, as it does in the Guryul Ravine section. The Selong section is very thin and can not be added to the high-resolution network by graphic means. However, the combination of biostratigraphic features just mentioned is sufficient to indicate that the *Otoceras latilobatum* Zone of the Tibetan section is largely equivalent to the Dorasham Beds and the Ali Bashi Formation of Transcaucasia. Equivalence to the *Otoceras woodwardi* Zone of the Guryul Ravine section is assumed from the common occurrence of *Otoceras*. Also, the upper two-fifths of the *Otoceras* beds at the latter locality yield *I.? parva* and *N. carinata*, but neither *N. changxingensis* nor *N. orientalis* has been reported.

Bhatt, Joshi & Arora (1981) have shown that *Neogondolella orientalis* and *N. subcarinata*, accompanied by several less diagnostic conodont species, also range through and to the top of the '*Otoceras* Bed' in the Lalung section of the Spiti District, Himalaya. *Isarcicella* has not been reported from the Lalung section, but occurrence of the two species of *Neogondolella* indicates that the '*Otoceras*' Bed' at Lalung is largely equivalent to the Dorasham Beds of Soviet Dzhulfa and the Ali Bashi Formation of nearby Iran. Again, equivalence of the *Otoceras* bed of Lalung and the *Otoceras woodwardi* Zone at Guryul Ravine is assumed from the common occurrence of *Otoceras*.

The *O. woodwardi* Zone in the Guryul Ravine section, Kashmir (= unit E2) is in the lower part of the Khunamuh Formation

(Nakazawa *et al.*, 1975), 100 to 105 m above the base of the section. Conodonts are rare in this part of the Guryul Ravine section and are referable to *Ellisonia triassica* Müller, *Hindeodus typicalis* (Sweet), *Isarcicella? parva* (Kozur & Pjatakova), *Neogondolella carinata* (Clark), and *Xaniognathus curvatus* Sweet. Specimens of *I. isarcica* (Huckriede) are reported by Matsuda (1981) from rocks just at the top of the *O. woodwardi* Zone, and my recent restudy of GRK conodont collections at Ohio State University has turned up two fragmentary specimens of *N. deflecta* Wang & Wang in a sample at the top of the *O. woodwardi* Zone. The combination of *I.? parva* and *N. deflecta*, in the same sample from the top of the *O. woodwardi* Zone indicates that the sample is from a level essentially equivalent to the top of the typical Changxing Limestone of South China. To be sure, the lower part of the *O. woodwardi* Zone of the Guryul Ravine section might, from conodont evidence alone, be as old as the Subcarinata Zone, but this seems unlikely because no physical evidence of discontinuity has been reported either at the top, within or at the base of unit E2. Furthermore, *I.? parva*, which appears about midway through unit E2, and *N. deflecta* are represented in beds with *Otoceras latilobatum* in Tibet (Yao & Li, 1987) and thus appear, in at least one other section, to be members of the conodont fauna of the '*Otoceras*' beds.

Equivalence of the Dorasham-Ali Bashi of Transcaucasia and the typical Changxingian of South China is also indicated by the fact that both include the distinctive *Neogondolella changxingensis* fauna and both lie immediately and conformably below strata with *Isarcicella? parva*. Thus, by almost syllogistic biostratigraphic reasoning it may be concluded that the Dorasham

Beds, Ali Bashi Formation, *Otoceras* beds, and all but the lower part of the typical Changxing Formation of South China are the same age. Stated in another way, the position of the *Otoceras* beds, in which Tozer (1988b) evidently includes all strata with *Otoceras*, regardless of species, is occupied at some localities by Dorasham or Ali Bashi or Changxing beds, which lack *Otoceras*, but are united with those that do by having a distinctive conodont fauna in common. Thus, in the absence of conclusive physical evidence, there is no need to postulate the existence of a hiatus, or unconformity, everywhere *Otoceras* is missing – the place of the *Otoceras* beds may simply be occupied in unbroken sections by beds with the *N. changxingensis* conodont fauna.

Tozer (1988b) suggests that the 'Dorashamian' condonts present in the *Otoceras* beds might have been reworked from older beds and that, if so, they are 'useless for precise correlation'. However, the presence of a common succession of conodonts through all the Permo-Triassic sections considered here argues forcefully against this suggestion. That is, it is difficult to imagine the sort of worldwide conditions that would cause conodonts to be reworked into younger deposits always and everywhere in the same order. A far simpler, if less imaginative, conclusion is that these conodonts represent the original succession and were members of an invertebrate fauna that from place to place may also have included various species of *Otoceras*.

In conclusion, the biostratigraphic framework built up by graphic correlation and recorrelation of sections in South China, India, Pakistan, and Iran, and summarized in Figs 11.7 and 11.8, invents nothing. It merely systematizes a scheme that is inherent in data on the physical stratigraphy of component sections and the stratigraphic ranges of conodonts and other invertebrates in them. Graphic assembly of the data makes it possible, however, to explore more rigorously the relationships between sections and to evaluate statistically the precision of correlation and questions of diachroneity.

Conodont zones and stages

The zonal sequence outlined along the right-hand side of Fig. 11.8 is similar to the one described by Ding elsewhere in this volume. Note, however, that in the scheme outlined by Ding the *Orientalis* Zone follows a *Neogondolella liangshanensis* Zone, which is best developed at localities in Shaanxi and Sichuan that can not at present be included in the graphic network. In sections that are considered here, the lower part of the Wuchiaping Formation is devoid of conodonts and *N. liangshanensis* is represented at Shangsi (Guangyuan County, Sichuan; SHA of this report) only in the upper part of the Wuchiaping, where it occurs with *N. bitteri, N. leveni, N. orientalis*, and *N. subcarinata* in strata that I assign to the *Subcarinata* Zone.

As indicated in Figs 11.7 and 11.8, the *Orientalis* and *Subcarinata* zones make up that part of the Akhura Suite at Dorasham between the Khachik Suite below and the Dorasham beds above. These two zones thus represent the Dzhulfian Stage. It is important to note, however, that the base of the *Orientalis* Zone has yet to be established in the graphic network. Although the first occurrence of *N. orientalis* is at or near the projected base of the Dzhulfian Stage in all sections considered here except the one at Nammal Nala, in the Salt Range of Pakistan, no conodonts

have been reported thus far from Wuchiapingian strata below the first occurrence of *N. orientalis* in the Shangsi section; the species is reported only at the base of the Changxing Limestone at Meishan; and very few diagnostic conodonts have been recovered from a considerable thickness of strata below the first occurrence of *N. orientalis* in Iran or Transcaucasia. In the Nammal Nala section of the Salt Range (Wardlaw, 1989, pers. comm.) *N. orientalis* has been recovered from a single sample from the lower part of the Chhidru Formation, and thus from a level in the Composite Standard Section described here that is more than 15 STUs earlier than its first occurrence elsewhere.

As noted by Ding elsewhere in this volume, the *Changxingensis* Zone is nearly coextensive with the Changxing Formation in South China. Overlying strata are generally assigned to the *Isarcicella*? *parva* Zone, which is succeeded by the *I. isarcica* Zone, and I have already emphasized that this zonal sequence is very widespread. In Fig. 11.8, however, the *I.*? *parva* Zone is included in the upper part of the *Changxingensis* Zone. There are two reasons for this. First, if the *Parva* Zone is defined as that part of the total range of the nominate species below the first occurrence of *I. isarcica*, it is less than one STU in extent and thus too short to be recognized in the graphic network at the 95% confidence level. Second, there is no consensus as to the generic assignment of *I.*? *parva*; questions about its phylogeny need to be resolved; and published illustrations suggest that elements representative of the species have not always been correctly identified. On the other hand, elements of *I. isarcica* (Huckriede) *sensu stricto* are very distinctive morphologically and thus are invariably identified correctly. In addition, of course, *I. isarcica* ranges through at least three STUs and a chronozone based on that range merits separate recognition.

Lower Triassic stages are founded on the succession of ammonoid faunas in sections at various places in the Canadian Arctic Archipelago (Tozer, 1965, 1967). Conodonts have been recovered from just a few beds in the stratotypes of these stages, hence they may be related only indirectly to the graphically assembled framework described in this report.

The lower two-thirds of the Akura Suite in the Dorasham II section of Soviet Dzhulfa may be regarded as stratotype of the Dzhulfian Stage, hence the extent of this chronostratigraphic unit in the graphic network is established beyond doubt. Stadial assignment of the *Changxingensis* and *Isarcica* zones is uncertain, however, but may be approximated indirectly. That is, Sweet (1988b) notes that the base of the *Kummeli–Cristagalli* Zone, which overlies the *Isarcica* Zone regionally, coincides in Salt Range sections with the level at which *Gyronites* first appears in those sections and hence with the base of the Dienerian Stage (Tozer, 1967). Thus the *Changxingensis* and *Isarcica* zones occupy the interval of the Griesbachian Stage. The *Changxingensis* Zone, which includes the *Otoceras* beds, is Lower Griesbachian (or Dorashamian), whereas the *Isarcica* Zone, with abundant *Claraias* is Upper Griesbachian.

In summary, strata in the Permo-Triassic boundary interval are conveniently divided into four conodont-based chronozones (Fig. 11.8) based on ranges of the nominate species in the CS assembled graphically in this report. The lower two of these zones (*Orientalis* and *Subcarinata*) make up the Dzhulfian Stage, whereas the upper two (*Changxingensis* and *Isarcica*) probably represent the Griesbachian Stage. The lower third of the latter, the *Changxingensis* Zone, is coextensive with the Dorashamian Stage of Transcaucasia, which is probably of too limited an extent to merit continued usage. The term Dorashamian, however, may be a better and more meaningful designation for this stratigraphic interval than 'Otoceras beds', particularly because *Otoceras* is not known everywhere from strata in this stratigraphic position.

References

Bhatt, D. K., Joshi, V. K. & Arora, R. K. (1981). Conodonts of the *Otoceras* bed of Spiti. *J. Palaeont. Soc. India*, **25**: 130–4.

Ellison, S. P., Jr. (1941). Revision of the Pennsylvanian conodonts. *J. Paleont.*, **15**: 107–43.

Iranian–Japanese Research Group (1981). The Permian and the Lower Triassic systems in Abadeh Region, central Iran. *Mem. Fac. Sci., Kyoto Univ., ser. Geol. & Min.*, **47**(2): 61–133.

Kapoor, H. H. *et al.* (1989). Report of Indian Working Committee on Upper Permian correlation and standard scale. *Permophiles*, **15**: 4–7.

Kozur, H. (1989). The Permian–Triassic boundary in marine and continental sediments. *Zbl. Geol. Paläont. Teil I*, **1988**(11/12): 1245–77.

Kozur, H., Leven, E. Ya., Lozovskiy, V. R. & Pyatakova, M. V. (1978). Subdivision of Permian–Triassic boundary beds in Transcaucasia on the basis of conodonts. *Bull. Obshch. Ispyateley Prirody, otdel. geol.*, **531**(5): 15–24. (Russian; English translation in *Int. Geol. Rev.* **22**(3): 361–8, 1980.)

Li Zishun & Yao Jianxin (1984). Biostratigraphic implications of *Otoceras* beds in China. *Scientific Papers on Geology for International Exchange, Prepared for 27th International Geological Congress*, pp. 75–86. Beijing: Geol. Publ. House (In Chinese, with English summary.)

Matsuda, T. (1981). Early Triassic conodonts from Kashmir, India. Part I: *Hindeodus* and *Isarcicella*. *J. Geosci. Osaka City Univ.*, **24**(3): 75–108.

Matsuda, T. (1982). Early Triassic conodonts from Kashmir, India. Part 2: *Neospathodus* I. *J. Geosci. Osaka City Univ.*, **25**(6): 87–103.

Matsuda, T. (1983). Early Triassic conodonts from Kashmir, India. Part 3: *Neospathodus* 2. *J. Geosci. Osaka City Univ.*, **26**(4): 87–110.

Matsuda, T. (1984). Early Triassic conodonts from Kashmir, India. Part 4: *Gondolella* and *Platyvillosus*. *J. Geosci. Osaka City Univ.*, **27**(4): 119–41.

Merrill, G. K. (1973). Pennsylvanian nonplatform conodont genera. I. *Spathognathodus. J. Paleont.*, **47**: 289–314.

Nakazawa, K., Kapoor, H. M., Ishii, K., Bando Y., Okimura, Y. & Tokuoka, T. (1975). The Upper Permian and Lower Triassic in Kashmir, India. *Mem. Fac. Sci., Kyoto Univ., ser. Geol. & Min.*, **42**(1): 1–106.

Pakistani–Japanese Research Group (1981). Stratigraphy and correlation of the marine Permian–Lower Triassic in the Surghar Range and the Salt Range, Pakistan. *Kyoto Univ.*, 25 pp., charts.

Pakistani–Japanese Research Group (1985). Permian and Triassic systems in the Salt Range and Surghar Range, Pakistan. In *The Tethys, Her Paleogeography and Paleobiogeography from Paleozoic to Mesozoic*, ed. K. Nakazawa & J. M. Dickins, pp. 219–312. Tokyo: Tokai Univ. Press.

Partowazar, H. (1989). Permian–Triassic boundary in Iran. *28th Int. Geol. Congress, Abstracts*, **2**: 576–7.

Rostovtsev, K. O. & Azaryan, N. R. (1973). The Permian–Triassic boundary in Transcaucasia. *Mem. Canadian Soc. Petrol. Geol.*, **2**: 89–99.

Ruzhentsev, V. E. & Sarycheva, T. G., (Eds) (1965). Razvitie i smena morskikh organizmov na rubezhe Paleozoya i Mezozoya. (Development and change of marine organisms at the Paleozoic-Mesozoic boundary.) *Trudy Paleont. Inst. Akad. Nauk SSSR*, **108**: 1–431.

Shaw, A. B. (1964). *Time in Stratigraphy*. New York: McGraw-Hill Book Co.

Sheng Jin-zhang, Chen Chu-zhen, Wang Yi-Gang, Rui Lin, Liao Zhuo-ting, Bando, Y., Ishii, K., Nakazawa, K. & Nakamura, K. (1984). Permian–Triassic boundary in middle and eastern Tethys. *J. Fac. Sci. Hokkaido Univ., Ser. IV*, **21**(1): 133–81.

Staesche, U. (1964). Conodonten aus dem Skyth von Südtirol. *N. Jb. Geol. Paläont.*, **119**(3): 247–306.

Stepanov, D. L., Golshani, F. & Stöcklin, J. (1969). Upper Permian and Permian–Triassic boundary in north Iran. *Geol. Surv. Iran, Rept.*, **12**: 1–72.

Sweet, W. C. (1970a). Permian and Triassic conodonts from a section at Guryul Ravine, Vihi District, Kashmir. *Paleont. Contr. Univ. Kansas*, **49**: 1–10.

Sweet, W. C. (1970b). Uppermost Permian and Lower Triassic conodonts of the Salt Range and Trans-Indus ranges, West Pakistan. *Univ. Kansas, Dept. Geology Spec. Publ.*, **4**: 207–75.

Sweet, W. C. (1979). Graphic correlation of Permo-Triassic rocks in Kashmir, Pakistan and Iran. *Geologica et Palaeontologica*, **13**: 239–48.

Sweet, W. C. (1988a). *The Conodonta*. Oxford Monographs on Geology and Geophysics, No. 10. New York: Oxford University Press.

Sweet, W. C. (1988b). A quantitative conodont biostratigraphy for the Lower Triassic. *Senckenbergiana leth.*, **69**(3/4): 253–73.

Teichert, C., Kummel, B. & Sweet, W. (1973). Permian-Triassic strata, Kuh-e-Ali Bashi, northwestern Iran. *Bull. Mus. Comp. Zool., Harvard Univ.*, **145**(8): 359–472.

Tozer, E. T. (1965). Lower Triassic stages and ammonoid zones of Arctic Canada. *Geol. Surv. Canada, Pap.*, **65**–12: 14 pp.

Tozer, E. T. (1967). A standard for Triassic time. *Bull. Geol. Surv. Canada*, **156**: 103 pp.

Tozer, E. T. (1988a). Definition of the Permian-Triassic (P–T) boundary: The question of the age of the *Otoceras* beds. *Mem. Soc. Geol. Ital.*, **34**: 291–301.

Tozer, E. T. (1988b). Towards a definition of the Permian–Triassic boundary. *Episodes*, **11**(3): 251–5.

Tozer, E. G. (1989). Permian–Triassic (P–T) correlation and boundary problems. *Permophiles*, **15**: 17–21.

Wang Cheng-yuan, & Wang Zhi-hao (1979). Permian conodonts from the Longtan Formation and Changhsing Formation of Changxing, Zhejiang and their stratigraphical and paleontological significance. *Select. Papers 1st Conv. Micropaleont. Soc. China*, pp. 114–20. (In Chinese.)

Yang Zunyi, Yin Hongfu, Wu Shunbao, Yang Fengqing, Ding Meihua & Xu Guirong (1987). Permian–Triassic boundary stratigraphy and fauna of South China. *Geol. Memoirs, ser. 2*, **6**: 379 pp. Beijing: Geol. Publ. House. (In Chinese, with English abstract.)

Yao Jianxin, & Li Zishun (1987). Permian-Triassic conodont faunas and the Permian-Triassic boundary at the Selong section in Nyalam County, Tibet, China. *Kexuo Tungbo*, **32**(22): 1556–60.

Yin Hongfu (1985). On the transitional bed and the Permian-Triassic boundary in South China. *Newsl. Stratig.*, **15**(1): 13–27.

Yin Hongfu, Yang Fengqing, Zhang Kexing & Yang Weiping (1988). A proposal to the biostragraphic criterion of Permian/Triassic boundary. *Mem. Soc. Geol. Ital.*, **34**: 329–44.

Zhang Jinghua, Dai Jinye & Tian Shugang (1984). Biostratigraphy of Late Permian and Early Triassic conodonts in Shangsi, Guangyuan County, Sichuan, China. *Scientific Papers on Geology for International Exchange, Prepared for 27th International Geological Congress*, 163–76. Beijing: Geol. Publ. House. (In Chinese with English summary.)

Ziegler, W., ed. (1973). *Catalogue of Conodonts*, vol. I. Stuttgart: E. Schweizerbart'sche Verlagsbuchhandlung.

Ziegler, W., ed. (1977). *Catalogue of Conodonts*, vol. III. Stuttgart: E. Schweizerbart'sche Verlagsbuchhandlung.

12 The palynofloral succession and palynological events in the Permo-Triassic boundary interval in Israel

YORAM ESHET

Introduction

The stratigraphy and nature of the Permo-Triassic boundary interval in the world have long been debated. Controversies have been fueled by incomplete sections, hiatuses, and by the provinciality of fossils. Although some fossil groups have been used to provide biozonations in the boundary interval, debates have centered on the question of whether these really represent the entire time span, or if there are unrecognized hiatuses. For example, Sokratov (1983) suggested that an additional, previously unrecognized ammonoid zone should be added between the highest Permian and the lowest Triassic biozones in the Caucasus region, where one of the most continuous sequences of boundary-interval strata is believed to occur.

Rock layers that represent the Permo-Triassic boundary interval are not exposed in Israel, and all available information comes from boreholes in the Negev in the southern part of the country (Fig. 12.1). The lithologic column in most boreholes, including the type section (Makhtesh Qatan 2 borehole: Figs 12.1, 12.2), represents what appears to be continuous sedimentation in shallow marine environments across the Permo-Triassic boundary (Weissbrod, 1981). However, in the Pleshet 1, Shezaf 1, and Zohar 8 boreholes, the boundary is within a thin layer of red clay of unknown origin.

In contrast with the apparent continuity in sedimentation, the palynologic assemblages exhibit a marked change within the boundary interval. As a result, assemblages thought to be Upper Permian can easily be distinguished from ones regarded as Lower Triassic. This fact enhances the usefulness of sporomorphs in defining the Permo-Triassic boundary in Israel.

In the past, various authors have discussed paleontologic aspects of the Permo-Triassic boundary in Israel. Using palynologic data from southern Israel, Horowitz (1973) placed the boundary within the Zafir Formation in the Advedat 1 borehole, and at the contact between the Yamin and Zafir formations in the Zohar 8 borehole (Fig. 12.1). R. Dunay (written communication, 1974) proposed that the boundary is in the upper part of the Arqov Formation in the Makhtesh Qatan 2 borehole, and in the lower part of the Yamin Formation in the Ramon 1 borehole. In the Zohar 8 borehole, he drew the boundary in the upper part of the Yamin Formation. Also using palynologic data, Eshet (1984) placed the boundary within the lower part of the Yamin Formation in the Makhtesh Qatan 2

borehole. Basing their decision on foraminifers and ostracodes, Derin & Gerry (1981) placed the Permo-Triassic boundary within the uppermost part of the Arqov Formation. Eshet's (1987) palynostratigraphic study indicated that the boundary occurs in the Pleshet 1 borehole at the same depth indicated by foraminifers and ostracodes (Derin & Gerry, written communication, 1986). These various placements of the systemic boundary in Israel reflect problems in the stratigraphic interpretation of the Permo-Triassic transitional interval.

In recent years, much attention has been paid to the sequence of events that occurred during the transition from Permian to Triassic. Of special importance is the work of Visscher & Brugman (1988), who attempted an ecologic explanation for the floral changes across the Permo-Triassic boundary in the Southern Alps.

The present report presents the results of a palynostratigraphic study of the Permo-Triassic boundary interval in Israel. In it I include a new stratigraphic interpretation based on both palyno- and lithostratigraphy. Since the palynologic succession is very similar to those in the Southern Alps (Visscher & Brugman, 1988) and Greenland (Balme, 1979), a comparison between those localities is also made.

Palynozonation of the Permo-Triassic boundary interval

Two interval zones have been defined in the Permo-Triassic boundary interval in Israel (Eshet, 1987). It is important to understand the relationship between these zones, as well as their stratigraphic position. The zones are:

(1) *Lueckisporites virkkiae* Zone. The top of this interval zone is defined by the highest occurrence of *Lueckisporites virkkiae*. A diverse assemblage of *Protohaploxypinus* (*P. varius*, *P. jacobii*, *P. microcorpus*, *P. hartii*, and *P. limpidus*) characterizes the uppermost part of the zone. *Klausipollenites schaubergeri*, *Plicatipollenites indicus*, *Guthoerlisporites cancellosus* and *Striatopodocarpites rarus* also make their last appearances at the top of the zone. As indicated in Figs 12.2–12.7, the top of the *L. virkkiae* Zone is marked by the disappearance of most of the significant species; almost none continues into overlying strata. Striate bisaccate sporomorphs are prominent and an abundance of fungal spores and tintinnids is a characteristic features of the top part of the zone.

The *L. virkkiae* Zone embraces the upper part of the Arqov Formation and the lower part of the Yamin Formation. Its

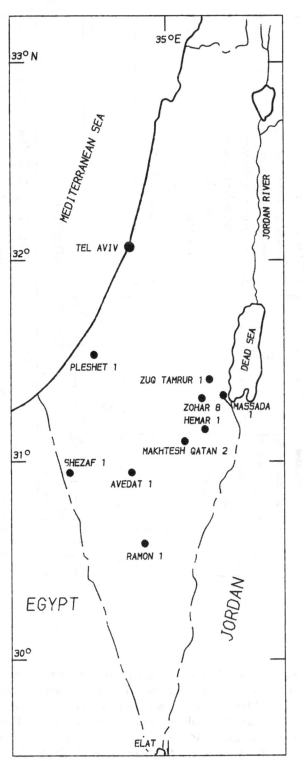

Fig. 12.1. Location of boreholes studied for this report.

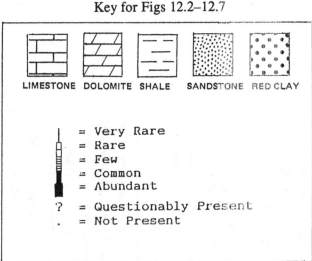

Key for Figs 12.2–12.7

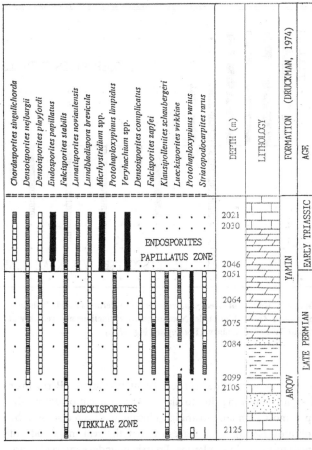

Fig. 12.2. Ranges of selected palynomorphs in Permo-Triassic boundary interval of Makhtesh Qatan 2 borehole. Note continuous sedimentary sequence and ranges at level of Permo-Triassic boundary.

upper boundary is in the lower part of the Yamin Formation in the Makhtesh Qatan 2 borehole (Eshet & Cousminer, 1986) and the Ramon 1 borehole (Eshet, 1987). In the Zohar 8 borehole, the boundary is in the upper part of the Yamin Formation, whereas in the Pleshet 1, Shezaf 1, and Zohar 8 boreholes, the top of the *L. virkkiae* Zone is marked by a thin, distinct layer of red clay.

The *L. virkkiae* Zone is assigned to the Late Permian as a result of comparison with assemblages in Europe (Clarke, 1965; Geiger & Hopping, 1968; Visscher, 1971; Visscher & Brugman, 1981), Pakistan (Balme, 1970), and Gabon (Jardine, 1974). In those localities, *L. virkkiae*, *K. schaubergeri*, and an abundance of *Protohaploxypinus* species are significant Upper Permian forms. Although the zone is undoubtedly of Late Permian age, it

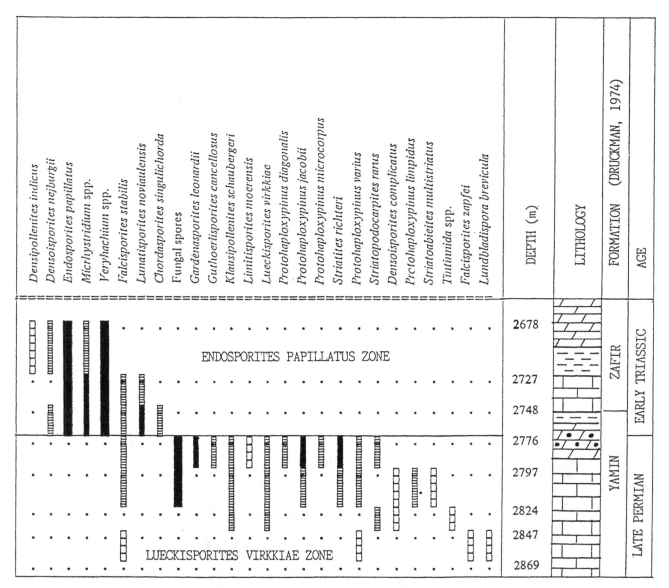

Fig. 12.3. Ranges of selected palynomorphs in Permo-Triassic boundary interval of Zohar 8 borehole. Note extensive extinctions and red clay at Permo-Triassic boundary. (For key see Fig. 12.2.)

is not yet possible to determine if it contains the latest Permian or if a part is missing due to hiatus. The problem of dating the uppermost Permian biostratigraphically has been discussed by Sokratov (1983), who suggested that the real magnitude of unconformities in the Permo-Triassic boundary interval is not always properly distinguished by conventional biozones.

(2) *Endosporites papillatus* Zone. The top of this interval zone is defined by the highest occurrence of *E. papillatus*, which is most common in the zone. *Lunatisporites noviaulensis* has its last appearance at the top of the zone in most sections. Species of *Kraeuselisporites* are common. The base of the zone is marked by the first appearance of types of acritarchs that are abundant throughout the Scythian-Anisian interval. At some places these comprise more than 70% of the palynologic assemblage. Striate bisaccates are less abundant in the *E. papillatus* Zone than in the subjacent *L. virkkiae* Zone.

In the boreholes studied, the *E. papillatus* Zone has been

identified in a lithostratigraphic interval that ranges from the Yamin Formation to the upper part of the Zafir Formation (Eshet, 1987). A Scythian age is assigned, based on a comparison of the zonal assemblage with equivalent ones in Europe (Orlowska-Zwolinska, 1977; Visscher, 1971; Visscher & Brugman, 1981) and North America (Jansonius, 1962), where the best Scythian guide is *E. papillatus*. In addition, Geiger & Hopping (1968) regarded *L. noviaulensis* as a Scythian form in North Sea sections. Balme (1979) and Visscher & Brugman (1981; 1988) reported that acritarchs begin to be abundant in the lower part of the European Triassic, although Balme noted that acritarchs were also numerous in the Upper Permian of East Greenland. A Scythian age for the *E. papillatus* Zone of Israel is also supported by *Glomospira* spp. and *Meandrospira julia* foraminifer zones, and by the '*Monoceratina* C.' ostracode zone of Derin & Gerry (1981). The chronostratigraphic resolution provided by the palynologic composition of the *E. papillatus*

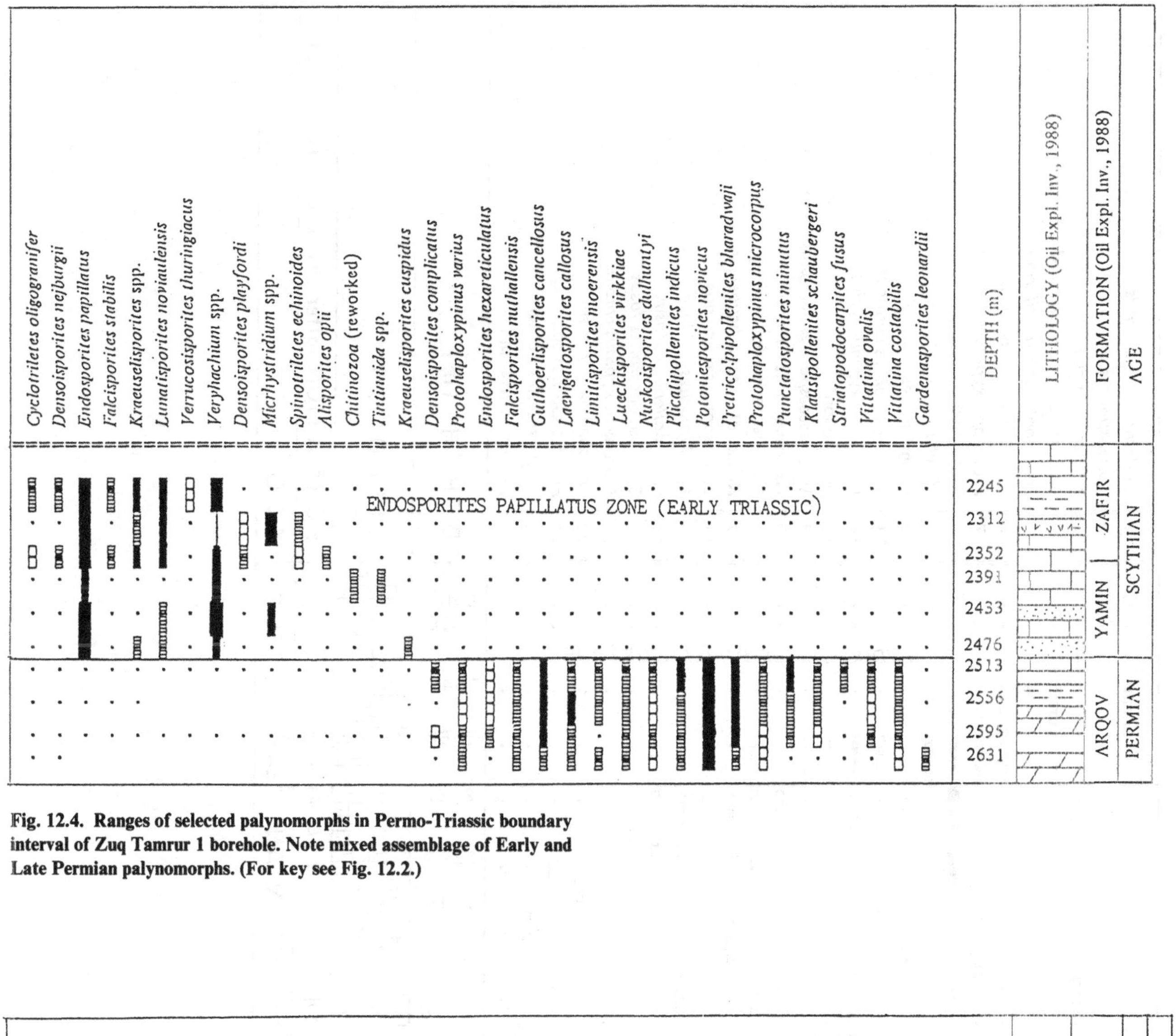

Fig. 12.4. Ranges of selected palynomorphs in Permo-Triassic boundary interval of Zuq Tamrur 1 borehole. Note mixed assemblage of Early and Late Permian palynomorphs. (For key see Fig. 12.2.)

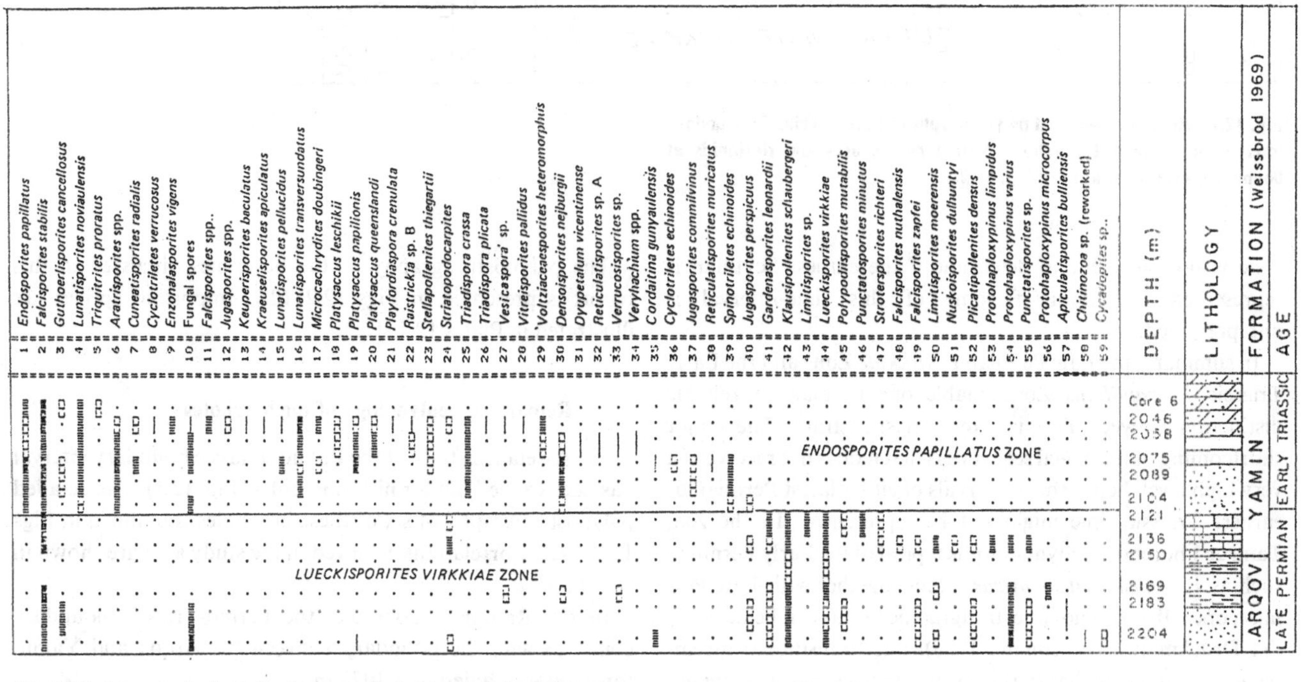

Fig. 12.5. Ranges of selected palynomorphs in Permo-Triassic boundary interval of Hemar 1 borehole. (For key see Fig. 12.2.)

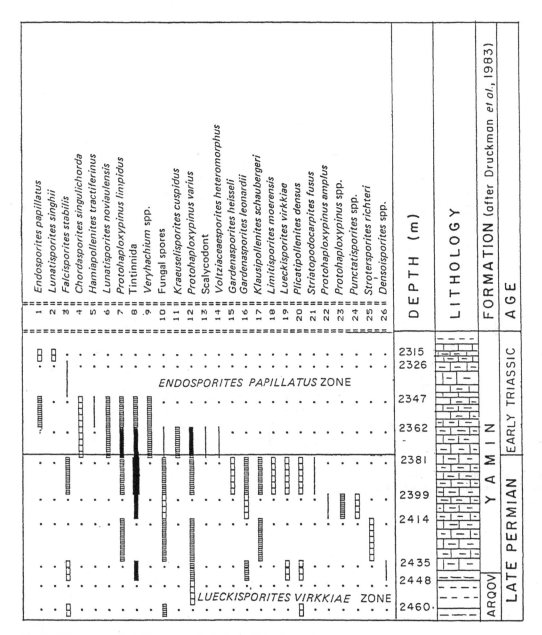

Fig. 12.6. Ranges of selected palynomorphs in Permo-Triassic boundary interval of Masada 1 borehole. Note fungal spores and tintinnids at boundary layer. (For key see Fig. 12.2.)

zone is not sufficient to determine if the earliest Scythian is represented in it, or if some of this time-interval is unrepresented.

In summary, the Late Permian *L. virkkiae* Zone and the Early Triassic *E. papillatus* Zone enable one to make a reliable distinction between Permian and Triassic strata. The major shortcoming of this zonal sequence is that it does not resolve at a very high level, hence short intervals of either latest Permian or earliest Triassic time might not be represented. In the Zuq Tamrur 1 borehole, palynomorphs typical of the Early Permian, such as *Potonieisporites novicus*, occur together with Late Permian ones. Because no lithostratigraphic hiatus can be detected (Y. Druckman, pers. comm., 1988), this 'transitional assemblage' may indicate that Early Permian species may have longer ranges in the Middle East than in Europe.

The palynostratigraphy of the boreholes discussed in this report is presented in Figs 12.2–12.7. Important species are illustrated in Plates 12.1–12.3.

Remarks on palynology of the boreholes

A detailed study of the Permo-Triassic boundary interval has been carried out for nine boreholes (Fig. 12.1). The detailed palynostratigraphy of six of these boreholes is outlined in Figs 12.2–12.7. Correlations achieved in the study area are shown in Fig. 12.8.

In the Ramon 1 borehole, the Permo-Triassic boundary coincides with the boundary between the Arqov and Yamin formations at a depth of 1070 m.

In the Avedat 1 borehole, the Permo-Triassic boundary is

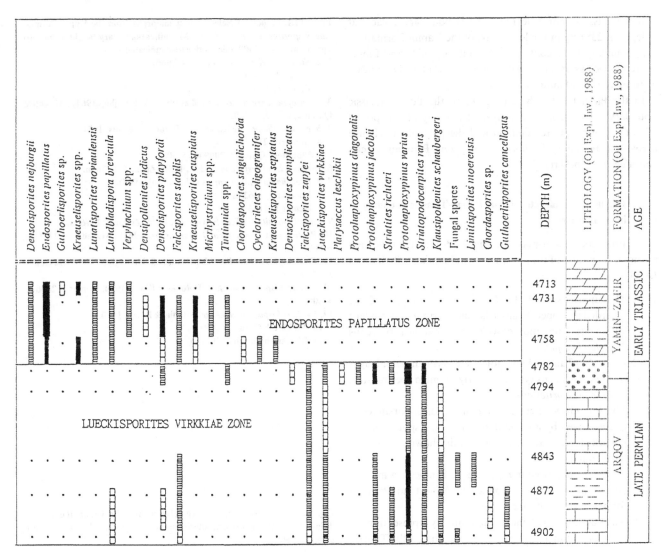

Fig. 12.7. Ranges of selected palynomorphs in Permo-Triassic boundary interval of Pleshet 1 borehole. Note red clay at boundary. (For key see Fig. 12.2.)

located provisionally at the top of core 5, in the lower Yamin Formation. The fossil assemblage contains abundant reworked terrestrial organic matter, fungal spores, and tintinnids, all recovered from a black shale.

In the Shezaf 1 borehole, the Permo-Triassic boundary is drawn provisionally in the upper part of the Arqov Formation, in a thin layer of red clay that has yielded fungal spores and tintinnids.

In the Makhtesh Qatan 2 borehole (Fig. 12.2), the Permo-Triassic boundary is at the top of core 9, in a black shale in the lower Yamin Formation (Eshet, 1984). Abundant reworked organic matter of terrestrial origin is typical of the transitional interval. There is palynologic continuity across the boundary, although there is an overall change in floral composition. Acritarchs are abundant at horizons immediately above the boundary.

In the Zohar 8 borehole (Fig. 12.3) the boundary is placed in the middle part of the Yamin Formation, within a red clay layer at a depth of 2776 m. Fungal spores are abundant in the boundary layer and acritarchs begin to appear in large quantities in strata above 2776 m.

In the Zuq Tamrur 1 borehole (Fig. 12.4), the Permo-Triassic boundary is located in the upper part of the Arqov Formation, at a depth of 2513 m. This boundary marks a drastic palynologic change. Below 2513 m the floral assemblage is a mixture of typical Early and Late Permian palynomorphs, all of which disappear at the boundary. Tintinnids are abundant at the boundary and reworked terrestrial organic matter is abundant. Acritarchs appear in large numbers immediately above the boundary layers, in strata assigned to the Scythian. There is no lithologic indication of hiatus at the boundary (Y. Druckman, pers. comm., 1988). This borehole exhibits the greatest palynologic change across the Permo-Triassic boundary in Israel.

In the Hemar 1 borehole (Fig. 12.5) the Permo-Triassic boundary is placed in the lower part of the Yamin Formation, at a depth of 2121 m. Fungal spores occur in the boundary layer; tintinnids are abundant; and reworked organic matter is prominent.

In the Masada 1 borehole (Fig. 12.6) the boundary is placed at a depth of 2381 m, in the lower part of the Yamin Formation. The boundary layer contains abundant tintinnids and fungal spores, as well as reworked palynomorphs, all of which indicate a strong terrestrial influence.

In the Pleshet 1 borehole (Fig. 12.7), the Permo-Triassic boundary is drawn in a thin layer of red clay at a depth of 4782 m. Tintinnids are common and the palynologic change across the boundary is drastic – only one Permian species continues into the Triassic.

Palynologic characters of the Permo-Triassic boundary interval in Israel

From the data displayed in Figs 12.2–12.8, the following are notable features of the transition from Permian to Triassic in Israel:

1 The upper part of the *Lueckisporites virkkiae* Zone, regarded as Upper Permian, is characterized by abundant striate bisaccates, most of which represent the form genus *Protohaploxypinus*.

2 In the Zuq Tamrur 1 borehole, Late Permian assemblages contain specimens typical of Early Permian species, such as *Potonieisporites novicus*.

3 The boundary interval in all the studied boreholes is characterized by an abundance of fungal spores and by reworked terrestrial particles, both showing a strong terrestrial influence.

4 Ranges summarized in Figs 12.2–12.8 indicate a marked change in flora at the boundary. Almost no Permian forms continue into the Triassic.

5 An abundance of acritarchs, presumably representing the phytoplankton, is typical of the beginning of the Triassic

Plate 12.1. (opposite) Selected palynomorphs from the Upper Permian *Lueckisporites virkkiae* Zone. Magnifications vary slightly but are approximately × 1000 unless otherwise indicated.

1, 2 *Strotersporites richteri* (Klaus, 1965).
   1 Zohar 8, 2941 m.
   2 Zohar 8, 2776 m.
3 *Striatopodocarpites rarus* (Bharadwaj & Salujha, 1964). Makhtesh Qatan 2, core 9, × 500.
4, 5 *Klausipollenites schaubergeri* (Potonie & Klaus, 1954).
   4 Hemar 1, 2121 m.
   5 Makhtesh Qatan 2, 2099 m, × 500.
6 *Jugasporites perspicuus* Leshik, 1956. Zohar 8, 2776 m.
7, 8 *Protohaploxypinus varius* (Bharadwaj, 1964).
   7 Masada 1, 2362 m.
   8 Hemar 1, 2183 m.
9 *Plicatipollenites indicus* Srivastava, 1970. Zohar 8, 2776 m.
10 *Protohaploxypinus jacobii* (Jansonius, 1962). Masada 1, 2381 m, × 800.
11 *Protohaploxypinus limpidus* (Balme & Hennelly, 1955). Masada 1, 2347 m.
12 cf. *Lueckisporites virkkiae* Potonie & Klaus, 1954. Hemar 1, 2136 m, × 800.
13, 14 *Lueckisporites virkkiae* Potonie & Klaus, 1954.
   13 Makhtesh Qatan 2, 2099 m.
   14 Makhtesh Qatan 2, 2021 m.

Fig. 12.8. Palynostratigraphic correlation of units within the Permo-Triassic boundary interval in subsurface of southern Israel. Sample depths given in meters.

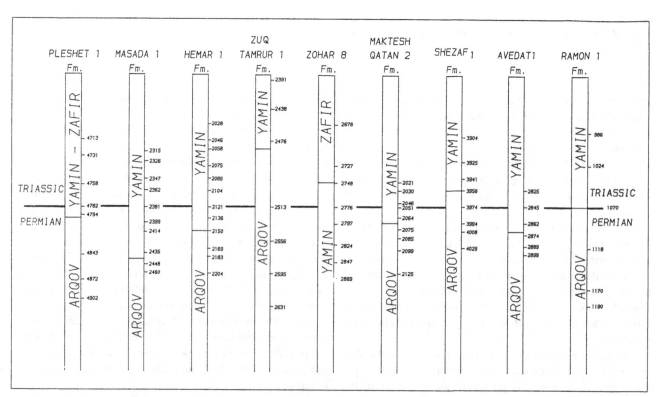

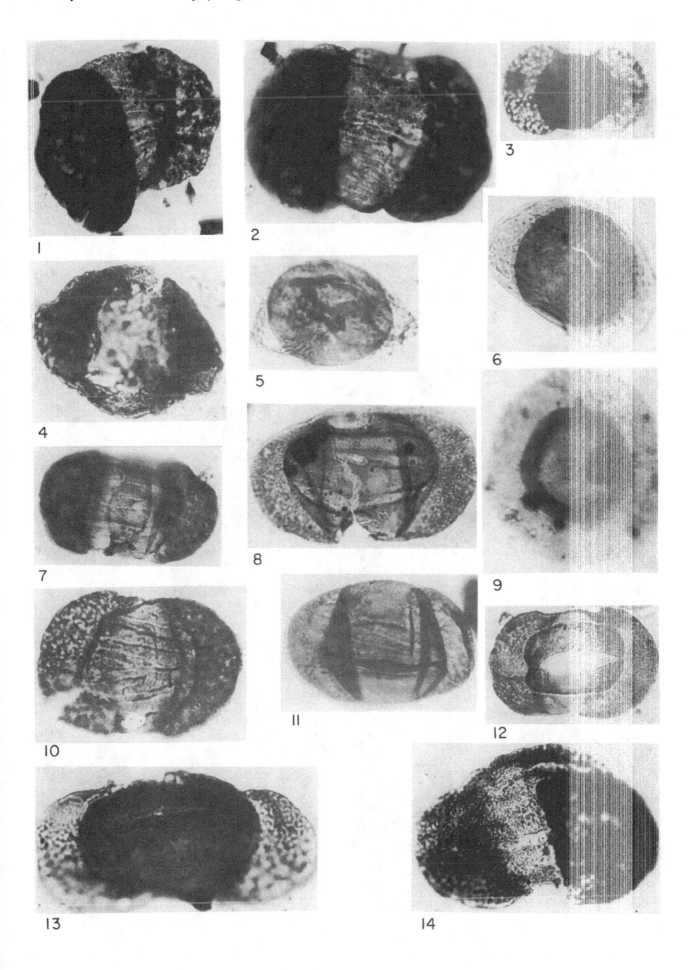

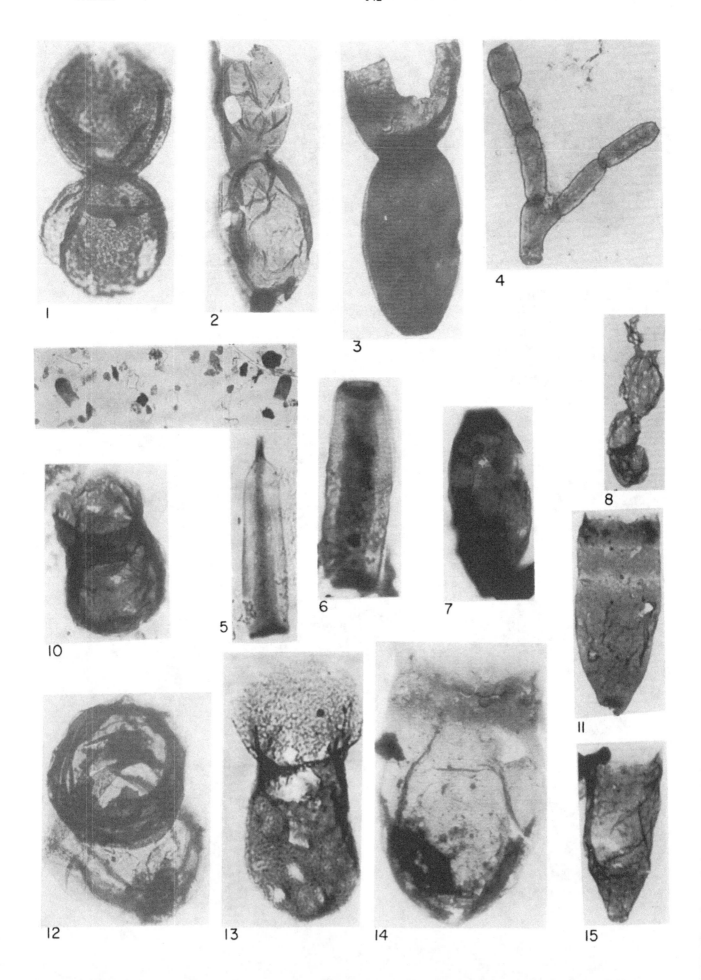

Plate 12.2. Selected fungal spores and tintinnids from Permo-Triassic boundary layers. Magnifications may vary slightly, but are approximately × 1000 unless otherwise indicated.
1–4, 8 Chains of fungal cells, generally assignable to *Tympanicysta* Balme, 1979.
   1 Masada 1, 2381 m, × 1250.
   2 Hemar 1, 2204 m.
   3 Hemar 1, 2183 m.
   4 Zohar 8, 2776 m.
   8 Masada 1, 2399 m.
5–7 Single fungal cells assignable to *Tympanicysta* Balme, 1979.
   5 Masada 1, 2362 m.
   6 Hemar 1, 2204 m.
   7 Zohar 8, 2776 m.
9–15 Various forms of tintinnids from Permo-Triassic transitional layers.
   9 Palynofacies with dominant tintinnids. Masada 1, 2381 m, × 80.
   10, 11, 13, 15 Masada 1, 2381 m, × 500.
   12 Masada 1, 2362 m.
   14 Zuq Tamrur 1, 2391 m.

and is found in most of the boreholes. In most boreholes, lycopode spores such as *Endosporites papillatus* and *Kraeuselisporites* spp. are found at levels slightly above the ones in which acritarchs first appear.

6 Although sedimentation across the boundary is considered to have been continuous (Weissbrod, 1981), a thin, distinct layer of red clay marks the boundary in the Pleshet 1, Zohar 8, and Shezaf 1 boreholes. The depositional environment represented by this layer has yet to be determined, but it is rich in terrestrial organic particles such as fungal spores and reworked tracheids and sporomorphs.

In southern Israel the Permo-Triassic boundary is situated stratigraphically at different levels between the upper part of the Arqov and the lower part of the Yamin formations (Fig. 12.8). This may indicate the existence of small, undetected hiatuses, or it may represent inaccurate definitions of the lithostratigraphic units.

Interpretation

In the interpretation of biostratigraphic information, one must consider the chronostratigraphy as well as the lithostratigraphic data. Available palynostratigraphic data (Figs 12.2–12.8) indicate that a complete sequence of palynozones may be recognized in the Permo-Triassic boundary interval. However, the extensive extinction of most Permian forms at the upper boundary of the *Lueckisporites virkkiae* Zone and the first occurrence of a new suite of palynomorphs at the base of the *Endosporites papillatus* Zone may suggest that a hiatus exists at that horizon. On the other hand, the lithostratigraphic succession is continuous and undisturbed across the zonal boundary (Y. Druckman, pers. comm., 1988).

An ecologic approach that does not require a hiatus might provide an alternative explanation for the drastic palynologic change across the *L. virkkiae–E. papillatus* zonal boundary, which is here regarded as the Permo-Triassic boundary. As mentioned previously, the zonal boundary is rich in the spores of fungi, which originated on land, and also contains a variety of

reworked terrestrial particles. This may suggest that harsh ecologic conditions existed on land during the transition from Permian to Triassic, and these conditions may have resulted in extinction of most of the Permian flora, represented mostly by bisaccate pollen, and expansion of fungi, which replaced the extinct vegetation. The nature of the harsh conditions cannot be reconstructed from available data, but the red clay at the boundary (in the Pleshet 1, Zohar 8, and Shezaf 1 boreholes) may be a product of terrestrial processes and might provide some clues.

The dominance of acritarchs at the level here regarded as the base of the Triassic can also be explained in ecologic terms. That is, the acritarchs probably represent an opportunistic phytoplankton group that conquered marine environments depleted of competitors during the ecologic break at the end of the Permian.

Correlation

Palynologic features of the Permo-Triassic boundary interval in Israel enable correlation with key sections in Europe and Asia. The dominance in rocks of Late Permian age of bisaccate pollen and other land-derived palynomorphs has been well documented by Balme (1970), and by Visscher & Brugman (1981, 1988). The appearance of fungal spores in thin horizons at the Permo-Triassic boundary in the Southern Alps of Italy has been noted by Visscher & Brugman (1988), and Balme (1979) has noted the same feature in Permo-Triassic boundary layers at Kap Stosch, East Greenland. Visscher & Brugman (1988) reported that the fungal remains appear in the Upper Permian, above the last assemblage of striate bisaccates. They regarded the fungal layer as highest Permian, and as a horizon that represents a great ecologic catastrophe. Because of the spacing of samples and problems of borehole-caving, no such resolution was possible in the study reported here. However, the fungal spores seem to appear at the same stratigraphic level in Israel as in Italy and East Greenland.

The abundance of acritarchs at the beginning of the Triassic is a worldwide phenomenon (Visscher & Brugman, 1981; Balme, 1970, 1979). This suggests that the acritarchs were the pioneering group that invaded Triassic oceans after the worldwide Permian regression. Also, the dominance of lycopode spores above the 'acritarch spike' is reported from Europe (Visscher & Brugman, 1981, 1988), Pakistan (Balme, 1970), and Israel (Eshet & Cousminer, 1986).

Conclusions

1 Palynologic assemblages undergo a drastic change within the Permo-Triassic boundary interval in Israel.

2 In the subsurface of Israel, the succession assigned to the Late Permian *Lueckisporites virkkiae* Zone is characterized by striate bisaccate pollen.

3 At the top of the Permian succession is a horizon rich in reworked palynomorphs and other land-derived organic debris. In most of the boreholes studied, there is a thin layer containing abundant fungal spores, associated with land-derived reworked organic particles. This layer is

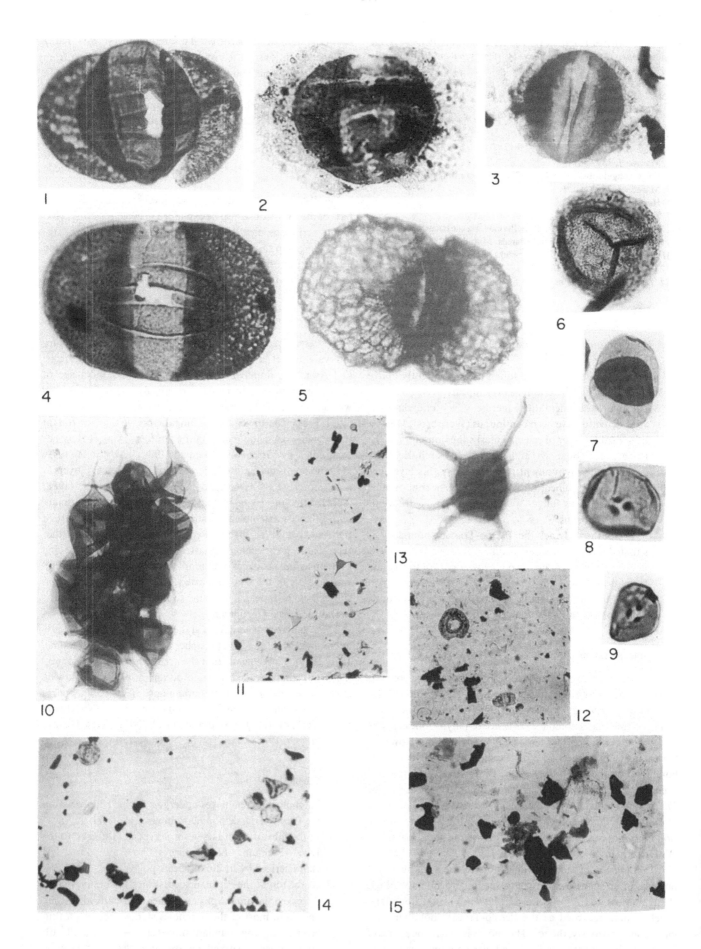

Plate 12.3. Selected palynomorphs from Lower Triassic beds. Magnifications may vary slightly, but are approximately × 100 unless otherwise indicated.

1, 2 *Lunatisporites noviaulensis* Leschik, 1956.
  1 Masada 1, 2347 m.
  2 Hemar 1, 2046 m.
3 *Falcisporites stabilis* Balme, 1970. Makhtesh Qatan 2, 1650 m, × 500.
4 *Lunatisporites* sp. Hemar 1, 2048 m.
5 *Platysaccus leschikii* Hart, 1960. Makhtesh Qatan 2, core 6.
6 *Densoisporites nejburgii* (Schulz). Makhtesh Qatan 2, 1550 m.
7–9 Inner body of *Endosporites papillatus* Jansonius, 1962. Makhtesh Qatan 2, core 2, Box A, × 1250.
10 A cluster of acritarchs assignable to *Veryhachium* (phytoplankton). Masada 1, 2347 m, × 800.
11 Palynofacies dominated by *Veryhachium* and *Micrhystridium* (acritarchs) at base of Triassic. Makhtesh Qatan 2, 1753 m, × 100.
12 Palynofacies at Permian-Triassic boundary dominated by terrestrial particles: pollen, spores, and inertinites. Avedat 1, core 5, × 100.
13 *Diexalophasis*, a reworked Silurian acritarch from the Permo-Triassic boundary layer. Hemar 1, 2183 m.
14 Palynofacies of land-derived palynomorphs at top of the Permian. Avedat 1, core 5, × 100.
15 Palynofacies of land-derived inertinites in the Permo-Triassic boundary layer. Avedat 1, core 5, × 100.

difficult to detect and its recognition requires detailed sampling, which has not always been possible in the material available for this study.

4 Acritarchs are abundant in layers in the lower part of the stratal sequence here regarded as Early Triassic.

5 Lycopodes, such as *Kraeuselisporites* and *Endosporites papillatus* are among the first representatives of land plants above the 'acritarch spike'.

6 Following Visscher & Brugman's (1988) ideas on the nature of the Permo-Triassic boundary in the Southern Alps, it is suggested here that palynologic associations in the Permo-Triassic boundary interval of Israel represent a vegetative succession that began with a massive extinction of the Late Permian flora and was followed by a proliferation of fungi, which could survive the ecologic stress. Later, at the beginning of the Triassic, acritarchs invaded marine niches and constituted the pioneering vegetation in the ocean after the break at the end of the Permian. The first Triassic land plants are represented by lycopodes, and are represented first in the record just above the 'acritarch spike'. This succession may be a model of a vegetative community recovering from a great ecologic shock.

References

Balme, B. E. (1970). Palynology of the Permian and Triassic strata in the Salt Range, Pakistan. In *Stratigraphic Boundary Problems: Permian and Triassic of West Pakistan*, ed. B. Kummel & C. Teichert, pp. 306–453. (Dept. Geology, Spec. Publ. 4.) Lawrence, Kansas: Univ. Kansas.

Balme, B. E. (1979). Palynology of Permian–Triassic boundary beds of Kap Stosch, East Greenland. *Medd. om Grønland*, 200(6): 1–37.

Clarke, R. F. A. (1965). British Permian saccate and monosulcate miospores. *Micropaleontology*, 8(2): 322–54.

Derin, B. & Gerry, E. (1981). Late Permian–Late Triassic stratigraphy in Israel and its significance to oil exploration in Israel. *Oil Exploration in Israel*, Israel Geol. Soc. Symp., 2–3 Dec., 1981, Abstracts, pp. 9–11.

Eshet, Y. (1984). A new palynologic evidence for the Permian–Triassic boundary in southern Israel. *Pollen et Spores*, 26(2): 285–92.

Eshet, Y. (1987). *Palynologic aspects of the Permo-Triassic sequence in the subsurface of Israel*. Ph.D. Dissertation, City University of New York, New York, 193 pp.

Eshet, Y., & Cousminer, H. L. (1986). Palynozonation and correlation of the Permo-Triassic succession in the Negev, Israel. *Micropaleontology*, 32(3): 193–214.

Geiger, M. E. & Hopping, C. A. (1968). Triassic stratigraphy of the southern North Sea basin. *Philos. Trans. Roy. Soc. London, Ser. B*, 254: 1–36.

Horowitz, A. (1973). Triassic miospores from southern Israel. *Rev. Palaeobot. Palynol.*, 16(3): 176–207.

Jansonius, J. (1962). The palynology of the Permian and Triassic sediments, Peace River, western Canada. *Palaeontographica, Abt. B*, 110: 35–98.

Jardiné, S. (1974). Microflores des formations du Gabon, atribuées au Karoo. *Rev. Palaeobot. Palynol.*, 17(1/2): 75–112.

Orlowska-Zwolinska, T. (1977). Palynological correlation of the Bunter and Muschelkalk in selected profiles from western Poland. *Acta Geol. Polonica*, 27(4): 417–30.

Sokratov, B. G. (1983). Oldest Triassic strata and the Permian–Triassic boundary in the Caucasus and the Middle East. *Internat. Geol. Rev.*, 25(4): 483–94.

Visscher, H. (1971). The Permian and Triassic of the Kingscourt Outlier, Ireland. *Ireland Geol. Surv., Spec. Pap.*, 1: 144 pp.

Visscher, H. & Brugman, W. A. (1981). Ranges of selected palynomorphs of the Alpine Triassic of Europe. *Rev. Palaeobot. Palynol.*, 34: 115–28.

Visscher, H. & Brugman, W. A. (1988). The Permian–Triassic boundary in the Southern Alps: A palynological approach. *Mem. Soc. Geol. Ital.*, 34: 121–8.

Weissbrod, T. (1981). The Paleozoic of Israel and adjacent countries. *Israel Geol. Surv. Rep.*, M. P. 600/81, 276 pp. (Hebrew, with English abstract.)

13 The effects of volcanism on the Permo-Triassic mass extinction in South China

YIN HONGFU, HUANG SIJI, ZHANG KEXING, HANSEN, H. J., YANG FENGQING, DING

MEIHUA AND BIE XIANMEI

Introduction

In this chapter we deal mainly with volcanism in the Permo-Triassic boundary interval in South China, with brief reference to other parts of the Tethyan realm. In much of South China a level commonly regarded as the Permo-Triassic boundary is marked by the so-called 'boundary clayrock', which is a layer of montmorillonite or interstratified montmorillonite and illite several centimeters to a few decimeters thick. The boundary clayrock is marine because it yields conodonts, ostracodes, foraminifera, and glauconite, and because its boron content is everywhere much higher than 100 ppm. It is at the level of the boundary clayrock that the strongest phase of Permo-Triassic mass extinction is recorded and it is also at this level that anomalous concentrations of $\delta^{13}C$, iridium, and other elements have been reported.

Permo-Triassic volcanism

Our ideas about volcanism in the Permo-Triassic boundary interval are based on the nature of the boundary clayrock, and on evidence from sphaerules in the boundary clayrock.

Nature of boundary clayrock

More than 30 occurrences of the boundary clayrock and other boundary rocks have been found to be of volcanic origin. The well-known boundary clayrock of the Meishan section, Changxing, Zhejiang Province, was first regarded as sedimentary. However, owing to discoveries in it of high-quartz (beta quartz, Plate 13.1, fig. 8) and other volcanic indicators (He *et al.*, 1987) this layer has been shown to be hydrolyzates of volcanic ash. The equally famous boundary clayrock in the Shangsi section, Guangyuan, Sichuan Province (bed 21 of Yang *et al.*, 1987, or bed 27b of Li *et al.*, 1986) is montmorillonitized tuff (Li *et al.*, 1986). Boundary clayrocks in the Meishan and Shangsi sections also exhibit porphyritic and blastotuffaceous textures, and yield automorphic plagioclase, zircon, apatite, and high-quartz crystals that lack transport or polishing scratches. Further, no terrigenous clasts have been found and there are a few clay balls (Fu Guoming, pers. comm.). At the Ermen locality, Huangshi, Hubei Province, which was studied in 1987 by a working group of IGCP Project 203, the interstratified montmorillonite-illite boundary clayrock (bed 34, Plate 13.1, fig. 13) includes melted quartz, high-quartz (Plate 13.1, figs 1, 2), a large quantity of sphaerules (Plate 13.3, figs 4, 6, 8, 11–14), and petalic zeolite druses that originated from alteration of volcanic glass in sea water (Plate 13.2, fig. 7). This bed is thus proved to be a montmorillonitized tuff. In bed 38 at the Ermen locality, several centimeters above the boundary clayrock, a large number of vitric shards have been discovered. These, together with oriented crystals, form a pseudoflowage structure.

Altogether at least 17 localities in South China are known to include Permo-Triassic boundary clayrocks of tuffaceous origin. At an additional 18 localities there are volcanic deposits at or near the Permo-Triassic boundary (Fig. 13.1). It would be too tedious to describe all of these. However, they have the following characteristics in common:

1 tuffaceous or blastotuffaceous, mainly porphyritic texture, with quartz and feldspar phenocrysts (Plate 13.2, figs 3–5), and vesiculate (Pl. 13.2, fig. 1) and pseudoflowage (Plate 13.2, fig. 6) structures. The latter display fluidal lamination, orientation of elongate shards and crystals, and plastic deformation of the shards along

Plate 13.1. (All figures except 14 are SEM micrographs)
1–3 High-quartz or β quartz paramorphs with characteristic hexagonal bipyramidal crystals that have altered into α quartz.
  1 bed 38, Ermen, Huangshi; lateral view, × 75.
  2 bed 34, Ermen, Huangshi; lateral view, × 140.
  3 bed ZCE-3, Meishan, Changxing; top view, × 100.
4 Zircon, bed 38, Ermen, Huangshi, × 400.
5 Zircon, bed 34, Ermen, Huangshi, × 180.
6 Zircon, bed ZCE-4, Meishan, Changxing, × 190.
7 Apatite, bed 38, Ermen, Huangshi, × 400.
8 Zircon, bed ZCE-3, Meishan, Changxing, × 180.
9 Gypsum, bed ZCE-3, Meishan, Changxing, × 130.
10, 11 Volcanic ash, bed 34, Ermen, Huangshi, × 300 and × 220, respectively.
12 Pyrite, bed ZCE-3, Meishan, Changxing, × 260.
13 Interstratified montmorillonite-illite microcrystals, bed 34, Ermen, Huangshi, × 5000
14 Clay ball, bed 38, Ermen, Huangshi, × 40.
15 Enlarged surface of clay ball, bed 38, Ermen, Huangshi, × 2000.
(Bed 34 of Ermen is the boundary clayrock; bed 38 of Ermen is a tuffaceous clay bed 115 cm above bed 34; beds ZCE-3 and 4 of Meishan are boundary clay beds; 3 is light colored, 4 is darker.)

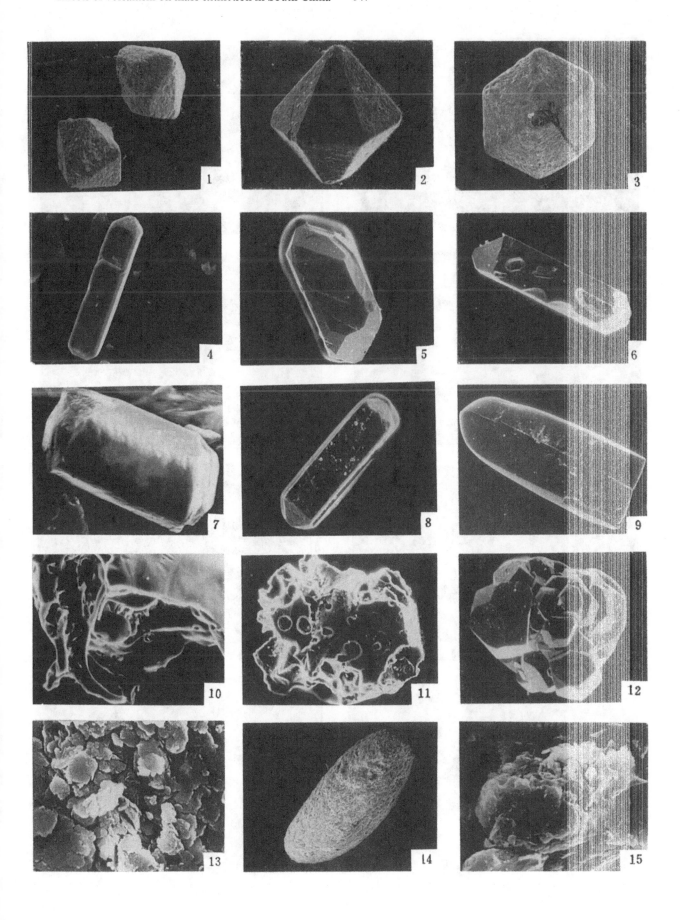

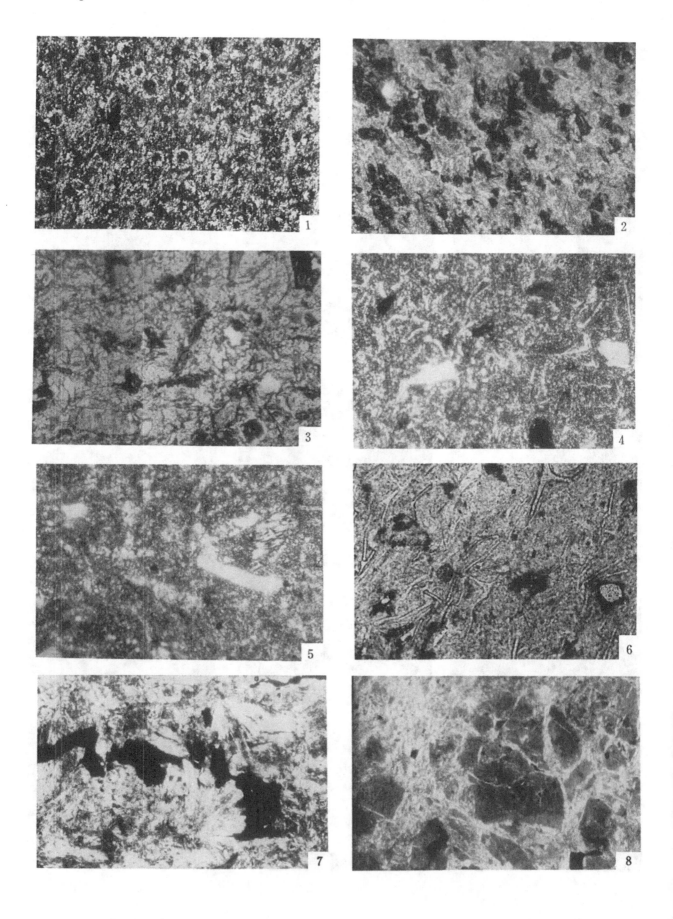

Plate 13.2. (opposite) **1** Montmorillonitized tuffite showing vesicles with siderite in center ringed by calcite. Bed 5, Huopu, Panxian; plane-polarized light, 10×3.2.

2 Montmorillonitized vitric tuff. Montmorillonite light colored; dark-colored lumps are residual vitric clasts. Bed 2, Xiaochehe, Guiyang; cross-polarized light, 4×2.5.

3 Silicified crystal-vitric tuff. Orthoclase crystals in lower left corner; quartz is light colored; biotite is short black strips. Matrix is volcanic vitrics with irregular cracks due to contraction during cooling. Bed 16, Heshan (Guangxi Province); cross-polarized light, 10×2.5.

4 Silicifed crystal-vitric tuff, showing light-colored angular quartz crystals and irregular shards. Partly silicified matrix of fine volcanic material shows glittering optical character. Bed 26, Paoshui, Laibin; cross-polarized light, 10×2.5.

5 Silicified crystal tuff. Quartz and orthoclase crystal strips on the right; darker parts of matrix are vitrics; glittering dots are microcrystal cherts. Bed 30, Paoshui, Laibin; cross-polarized light, 10×2.5.

6 Montmorillonitized vitric tuff showing glass shards forming pseudoflow-age structure. Bed 7, Xiaochehe, Guiyang; plane-polarized light, 10×3.2.

7 Interstratified montmorillonite-illite clayrock with petalic zeolite druses radiating toward a relatively large crack. Bed 34, Ermen, Huangshi; cross-polarized light, 4×2.5.

8 Montmorillonitized tuff with silicified ash lumps (darker) separated by thin montmorillonite. Bed 42, Kejiawan, Huangshi; cross-polarized light, 4×2.5.

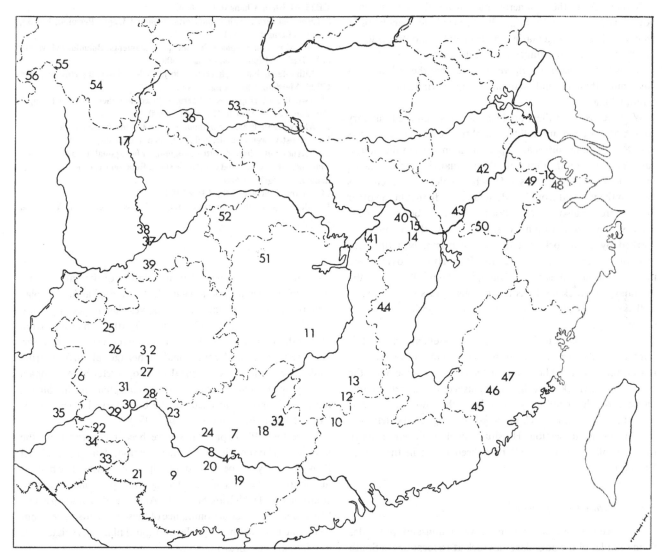

Fig. 13.1. Distribution of volcanic and related deposits at the Permo-Triassic boundary, South China.

1–17: Tuffaceous Permo-Triassic boundary rocks. 1, Xiaochehe, Guiyang; 2, Wengjingkou, Guiyang; 3, Liuchang, Qingzhen; 4, Hongshuihe, Laibin; 5, Paoshui, Laibin; 6, Hopu, Panxian; 7, Liuzhou; 8, Heshan; 9, Pinggo; 10, Lianxian; 11, Shaodong; 12, Yizhang; 13, Chenxian; 14, Ermen, Huangshi; 15, Kejiawan, Huangshi; 16, Meishan, Changxing; 17, Shangsi, Guangyuan.

18–35: Volcanic deposits near or at Permo-Triassic boundary. 18, Lipu; 19, Guixian; 20, Shangling; 21, Jingxi; 22, Xiling; 23, Nandan; 24, Yishan; 25, Dafang; 26, Zhijing; 27, Huishui; 28, Lodian; 29, Ceheng; 30, Wangmo; 31, Houchang, Ziyun; 32, Pingle; 33, Funing; 34, Guangnan; 35, Sizhong.

36–51: Permo-Triassic boundary clayrock (montmorillonite or montmorillonite-illite). 36, Xixiang; 37, Lianfengya, Chongqing; 38, Hechuan; 39, Qijiang; 40, Daye; 41, Puqi; 42, Caoxian; 43, Huaining; 44, Yichun; 45, Yanshi; 46, Zhangping; 47, Datian; 48, Huangzishan, Huzhou; 49, Niutoushan, Guangde; 50, Xiancha, Jingdezhen; 51, Shangzhi.

52–56: Sphaerules at Permo-Triassic boundary. 52, Lichuan; 53, Zhenan; 54, Dangchang; 55, Diebu; 56, Luqu. Sphaerules also occur at Permo-Triassic boundary at 4, Laibin; 14, 15, Kejiawan and Ermen, Huangshi; 16, Meishan; 30, Wangmo; 31, Houchang, Ziyun; 42, Caoxian; 47, Datian; 49, Guangde.

the orientation. Clay balls (Plate 13.1, figs 14, 15) may have been formed by rolling together of volcanic ash or by infilling of vesicles in the tuff.

  2 index minerals of tuffaceous rocks, mainly high-quartz (Plate 13.2, figs 1–3) and zeolites. High-quartz crystallizes between 573 and 867°C and is a typical mineral of rapidly cooling volcanic or hyperbyssal rocks.

  3 volcanic ash (Plate 13.1, figs 10–11), vitrics and crystals. Chicken-bone shaped shards and other vitrics are typical clasts in tuff (Plate 13.2, figs 2–6). Common crystals are automorphic to hypautomorphic plagioclase, apatite, zircon, pyrite, and gypsum (Plate 13.1, figs 4–8, 10).

Judging from their mineral and chemical content, most boundary clayrocks were originally intermediate to acidic tuffs, which readily alter in a marine environment to montmorillonite or interstratified montmorillonite-illite clay. Origin of these rocks as tuffs explains the interesting fact that these thin layers cover more than one million square kilometers in 12 provinces in South China.

We note also that there are localities at which the boundary rocks are unaltered tuff, tuffite, ignimbrite, or tuffaceous clastics and lava. Such localities are especially numerous in the Southeast Yunnan–South Guizhou–West Guanxi Permo-Triassic fault block (Plate 13.2, figs 1–5), where volcanism reached its acme in the interval from late Permian to Induan. On the other hand, there are also a few localities at which the boundary beds are not of volcanic origin. At some of these the boundary is marked by a clayrock, but it was not originally a tuff. For example, in the Guanyinshan section, Puqi, Hubei Province, and in the Niutoushan section, Guangde, Anhui Province, the boundary clayrock contains terrigenous sand and pebbles, as well as reworked fossils, and is thus considered to be sedimentary.

At many localities in South China, the boundary clayrock is not the only clay bed. Two or more such beds may occur within a few decimeters of the boundary. At some of these localities the boundary clayrock may be distinguished by its remarkable biotic and lithic characters. At others, the lowermost bed is taken to be the boundary clay if no great change can be detected. In the Duanshan section, Huishui, Guizhou Province, more than 10 tuffaceous clay beds have been found in the Upper Permian.

Evidence from sphaerules

Since 1984, we have obtained several thousand sphaerules from samples of the boundary clayrock at 16 localities in South China (Fig. 13.1). Similar discoveries have been reported by other authors (He, 1985; Gao *et al.*, 1987).

In the Meishan section, Changxing, Zhejiang Province, 800 magnetic sphaerules have been found in the boundary clayrock, and 1000 have been recovered from the same unit in the Pingdingshan section, Caoxian, Anhui Province. Energy-spectrum analyses show that the majority of these are ferruginous (Plate 13.3, fig. 1) or lithic-ferruginous (Plate 13.3, figs 5, 7, 9, 10). Some have contraction wrinkles or spiral filaments on the surface (Plate 13.3, fig. 9). The wrinkles may have formed by contraction of the surface of molten sphaerules, and the spiral

Plate 13.3. (All figures are SEM micrographs)
1 Ferruginous sphaerule. Boundary clayrock, bed CB25, Meishan, Changxing, × 100.
2 Chrome spinel sphaerule with jet hole. Base of Triassic, bed 2, Yidaohe, Zhengan, × 100.
3 Apatite sphaerule. Boundary clayrock, bed 16, Niutoushan, Guande, × 180.
4 Lithic-ferruginous sphaerule with wrinkles on surface produced by contraction during cooling. Boundary clayrock, bed 34–1, Ermen, Huangshi, × 1300.
5 Lithic-ferruginous sphaerule with shape of tapering droplet. Boundary clayrock, bed 41, Pingdingshan, Caoxian, × 660.
6 Lithic-ferruginous sphaerule; tapering droplet with spiral filaments. Topmost Upper Permian, bed 33, Ermen, Huangshi, × 550.
7 Lithic-ferruginous sphaerule with impact pits. Boundary clayrock, bed CB25, Meishan, Changxing, × 660.
8 Lithic sphaerule with impact pits. Topmost Upper Permian, bed 33, Ermen, Huangshi, × 500.
9 Lithic-ferruginous sphaerule with spiral filaments. Boundary clayrock, bed CB25, Meishan, Changxing, × 600.
10 Lithic-ferruginous sphaerule with jet holes. Boundary clayrock, bed CB25, Meishan, Changxing, × 500.
11 Two lithic-ferruginous sphaerules fused together. Topmost Upper Permian, bed 33, Ermen, Huangshi, × 600.
12 Many small lithic-ferruginous sphaerules fused onto a larger one. Lowermost Lower Triassic, bed 39, Ermen, Huangshi, × 430.
13 Ferruginous sphaerule with pentagonal or hexagonal surface structure, a microcrystal center, and concentric rings. Boundary clayrock, bed 34–1, Ermen, Huangshi, × 500.
14 Enlarged surface of fig. 13, × 2100.
15 Glassy sphaerule. Lowermost Lower Triassic, bed 14–1, Yewagou, Diebu, × 300.

filaments may have resulted from rotation of the half-consolidated sphaerules and friction with dust. Some sphaerules display structures such as fusions, jet holes, vesicles, impact pits, and tapering droplets (Plate 13.3, figs 5, 7, 10). Sphaerules from Huangshi (Plate 13.3, figs 13, 14) have a microcrystalline center, concentric rings, and pentagonal or hexagonal surface structures. Such features are typical of alloy surfaces that cooled rapidly by quenching. Similar but less conspicuous phenomena have been reported to occur in the boundary-clay sphaerules from Shangsi, Sichuan (Gao *et al.*, 1987).

More than 1000 sphaerules have been recovered from the boundary clayrock in the Ermen section, Huangshi, Hubei Province. They can be divided into three categories. The first, or major group, consists of lithic-ferruginous sphaerules, the main components of which are Si, Fe, Ti, Al, Ca, and K, and, in a few, Mn and S. The order of dominance of these elements varies from sphaerule to sphaerule. Vesicles and spiral filaments (Plate 13.3, figs 4, 6) are also present.

The second group of sphaerules from the Ermen boundary clay accounts for 5% of the total. These are siliceous sphaerules. Some show irregular pits and obscure spiral filaments (Plate 13.3, fig. 8). In form and composition this group of sphaerules is comparable with the silicate sphaerules intercepted in the plume of the volcano Etna (Lefèvre, Gaudichet & Billon-Galland, 1986).

Sphaerules of the third group, found in the Kejiawan section, near Ermen, are partly filled with a rust-colored material, which is reminiscent in its chemical composition of goethite and chamosite. These sphaerules are all coated with an organic layer

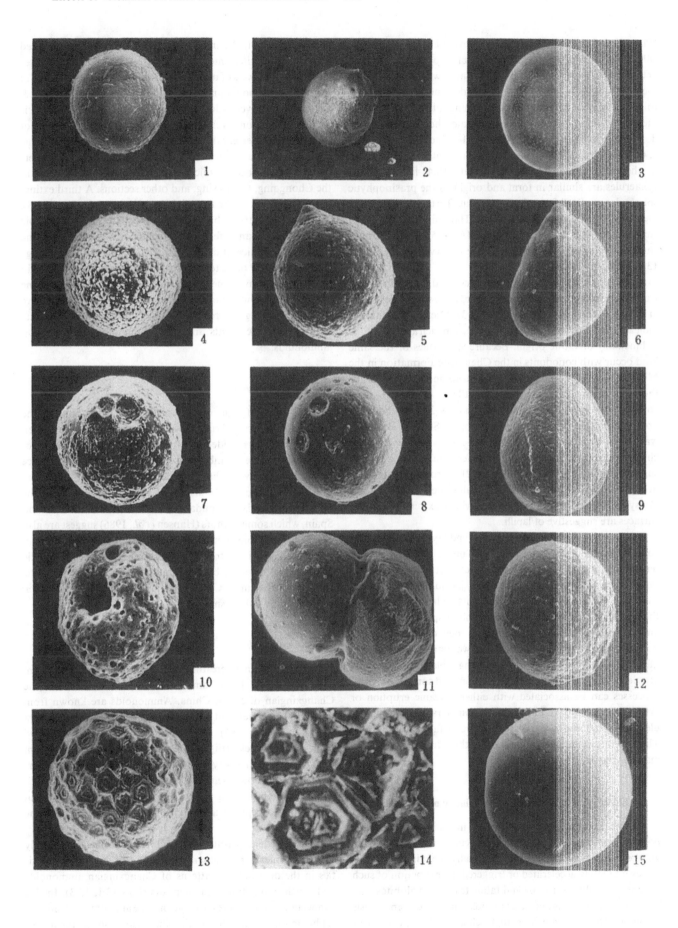

(Plate 13.4, figs 3, 4), which can be recovered by processing the sphaerules individually in a palynological preparation. Some are actually preserved without infill, so that it was possible to fix them biologically with glutaraldehyde and osmium acid, followed by embedding and ultrasectioning (Plate 13.4, fig. 5). In the SEM, dried specimens demonstrate the distribution of double or triple perforations of the algal surface (Plate 13.4, figs 5, 6). Structure of the organic material indicates that it belongs to the prasinophyte algal group (Plate 13.4, fig. 2). These sphaerules are similar in form and origin to the prasinophytic sphaerules reported from the Cretaceous–Tertiary boundary in Denmark, New Zealand, and Spain (Hansen *et al.*, 1986). In chips of the black boundary clayrock (AG–253) from Shangsi, clusters of infilled prasinophytes were observed *in situ* (Plate 13.4, fig. 1).

A few sphaerules from other localities are different in nature from those just mentioned. For example, some sphaerules found at Guande are apatitic (Plate 13.3, fig. 3). Microprobe analysis indicates 38.99% P and 56.92% Ca. Calcium-rich apatites are of sedimentary or biotic origin. Because sphaerules of the same kind occur with conodonts in the Changxing Formation in the Meishan section, it is possible that they are conodont 'pearls', or they may have originated from nautiloids (P. Ward, pers. comm.). One sphaerule from a horizon 0.78–3.27 m above the boundary in the Yidaohe section, Zhengan, Shaanxi Province, was found by energy-dispersive X-ray analysis to be made of chrome spinel, or chromite. This sphaerule also has a papilla suggestive of a jet hole (Plate 13.3, fig. 2). Chrome spinels occur in ultrabasic rocks, or meteorites. Finally, there are also a few glassy sphaerules (Plate 13.3, fig. 15) whose extremely smooth surfaces are suggestive of lapilli.

In summary, the boundary clayrock and layers adjacent to it contain a myriad of sphaerules, the majority of which are ferruginous or lithic-ferruginous and exhibit features such as impact pits, fusions, tapering droplets, jet holes, vesicles, contraction wrinkles, and spiral filaments. These features are thought to have formed as falling, molten sphaerules rotated and collided with one another during their descent and then quickly cooled under water, where they were subjected to a series of physicochemical processes including ablation, degassing, droplet cooling, quenching, condensation, and concretion. Such processes can be associated with either volcanic eruption or extraterrestrial impact, although the diameters of some sphaerules (0.37–0.60 mm) are too great for freely falling cosmic ash. The siliceous sphaerules, reminiscent of those in the Etna plume, may be volcanic.

Effects of volcanism on Permo-Triassic mass extinction

The mass extinction recorded in the Permo-Triassic boundary interval has two basic aspects. First, not all fossil groups experienced mass extinction equally. That is, the boundary event merely accelerated or triggered the extinction of such groups as fusulinids, rugose and tabulate corals, trilobites, and Paleozoic ammonoids and brachiopods, which had been in crisis through at least the Late Permian. Decline and disappearance of these groups seems to have been evolutionarily inevitable and was controlled mainly by long-term, worldwide environmental

changes. Second, the mass extinction itself was concentrated within a short time interval. In South China, there are three closely spaced levels of mass extinction. The strongest phase is at the level of the boundary clayrock (data reported in Yin, Xu & Ding, 1984). Several decimeters below this level, however, is another extinction level, which is marked by disappearance of major foraminifer and brachiopod taxa and, in many sections, corresponds to a change from crystalline limestone with chert nodules to thin-bedded micrites. This level may be observed in the Chongqing, Changxing, and other sections. A third extinction horizon lies at the top of the transitional bed (Yin, 1985). This level is marked by the extinction of residual Paleozoic ammonoids, foraminifers, and brachiopods, as well as by the radiation of ophiceratids and claraiids. In this way, the change from a Paleozoic biota to a Mesozoic biota was accomplished in three steps within a time interval recorded by deposits only one meter thick. Although a series of interrelated events must have transpired during this relatively short interval, we concentrate here on volcanic events and discuss the effect on algae and ammonoids.

Prasinophytes

Clusters of prasinophyte algae have been found in the volcanogenic Permo-Triassic boundary clayrock of the Ermen section, which has yielded no other microphytes. Abundance of such algae may be a symbol of disaster. That is, such abundance may imply an event that was destructive of microplankton. Prasinophyte algae have also been found in clayrocks at the Cretaceous–Tertiary boundary in Denmark (Stevns Klint) and Spain, which some authors (Hansen *et al.*, 1986) suggest are also volcanogenic. Coincidentally, other microphytes are lacking at these horizons. Disastrous circumstances such as the anoxic environment detected by Cai Zhifang (pers. comm.) and the sudden decrease in $\delta^{13}C$ values discussed below may have been introduced by volcanism.

Ammonoids

There is an interesting relationship between volcanism and the rise and fall of ammonoids, which flourished in the Changxingian of South China. Ammonoids are known from nearly 100 localities, and represent 152 species of some 32 genera (Yang & Zhang, 1986). Most ammonoids have been collected from siliceous rocks (Fig. 13.2) and their period of maximum abundance and diversity was coincident with a period of chert deposition. Cherts essentially ceased to accumulate in South China at the beginning of the Triassic, and all Changxingian ammonoids except for a few relict *Pseudogastrioceras* became extinct.

It has been demonstrated that Changxingian cherts are closely related to volcanism (Yang *et al.*, 1987), and there are similarities in the distribution patterns of Changxingian ammonoids and synchronous tuffogenic deposits (Figs 13.1, 13.3). In this connection, it is of interest to note that the abundance of modern cephalopods around the Philippines may also be related to active volcanism there. The postulation is that submarine volcanism may have provided a favorable hydrochemical environ-

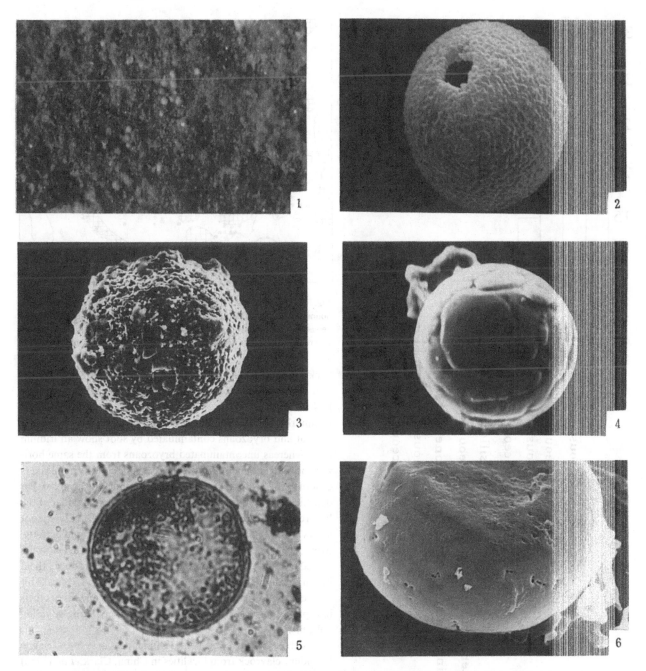

Plate 13.4. **1** Detail of black boundary clayrock (bed AG92) from Shangsi, Guangyuan, showing yellowish-red sphaerules *in situ*. Diameter of single sphaerule 40 microns.
2 Ferruginous sphaerule showing phylome, infilled prasinophyte alga. Brecciated black boundary clayrock, bed 34–2, Ermen, Huangshi, × 1000.
3, 4 Ferruginous sphaerules. In the fluorescence microscope with ultraviolet illumination these two specimens showed distinct fluorescence that indicates a surface cover of organic material. Brecciated black boundary

clayrock, bed 34–2, Ermen, Huangshi, × 1000.
5 Light micrograph of organic sphaerule (prasinophyte) left after dissolution in HC1 and HF of reddish, mineralized sphaerule. Note double and multiple pattern of pores in the wall. Boundary clayrock (bentonite), bed 42, kejiawan, Huangshi, × 1100.
6 Unmineralized prasinophyte alga. Note pore pattern (compare fig. **5**). Alga is slightly compressed. Boundary clayrock (bentonite), bed 42, Kejiawan, Huangshi, × 2000.

ment for Changxingian ammonoids, and that a change in that environment caused ammonoid extinction.

Anomalies in occurrence of iridium and other trace elements and their possible relationship to volcanism

Iridium anomalies reported from the boundary clayrock of Changxing (Sun *et al.*, 1984; Chai *et al.*, 1986) and Guangyuan (Xu *et al.*, 1985) have been taken as strong evidence of impact by an extraterrestrial body. However, this interpretation must be reconsidered.

Thus far, 11 analyses have been made of rocks at the Permo-Triassic boundary in South China (Table 13.1). Five of these have been made at Changxing and two at Guangyuan. However, only three of these analyses (two at Changxing, one at Guangyuan) have detected anomalously high values of iridium. This indicates that even in these two sections iridium anomalies exist in only selected materials or at certain levels. Analyses of the

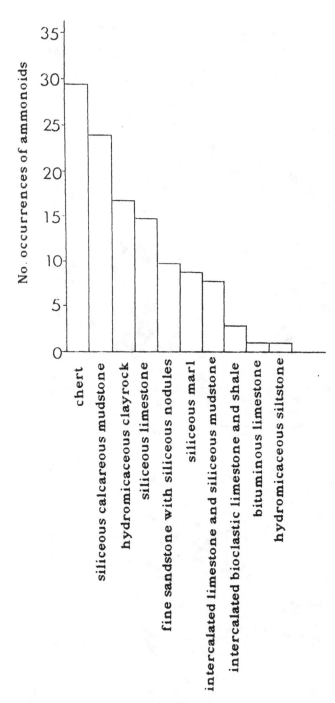

Fig. 13.2. **Relationship between rock type and Changxingian ammonoids in South China.**

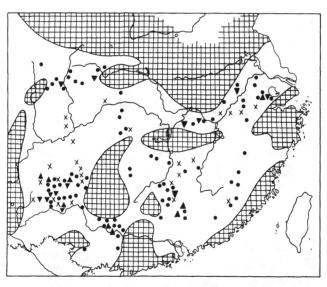

Fig. 13.3. **Geographic distribution of Changxingian ammonoids and boundary clayrocks. Dots = Changxingian ammonoids; triangles = tuffaceous boundary clayrocks; inverted triangles = montmorillonite boundary clayrocks; Xs = other boundary clayrocks; trellis pattern = oldlands.**

boundary clayrock in Transcaucasia have failed to establish an iridium anomaly. Brandner *et al.* (1986) reported an iridium anomaly (3 + 1 ppb) at the Permo-Triassic boundary in the Southern Alps, Italy, but Oddone & Vannucci (1988) reported no iridium anomaly there. Thus, the selective existence of iridium anomalies in time and space may indicate causes other than bolide impact.

It is notable that iridium highs at the Chinese boundary are mostly in a black layer, 1–7 cm thick, that immediately overlies the boundary clayrock, not in the white clayrock itself. One of the anomalies (0.6 ppb) is only one degree higher than the

background value, and could readily have resulted from adsorption of iridium by the organic carbon in the black layer. In the Fish Clay at the Cretaceous–Tertiary boundary of Stevns Klint, Denmark, iridium is highly absorbed by free-state carbon black or soot and bryozoans contaminated by soot show an iridium high, whereas uncontaminated bryozoans from the same horizon show no anomalous iridium values. The concentrated layers of soot in the Fish Clay, namely the black and rusty layers of that deposit, display higher iridium values than the remainder of the bed, reaching 3 ppm in one test. According to $\delta^{13}C$ analysis, soot layers in the Fish Clay are the products of volcanic activity. Moreover, such layers are known from as much as 1 m below the boundary, and this suggests that the events that produced them began at least 60,000 years before the end of the Cretaceous. This excludes their origination from a single impact, and favors their interpretation as the results of relatively long-standing volcanism.

Based on a survey of elemental abundances in Permo-Triassic boundary clayrock from localities in China, Clark *et al.* (1986) noted a strong enhancement of K, Cs, Hf, Ta, and Th, and depletion of Na, V, Cr, Mn, Co, and Ir, relative to average crustal abundances. From this, they concluded that the most likely source of the boundary clayrock material was from a massive volcanic eruption. The extremely low ratios in the boundary clay of TiO to Al O (0.014 to 0.017) are compatible with highly silicic ash. After study of the geochemistry and mineralogy of some Permo-Triassic boundary sections in South China, Zhou *et al.* (1987) concluded that the boundary clayrock might have formed from acidic to intermediate volcanic ash. Asaro (pers. comm.) has expressed a similar opinion.

Stable carbon anomaly and its possible relationship to volcanism

Thus far, some 26 analyses of the stable isotope of carbon have been made across the Permo-Triassic boundary along the

Table 13.1. *Iridium abundance in Permo-Triassic boundary clayrock of South China*

Bed no.	Meishan, Changxing							Shangsi, Guangyuan, Sichuan		Ermen, Huangshi, Hubei			Yanshi, Fujian	Lichuan, Hubei	Guiyang, Guizhou	earth back-ground
	*1	Cy-1	AG-91	AG-92	B3-AG91-92	AG-91	AG-92	AG-252	AG-253	HE-34-1	HE-34-2	HE-34-1, 2	FY-42-A-8			
Ir (ppb)	≤ 0.40	---- *2	8.0	5.0	0.002	0.12	0.60	----	2.48	< 0.82	< 4.58	< 0.05	< 4.72	0.0012	≤ 0.50	0.005-0.080
Ref.	Asaro et al., 1982	Zhang et al., 1983	Sun et al., 1984			Clark et al., 1986	Chai et al., 1986		Xu et al., 1985		*3	*4	*3		Clark et al., 1986	Asaro et al., 1982	Clark et al., 1986

*1, AG-91 and AG-252 are white clayrocks; AG-92 and AG-253 are black shales; blank if bed nos. are unknown.
*2, ---- = undetected.
*3, Data kindly provided by Chai Zhifang (Inst. High Energy Physics, Acad. Sinica).
*4, Hansen and co-workers, this chapter.

Tethys and in South China (Yan et al., 1987; Baud, Holser & Magaritz, 1986; Chen et al., 1984). All results show a sudden drop of 2 to 5% in the $\delta^{13}C$ value and, in a majority of the sections, the fall-off interval is no more than several decimeters from the boundary. Thus, just as at the Cretaceous–Tertiary boundary, a drop in values of $\delta^{13}C$ at the Permo-Triassic boundary has now been firmly established.

One possible cause of the rapid decrease in $\delta^{13}C$ values at the Cretaceous–Tertiary boundary is the mass extinction of phytoplankton recorded in the fossil record. The diminished marine biomass would absorb much less light carbon (^{12}C) from sea water, and a large percentage of the light carbon isotope remaining in the water body would, in turn, dramatically reduce the $\delta^{13}C$ value. Another possibility is tied to the decline in terrestrial plants also documented by the fossil record. The massive burn-out of forests suggested by the soot layers would provide a large quantity of ^{12}C to sea water through surface run-off, thus initiating a $\delta^{13}C$ fall-off. Both possibilities may be connected with volcanism.

One of us (Hansen et al., 1987) has made a series of measurements of the stable isotopes of carbon at the Cretaceous–Tertiary boundary Fish Clay in Denmark. Elemental carbon values are surprisingly light while charcoal values are heavier. A parallel study was made of the section at Shangsi, since this outcrop represents a tectonically undisturbed Permo-Triassic transition. The preparation of acids follows the method described in Hansen et al. (1987).

The black boundary clay (AG253) was processed and the acid residue was filtered into two fractions, of which the one larger than 10 microns in aliquots was found to be composed mainly of marine kerogen and some pieces of charcoal, whereas the fractions smaller than 10 microns were composed mainly of very small carbon particles in the size range of 1 micron or smaller. Carbon isotope values obtained are in good agreement with those reported from the Cretaceous–Tertiary boundary of Denmark.

The smaller fraction had a $\delta^{13}C$ value of $-27.88 + 0.02$(PDB) while the kerogen-dominated fraction larger than 10 microns had a value of $-26.25 + 0.02$(PDB). The value of the Upper Permian coal was $-23.18 + 0.02$(PDB).

Hansen et al. (1987) reported $\delta^{13}C$ of elemental carbon from the uppermost Maastrichtian in Denmark of $-27.81 + 0.05$, charcoal values in the range of -23 to -25, and a marine background value of $-26.62 + 0.05$.

The interpretation of these values implies that the elemental carbon value corresponds to that of igneous carbon and not to meteoritic carbon or carbon from the burning of wood. Thus we suggest that the carbon at the Permo-Triassic boundary is of igneous (i.e., volcanic) origin.

On the possible relations between volcanism and mass extinction

Thus far, two possible relationships have been suggested between volcanism and mass extinction. McLean (1985) suggested mantle degassing of CO_2 through basic volcanic activities. According to his calculations, eruption of the Deccan Trap would have been sufficient to induce such effects at the Cretaceous–Tertiary boundary. Tremendous basic eruptions also occurred during the transition between Permian and Triassic. The Permo-Triassic Tungus Trap in Siberia alone has an areal extent of 1.5 million km^2, three times that of the Deccan Trap (ca. 0.5 million km^2). There were also large-scale basaltic eruptions in west China, mainly in the Late Permian, but ranging in age from Early Permian to Early Triassic. The Emeishan Basalt by itself is estimated to cover an area of about 0.3 million km^2 and is 2000 m thick in the Nanpanjiang drainage area.

Another possible relationship between volcanism and mass extinction involves massive volcanic explosion, which may produce effects such as dusting, overshadowing, and toxification, similar to those of asteroid impact. Budyko & Pivivariva (1967) estimated that 10 episodes of Krakotoa-scale volcanism within a period of 10 generations would decrease solar radiation by 80 to 90% and thus seriously affect temperature and photosynthesis. Volcanic activities would differ from impact in being multiphase and operating over a longer time span. Such a scenario seems compatible with the existence in China of several clay beds, a three-phase extinction, and the relatively long duration of $\delta^{13}C$ fall-off at the Permo-Triassic boundary. Also, as mentioned previously, intermediate to acidic volcanics were widespread in South China during that interval of time.

Other regions

Except in the Tethyan realm of Eurasia, there was erosion and thus a major gap between the Permian and Triassic elsewhere in the world. Therefore, it can not be determined if ash layers accumulated during this interval in the rest of the world. The same situation applies in most of the Gondwanan portion of the Tethyan realm. Even the well-known Guryul Ravine section of Kashmir may have undergone end-Permian erosion (Gupta & Brookfield, 1987). We have been informed by Li Zishun, however, that a boundary clay exists in the Nammal section of the Salt Range, in Pakistan.

The Eurasian margin of the Tethyan realm, on the other hand, records continuous or nearly continuous deposition through the Permo-Triassic interval and clay beds are widespread at the Permo-Triassic boundary. In Iran, a clay bed has been reported at the supposed Permo-Triassic boundary in the Abadeh section (Iranian–Japanese Research Group, 1981), and similar beds occur in the Gheshlagh section of the eastern Elburz (Altiner *et al.*, 1979) and in the Kuh-e-Ali Bashi section of northwest Iran (Teichert, Kummel & Sweet, 1973). Such clay beds also exist in the Dorasham section in Soviet Transcaucasia (Ruzhentsev & Sarycheva, 1965). In the Southern Alps, a clay bed has been discovered between the Bellerophon and Werfen Formations in the Casera Federata section and also in the lower part of the Tesero Member. However, further study will be required to determine if any of these clay beds is related to volcanism.

Summary

Clayrocks at the Permo-Triassic boundary in South China are mainly tuffaceous, blastotuffaceous, or volcanogenic rocks of other types. Mass extinctions in the boundary interval of phytoplankton, for example, may have been related to the volcanism, which may also have produced anomalies in the distribution of $\delta^{13}C$, Ir, and other elements. Frequent and widespread volcanic activity took place in South China in the Late Permian and contributed to formation of large-scale siliceous deposits. Environmental changes induced by Late Permian volcanism may have influenced the distribution and evolutionary development of fossil groups such as the ammonoids.

Permo-Triassic basic volcanism was widespread, and intermediate to acidic volcanism was intensive in South China. Although the fact that extensive Permo-Triassic volcanism has not been reported from other parts of the world might be a major argument against extending the South China model worldwide, we suggest that because the primary effects on the biota would have been produced by dusting, overshadowing, and similar effects (rather than by the volcanism itself) intensity of volcanism would have been much more important than its ubiquity. In addition, clay beds are widespread all along the Eurasian margin of the Tethyan realm. Some might be volcanogenic, although their origin and nature need further investigation.

References

Altiner, D., Baud, A., Guex, J. & Stampfli, G. (1979). La limite Permien-Trias dans quelques localités du Moyen-Orient: Recherches stratigraphiques et micropaléontologiques. *Riv. Ital. Paleont.*, **85**(3–4): 683–714.

Asaro, F., Alvarez, L. W., Alvarez, W. & Michel, H. V. (1982). Geochemical anomalies near the Eocene/Oligocene and Permian/Triassic boundaries. *Geol. Soc. Amer. Spec. Paper*, **190**: 517–28.

Baud, A., Holser, W. T. & Magaritz, M. (1986). Carbon-isotope profiles in the Permian-Triassic of the Tethys from the Alps to the Himalayas [Abstract]. In *Field Conference on Permian and Permian-Triassic Boundary in the South-Alpine Segment of the Western Tethys, and Additional Regional Reports*, Abstracts vol., p. 14. Pavia: Soc. Geol. Ital. and IGCP Project 203.

Brandner, R., Donofrio, D. A., Krainer, K., Mostler, H., Nazarow, M. A., Resch, W., Stingl, V. & Weissert H. (1986). Events at the Permian–Triassic boundary in the southern and northern Alps [Abstract]. In *Field Conference on Permian and Permian–Triassic Boundary in the South-Alpine Segment of the Western Tethys, and Additional Regional Reports*, Abstracts vol., p. 15. Pavia: Soc. Geol. Ital. and IGCP Project 203.

Budyko, M. I. & Pivivariva, Z. I. (1967). The influence of volcanic eruptions on solar radiation incoming to the Earth's surface. *Meteorologiya i Gidrologiya*, **10**: 3–7.

Chai Chifang, Ma Shulan, Mao Xueying, Sun Yiying, Xu Daoyi, Zhang Qinwen & Yang Zhengzong (1986). Elemental geochemical characters at the Permian–Triassic boundary section in Changxing, Zhejiang, China. *Acta Geologica Sinica*, **60**(2): 140–50. (Chinese, with English abstract.)

Chen Jinshi, Shao Maorong, Huo Weiguo & Yao Yuyuan (1984). Carbon isotopes of carbonate strata at Permian–Triassic boundary in Changxing, Zhejiang. *Scientia Geologica Sinica*, **1984**(1): 88–93. (Chinese, with English abstract.)

Clark, D. L., Wang Chengyuan, Orth, C. J. & Gilmore, J. S. (1986). Conodont survival and low Iridium abundances across the Permian–Triassic boundary in South China. *Science*, **233**: 984–6.

Gao Zhengang, Xu Daoyi, Zhang Qinwen & Sun Yiyin (1987). Discovery and study of microspherules at the Permian–Triassic boundary of the Shangsi section, Guangyuan, Sichuan. *Geol. Review*, **33**(3): 203–11. (Chinese, with English abstract.)

Gupta, V. J. & Brookfield, M. E. (1987). Preliminary observations on a possibly complete Permian–Triassic boundary section at Pahlgam, Kashmir, India. *Newsl. Stratig.*, **17**(1): 29–35.

Hansen, H. J., Gwozdz, R., Bromley, R. G., Rasmussen, K. L., Vogensen, E. & Rasmussen, K. R. (1986). Cretaceous–Tertiary boundary spherules from Denmark, New Zealand and Spain. *Bull. Geol. Soc. Denmark*, **35**: 75–82.

Hansen, H. J., Rasmussen, K. L., Gwozdz, R. & Kunzendorf, H. (1987). Iridium-bearing carbon black at the Cretaceous–Tertiary boundary. *Bull. Geol. Soc. Denmark*, **36**: 305–14.

He Jinwen (1985). Discovery of microsphaerules from the Permian–Triassic mixed fauna bed No. 1 of Meishan in Changxing, Zhejiang and its significance. *J. Stratigr.*, **9**(4): 293–7. (Chinese, with English abstract.)

He Jinwen, Rui Lin, Chai Chifang & Ma Shulan (1987). The latest Permian and earliest Triassic volcanic activities in the Meishan area of Changxing, Zhejiang. *J. Stratigr.*, **11**(3): 194–9. (Chinese, with English abstract.)

Iranian–Japanese Research Group (1981). The Permian and the Lower Triassic systems in Abadeh region, central Iran. *Mem. Fac. Sci., Kyoto Univ., Ser. Geol. Min.*, **47**(2): 61–133.

Lefèvre, R., Gaudichet, A., & Billon-Galland, M. A. (1986). Silicate microspherules intercepted in the plume of Etna volcano. *Nature*, **322**: 817–20.

Li Zishun, Zhan Lipei, Zhu Xinfang, Zhang Jinhua, Jin Ruogu, Liu Guifang, Sheng Huaibien, Shen Guimei, Dai Jinye, Huang Hengquan, Xie Longchun & Yan Zheng (1986). Mass extinction and geological events between Palaeozoic and Mesozoic era. *Acta Geologica Sinica*, **60**(1): 1–15. (Chinese, with English abstract.)

McLean, D. M. (1981). Terminal Cretaceous extinctions and volcanism [Abstract]. *Am. Assoc. Adv. Sci., 147th Ann. Meeting, Abstracts of Papers*, p. 128.

McLean, D. M. (1985). Deccan Trap's mantle degassing in the terminal Cretaceous marine extinctions. *Cretaceous Research*, **6**: 235–59.

Oddone, M. & Vannucci, R. (1988). PGE and REE geochemistry at the B–W boundary in the Carnic and Dolomite Alps (Italy). *Mem. Soc. Geol. Italiana*, **34**: 129–40.

Ruzhentsev, V. E. & Sarycheva, T. G. (Eds) (1965). Razvitie i smena morskikh organismov na rubezhe Paleozoya i Mezozoya. *Trudy, Akad. Nauk SSSR, Paleont. Inst.*, **108**: 431 pp. (Russian.)

Sun Yiyin, Xu Daoyi, Zhang Qinwen, Yang Zhengzhong, Sheng Jinzhang, Chen Chuzhen, Rui Lin, Liang Xiluo, Zhao Jiaming & He Jinwen (1984). The discovery of Iridium anomaly in the Permian–Triassic boundary clay in Changxing, Zhejiang, China and its significance. In *Developments in Geoscience: Contribution to 27th International Geological Congress*, ed. Tu Guangzhi, pp. 235–45. Beijing: Science Press. (Chinese with English abstract.)

Teichert, C., Kummel, B. & Sweet, W. C. (1973). Permian–Triassic strata, Kuh-e-Ali Bashi, northwestern Iran. *Bull. Mus. Comp. Zool.*, **145**(8): 359–472.

Xu Daoyi, Ma Shulan, Chai Zhifang, Mao Xuoying, Sun Yiying, Zhang Qinweng & Yang Zhengzhong (1985). Abundance variation of Iridium and trace elements at the Permian/Triassic boundary at Shangsi in China. *Nature*, **314**: 154–6.

Yan Zheng, Xu Daoyi, Zhang Qinwen, Sun Yiyin & Ye Lianfang (1987). Some stable isotope anomalous events across the P/T boundary. *Abstracts, IGCP Project 199, 'Rare Events in Geology', 2d session, 3–5 March, 1987, Beijing*.

Yang Fengqing & Zhang Yijie (1986). Characters of distribution and evolution of Changxingian ammonoid fauna of South China. *Acta Geologica Sinica*, **60**(4): 320–8. (Chinese, with English abstract.)

Yang Zunyi, Yin Hongfu, Wu Shunbao, Yang Fengqing, Ding Meihua & Xu Guirong (1987). Permian–Triassic boundary stratigraphy and fauna of South China. *Geol. Memoirs, ser. 2*, **6**: 379 pp. Beijing: Geol. Publ. House. (Chinese, with English abstract.)

Yin Hongfu (1985). On the Transitional Bed and the Permian–Triassic boundary in South China. *Newsl. Stratig.*, **15**(1): 13–27.

Yin Hongfu, Xu Guirong, & Ding Meihua (1984). Palaeozoic–Mesozoic alternation of marine biota in South China. *Scientific Papers on Geology for International Exchange, Prepared for 27th International Geological Congress*, **1**: 195–204. (Chinese, with English abstract.)

Zhang Jinghua, Zhang Yuanji, Wang Yuqi, Chen Bingru & Sun Hingxin (1983). Features of rare earth elements of Permian-Triassic boundary clay rocks in South China, with its stratigraphic significance. *Acta Petro. Min. Analytica*, **2**(2): 81–6. (Chinese, with English abstract.)

Zhou Yaoqi, Chai Zhifeng, Mao Xueying & Ma Shulan (1987). On the catastrophic environments of the P/Tr boundary in South China [Abstract]. *Abstracts of the Final Meeting on Permo-Triassic Events of East Tethys Region and Their Intercontinental Correlations*, p. 38. Beijing: 5–20 Sept., 1987.

14 Geochemical constraints on the Permo-Triassic boundary event in South China

CHAI CHIFANG, ZHOU YAOQI, MAO XUEYING, MA SHULAN, MA JIANGUO, KONG

PING, AND HE JINGWEN

Introduction

In the 570 Ma since the beginning of the Phanerozoic, five great extinctions of organisms have been recorded, of which the one at the end of the Permian was the most serious. Thus, during the last decade, there has been a growing interest in study of the Permo-Triassic event (Asaro *et al.*, 1982; Alekseev *et al.*, 1983; Sun *et al.*, 1984; Xu *et al.*, 1985; Brandner *et al.*, 1986; Oddone & Vannucii, 1986, 1988; Chai *et al* 1987; Clark *et al.*, 1986; He *et al.*, 1987, 1988; Zhou *et al.*, 1987a,b, 1988). Much work has been conducted in the fields of mineralogy, petrology, and isotopic and elemental geochemistry, and on microspherules and shock-metamorphosed quartz. Even so, the nature of the mass extinction at the end of the Permian is still quite a controversial issue.

Authors have put forward various models to interpret the Permo-Triassic boundary (P/T) event, among which those favoring volcanic eruption and ones advocating extraterrestrial impact are dominant. Each model, of course, is supported by its own evidence.

Observed and experimentally derived facts that favor the volcanic-eruption model are:

1 Clays at the P/T boundary exhibit a residual rhyotaxitic structure and contain high-temperature beta-quartz, zircon, and apatite, which are common minerals in acidic-intermediate volcanic rocks.

2 Multilayered clays present above and below the P/T boundary are similar to the boundary clay. Their elemental compositions resemble those of the volcanic ashes in the Triassic of South China.

Lines of evidence that favor an extraterrestrial event are:

1 An anomalous abundance of siderophile elements, including a moderate enrichment of iridium at, or right above, the P/T boundary.

2 Occurrence of a small amount of metamorphosed upper-crust debris and quartz; and microspherules consisting of diopside and chrome-bearing spinel, all likely related to impact.

3 Good stability of the P/T boundary clay layers, which are comparable with each other but considerably different from the non-boundary clays.

In order to decipher the mechanism of the P/T boundary event, we analyzed nine P/T sections in South China. These are:

Changxing, Zhejiang; Guangyuan, Sichuan; Liangfengya, Sichuan; Heshan, Guangxi; Fusui, Guangxi; Zhuzang, Guizhou; Duansan, Guizhou; Longyan, Fujian; and Huangshi, Hubei (Fig. 14.1). In addition, some continental P/T sections were also studied. These are: Yujiazhuang, Sandong; Xingxing, Henan; and a section at Nammal, Pakistan. Results of studies reported in the literature are also cited, e.g., Lichuan, Hubei (Clark *et al.*, 1986), Transcaucasia (Alekseev *et al.*, 1983), and some sections in the Southern Alps (Brandner *et al.*, 1986; Oddone & Vannucci, 1986).

Stratigraphy, petrology, and mineralogy

The biostratigraphy of P/T sections in South China has been thoroughly studied (Sheng *et al.*, 1984; Yang *et al.*, 1987). The Meishan (Changxing, Zhejiang), and Shangshi (Guangyuan, Sichuan) sections are the best of the sections studied and are candidates for boundary stratotype. All other P/T sections in South China are comparable with them (Fig. 14.2).

Generally, upper Permian rocks in South China are mainly limestone and siliceous limestones in which there are many fossils. The lower Triassic consists mainly of shale, sandy shale, clay, and dolomite.

There is a clay layer 2–25 cm thick at the P/T boundary at a

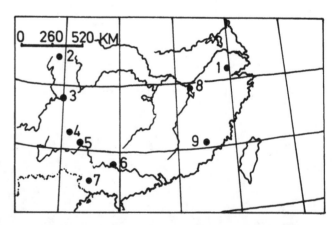

Fig. 14.1. Map of South China showing location of Permo-Triassic boundary sections. 1, Changxing, Zhejiang; 2, Guangyuan, Sichuan; 3, Liangfengya, Sichuan; 4, Zhuzang, Guizhou; 5, Duansan, Guizhou; 6, Heshan, Guangxi; 7, Fusui, Guangxi; 8, Huangshi, Hubei; 9, Longyan, Fujian.

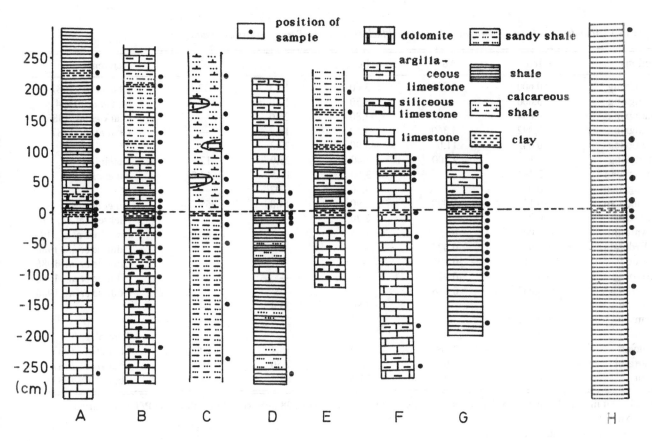

Fig. 14.2. Comparison of Permo-Triassic boundary sections in South China. A, Changxing, Zhejiang; B, Guangyuan, Sichuan; C, Longyan, Fujian; D, Huangshi, Hubei; E, Heshan, Guangxi; F, Liangfengya, Sichuan; G, Fusui, Guangxi; H, Duansan, Guizhou.

majority of localities in South China. At a few localities this clay layer is overlain by a black shale layer, 3–20 cm thick. In addition to the boundary clay, there are also two to six clay layers near the P/T boundary in South Chinese sections.

Clays near the P/T boundary have a volcanic rhyotaxitic structure and include greater or lesser amounts of beta quartz. Hence the formation of those clays may have been related to volcanic activity. However, the boundary clays seem to be more complex than the non-boundary clays. That is, in addition to the high-temperature beta-quartz and rhyotaxitic structure, the boundary clays also include a large number of microspherules and sedimentary or metamorphic clastic debris (Zhou *et al.*, 1987). The microspherules contain ferric chromeceylonite, diopside and apatite. The chromeceylonite and diopside were thought to be related to an impact event (Xu *et al.*, 1988), and apatitic microspherules were considered as biogenic products (Yin *et al.*, 1988). Thus, the ferric microspherules may be related not only to impact, but also to volcanogenic events (Zhou *et al.*, 1988).

Elemental geochemistry

Ir, Au, Co, Ni

Our results (Table 14.1) indicate that, in general, elemental geochemical anomalies occur in the black shale layer just

above the boundary clay. The average values of those elements, calculated from the data in Table 14.1, are Ir (123 ppt) (all Ir data above 1 ppb were discarded on the assumption they may have been caused by accidental factors); Au (1.8 ppb); Co (22.6 ppm); and Ni (126 ppm). These values may serve roughly as an anomaly index for the end of the Permian. Although this anomaly is much less than the one at the end of the Cretaceous, it is still much higher than local background values in upper Permian and lower Triassic rocks.

It is worth mentioning that, unlike the situation at the Cretaceous–Tertiary boundary, the siderophile-element anomalies at the P/T boundary do not occur in the boundary clays but in the black shale layers just above them (Fig. 14.3). This implies a different origin of the siderophile-element anomaly. The view of Clark *et al.* (1986) that the siderophile elements were enriched by black shale seems valid. However, the mechanism and scale of the enrichment are now unclear. Reasonably, the enrichment should be related to the amount of organic matter in the black shale and to background values of trace elements in the marine system. But, after careful analysis of the black shales just above the P/T boundary, we found that enrichment of siderophile elements by organic matter does not appear to be synchronous (Fig. 14.4).

From Fig. 14.4 and Table 14.1 it is evident that Ir and Ni are much more enriched than other siderophile elements. There may be two reasons for this: (1) Ir and Ni are easily absorbed by

Table 14.1. *Abundances of some siderophile elements in black shale at the Permo-Triassic boundary*

Sections	Ir	Au	Fe	Co	Ni	References
Changxing	0.600	2.700	3.60	13.3	57.1	Chai *et al.*, 1987
Changxing	5.000	—	3.30	12.9	70.2	Sun *et al.*, 1984
Changxing	0.034	6.00	3.80	26.0	—	Clark *et al.*, 1986
Changxing	0.092	4.360	3.50	14.4	—	Chai *et al.*, 1987
Changxing	0.042	11.0	3.90	22.3	—	**
Changxing	0.150	9.100	3.40	25.0	124	'herein'
Guangyuan	2.480	—	3.60	17.5	63.2	Xu *et al.*, 1985
Guangyuan	0.027	—	4.50	9.92	—	**
Guangyuan	0.125	1.090	2.90	14.8	59.2	Xu *et al.*, 1985
Transcauc.	0.039	—	2.70	—	—	Alekseev *et al.*, 1983
S. Antonio	3.000	—	—	—	60.0	Brandner *et al.*, 1986
C. Federata	0.077	0.022	—	—	—	Oddone *et al.*, 1986
Sass de P.	0.019	0.005	—	—	—	Oddone *et al.*, 1986
Tesero	0.037	0.010	—	—	—	Oddone *et al.*, 1986
Butterloch	0.057	0.016	—	—	—	Oddone *et al.*, 1986
Zhuzang	—	—	7.50	33.4	—	**
Duansan	0.052	—	3.30	21.0	—	**
Nammal	0.366	—	—	—	—	'herein'
Longyan	—	2.460	4.90	16.4	298	Chai *et al.*, 1987
Huangshi	—	0.690	3.00	7.50	56.5	Chai *et al.*, 1987
Fusui	—	5.90	13.7	82.5	347	Chai *et al.*, 1987

Unit: Ir, Au in ppb; Fe in %; Co, Ni in ppm. We also detected 5.2 ppb Os, 38 ppb Re, 48 ppb Pt at the P/T boundary at Changxing (Chai *et al.*, 1987).
** Data from Los Alamos laboratory, USA.

organic matter; and (2) there may have been another Ir reservoir, in addition to the one supplied by seawater.

The ratio of Ir to Au is often used to judge the origin of an Ir anomaly. Data in Table 14.1 show that, at Changxing, South China, the ratio is 0.22, which is lower than the cosmic one (3.3) by a factor of 16, but higher than the crustal one (0.02) by almost the same factor. In the Casera Federata, Sass de Putia, Tesero, and Butterloch sections of the Southern Alps (Oddone & Vannucci, 1986, 1988), average Ir/Au ratios for the samples studied are 3.5, 3.8, 3.7, and 3.6, respectively, and are thus close to the cosmic ratio. Although this might suggest that the iridium anomaly at the P/T boundary in the Southern Alps had an extraterrestrial cause, Oddone & Vannucci (1988) are of the opinion that there is at present no evidence to support an impact event there. Ir to Au ratios in South China, on the other hand, indicate a certain mixing effect.

In addition, we have determined the content of Os, Pt, and Re in a few P/T sections in South China (Chai *et al.*, 1987).

As, Se, Sb, and Mo

At the P/T boundary localities we have studied for this report, almost all chalcophile elements are enriched. Generally speaking, the enrichment is caused by an anoxic-reducing environment. As with the siderophile elements, chalcophile elements are also enriched in the black-shale layers, a result undoubtedly of the absorption of trace elements in seawater by the organic matter contained in the black shale. However, it should be noted that layers with the highest abundances of chalcophile elements are at the base of the boundary clays, rather than in the black-shale layers (Fig. 14.5). This phenomenon is related to the presence of a large amount of pyrite at the base of the boundary clay. Mössbauer spectrometry of samples from the base of the boundary clay indicates that more than 99% of the iron is present in the Fe^{++} form, which implies a strongly reducing environment during formation of the boundary clays. Enrichment of As, Se, and Sb is likely to have resulted from their substitution for an S atom in FeS, whereas Mo may replace Fe^{++} in FeS. The reducing environment may have been a result of the mass extinction of organisms in the marine system.

Although chalcophile elements are enriched with siderophile elements in black shale, they do not show a significant correlation (see Fig. 14.6).

Na, K, Rb, Cs, Mg, Ca, Sr, Ba, Sc, Al, Ti, Th, U, Hf, and Ta

Abundances of almost all lithophile elements rise considerably at the P/T boundary, as a result of an increase in the supply of continental debris following the P/T event(s). This rise indicates that a severe regression event took place at the end of the Permian, in contrast to the marine transgression deduced by Alekseev *et al.* (1983) in their study of the P/T section in Sovetashan, Transcaucasia.

In Fig. 14.7 it can be seen that abundance spikes of some lithophile elements appear at the boundary clay layers, whereas

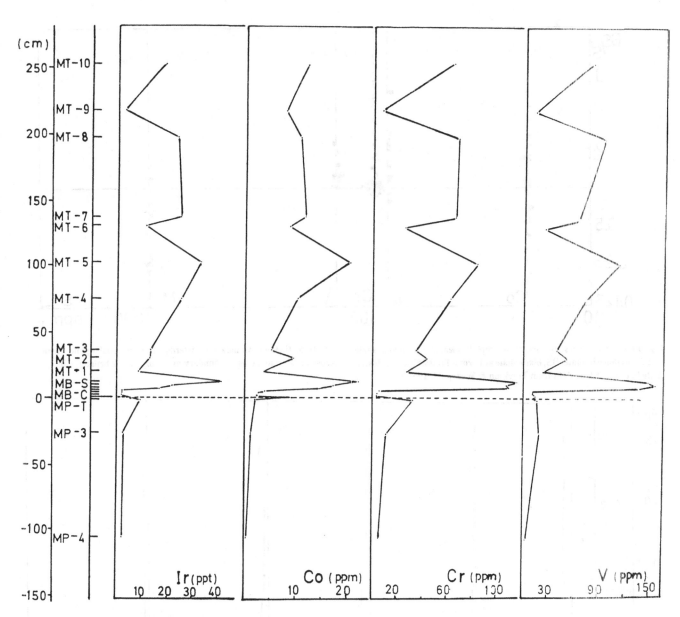

Fig. 14.3. Abundance patterns for some siderophile elements across the Permo-Triassic boundary at Changxing, Zhejiang, China. Ir data are from one of the authors, Mao Xueying, with C. J. Orth, Los Alamos, NM, USA. Other Ir data sets show similar variation, but with different absolute magnitudes.

the content of these elements in overlying strata other than some non-boundary clay layers, ranges between that of the boundary-clay layer and marine-facies limestones of the Permian. Thus we conclude that the boundary-clay layers really originated from continental debris, whereas Triassic rocks are products of marine sedimentation involving a large amount of continental debris.

Formation of the boundary-clay layers was clearly a rapid process because the sedimentation rate of normal clay is so low that contents of trace elements from the marine system become higher. In fact, variations in the Mn/Ti ratios along the clay layers shown in Fig. 14.8 are the same as expected. It is known that Mn represents marine deposition, whereas Ti is a typical indicator of continental rocks. Thus, the higher the Mn/Ti ratio,

the slower the sedimentation rate. At the boundary, or in non-boundary clays, the Mn/Ti ratio is generally below 0.1. Such a low ratio indicates that a clay layer 5 cm thick might have been deposited in just a few years. This rapid sedimentation could have been completed only by turbidity flow, volcanic ashfall, or by impact-related fallout. Geological records of the sections studied do not support deposition from turbidity currents. However, the occurrence in the clays of rhyotaxitic structure and a large amount of minerals characteristic of acidic-intermediate volcanic rocks favors a volcanogenic origin for the clay layers.

Aluminum is also a typical continental element, and the Mn/Al ratio reflects variations in sedimentation rate. If Ir abundances are normalized to the sedimentation rate, there appears to be a maximum of Ir at the top of the black shale (Fig. 14.9).

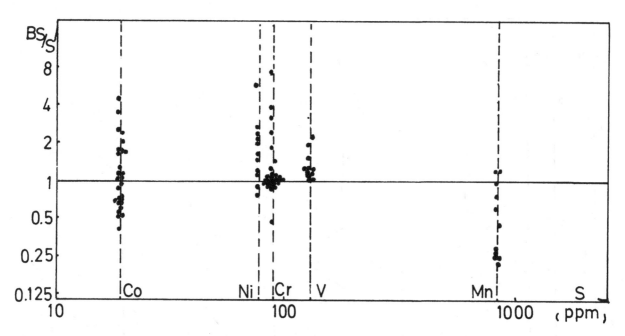

Fig. 14.4. Abundances of various siderophile elements in black shale relative to those in shale. BS, black shale at Permo-Triassic boundaries; S, shale in crust. (Shale data from Turekian & Wedepohl, 1961.)

Fig. 14.5. (below) Abundance patterns for some chalcophile elements across the Permo-Triassic boundary at Changxing, Zhejiang, China.

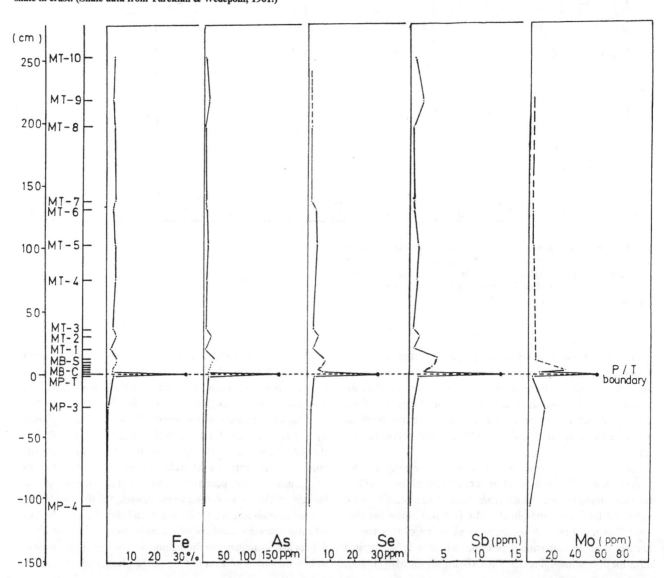

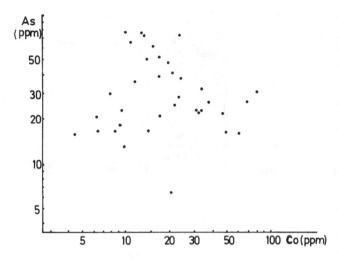

Fig. 14.6. Correlation of As and Co in the boundary black shale.

Clearly, the black shale is able to absorb Ir above the normal background.

The Rb/K ratio is an indicator of seawater salinity. Fig. 14.9 shows that, from the end of the Permian, salinity of seawater suddenly became high. This may have been caused by a high-temperature, evaporating environment. Indeed, at the base of the boundary-clay layers we found a large amount of gypsum debris, which indicates strong evaporation.

Th/U, K/Na, and Hf/Ta ratios may also be used to determine the origin of sediments. In general, the larger Th/U and K/Na ratios are, the more continental matter there is in sedimentary rocks. Hf/Ta ratios, however, are relatively constant and different rocks generally have the corresponding range in Hf/Ta ratios. Figs 14.10 and 14.11 show the relationship between Hf/Ta and Th/U in the boundary clays and the black shales. In the boundary-clay layers, it appears that the Hf/Ta ratio has a weak negative correlation with the Th/U ratio. However, in the black shales there is no correlation between these ratios.

In contrast with the boundary clays, Hf/Ta is positively correlated with Th/U in non-boundary clays near the P/T boundary (Fig. 14.12). Thus the boundary clays differ from the non-boundary clays in mechanism of formation.

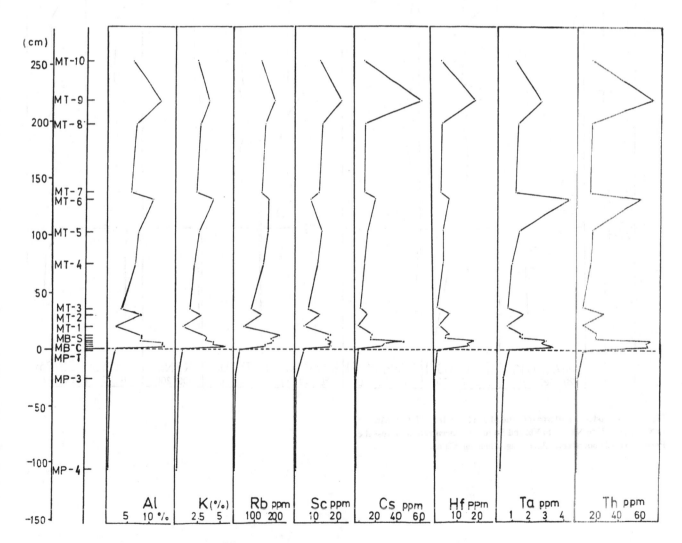

Fig. 14.7. Abundance patterns for some lithophile elements across the Permo-Triassic boundary at Changxing, Zhejiang, China.

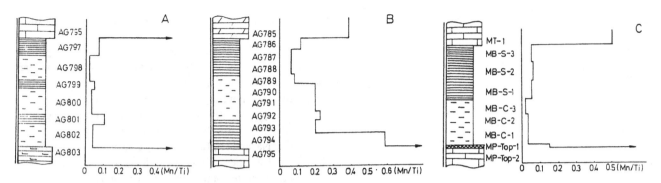

Fig. 14.8. Abundance patterns for ratio of Mn to Ti concentrations through the clays at and near the Permo-Triassic boundary of South China. A, GS 25 clay of Guanyuan section, Sichuan; B, boundary clay in Guangyuan section; C, boundary clay in Changxing section.

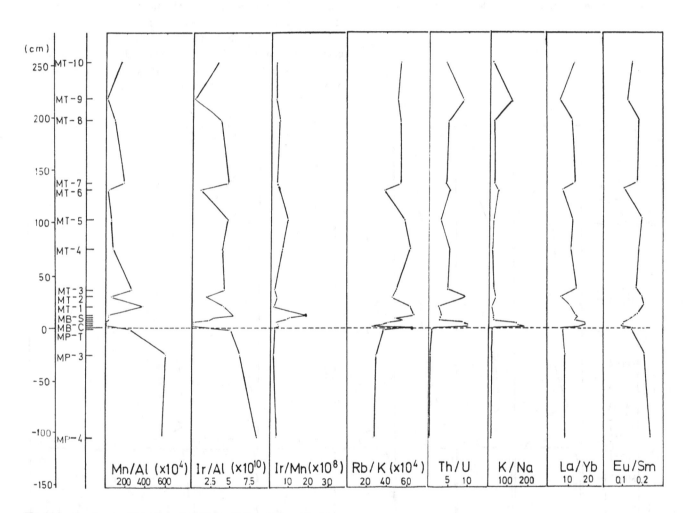

Fig. 14.9. Abundance patterns for ratio of Mn to Al, Ir to Al, Ir to Mn, Rb to K, Th to U, K to Na, La to Yb, and Eu to Sm concentrations across the Permo-Triassic boundary, Changxing, Zhejiang, China.

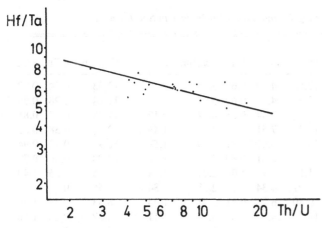

Fig. 14.10. Correlation of Th/U and Hf/Ta ratios in Permo-Triassic boundary clays of South China.

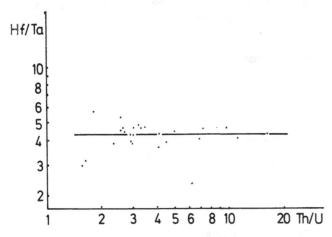

Fig. 14.11. Correlation of Th/U and Hf/Ta ratios in black shales above the Permo-Triassic boundary in South China.

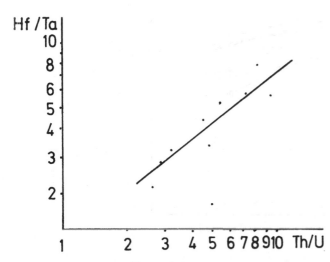

Fig. 14.12. Correlation of Th/U and Hf/Ta ratios in non-boundary clays near the Permo-Triassic boundary in South China.

Rare-earth elements

Abundance patterns of rare-earth elements (REE) can play important roles in determining the origins of sediments. Among the various rocks in the P/T sections we have studied, only the origin of the clay layers at and near the boundary is a controversial issue. REE abundances at the P/T boundary in sections in South China are listed in Table 14.2 and REE abundances in the Cretaceous–Tertiary (K/T) boundary clay and the Triassic volcanic ash of South China are included for comparison.

Figs 14.13 and 14.14 show the REE patterns of clays at and near the P/T boundary in South China. In these figures we note the following differences in REE patterns between the boundary and the non-boundary clays:

1 compared with the non-boundary clays, the boundary clays are more enriched in the light REEs;
2 slope of the REE patterns in the boundary clays is steeper than that of the non-boundary clays, e.g., the La/Yb ratio of 13 and 5.7, respectively;
3 the Eu negative anomaly index of the boundary clay (δ Eu 0.39) is less than that of the non-boundary clay (δ Eu 0.28);
4 boundary clays show a negative Ce anomaly (δ Ce = 0.88), whereas non-boundary clays show no such anomaly or may even have a small positive Ce anomaly (δ Ce = 1.15).

In Table 14.2 and Fig. 14.15 it can be seen that, compared with the boundary clays, the REE patterns of the non-boundary clays are more similar to those of the Triassic ash. Boundary clays have smaller negative Eu and Ce anomalies and larger La/Yb, La/Sm, and Tb/Yb ratios. Thus, although formation of the boundary clay was undoubtedly related to volcanic activity, that was not the only factor affecting formation of the clays. Non-boundary clays, on the other hand, are made mainly of volcanic alteration products because they include volcanic minerals, have a rhyotaxitic structure, and have REE patterns that resemble that of the volcanic ash.

In general, clay layers at the K/T boundary originated from alteration of matter spattered by impact. Fig. 14.15 shows that the REE pattern at the K/T boundary is more or less similar to that of the P/T boundary clays; that is, there is a negative Ce and Eu anomaly and a large La/Yb ratio. However, there is also a significant difference in the total REE amount and in the magnitude of the Ce and Eu anomalies. The low total REE amounts can be interpreted to mean a different level of enrichment during the alteration process, which generally would not lead to REE fractionation. Thus the negative Ce and Eu anomalies would be characteristic of their precursors or the diagenetic process.

We conclude from these observations that the REE pattern of the boundary clay in South China may result from a mixture of acidic-intermediate volcanic ash and upper crust matter spattered by an impact event.

In order to estimate mixing ratios of the two components, we made two hypotheses

1 that the Eu index depends simply on the mixing proportions, and

Table 14.2. *Concentrations of rare-earth elements in clays at and near the Permo-Triassic boundary in South China**

Section	Distance (cm)	La	Ce	Nd	Sm	Eu	Tb	Dy	Yb	Lu	(La/Sm)	(Tb/Yb)	(La/Yb)	δ Eu	δ Ce
Changxing	0	82.3	203	79.6	15.01	1.46	2.14	9.58	4.50	0.68	3.45	2.10	12.33	0.30	0.92
Guangyuan	0	76.5	163.7	73.5	12.34	1.05	1.42	8.51	4.59	0.60	3.90	1.37	11.24	0.28	0.99
Liangfengya	0	90.7	152.0	67.4	12.3	2.28	1.71	—	3.44	0.56	4.64	2.19	17.78	0.57	0.81
Duansan	0	141.8	243.0	106.0	26.3	2.95	2.45	16.2	7.51	0.75	3.39	1.44	12.73	0.38	0.83
Zhuzang	0	87.7	185.7	102.0	14.56	1.95	1.95	10.3	5.41	0.81	3.79	1.59	10.93	0.42	0.94
Heshan	0	127.3	208.3	93.4	16.8	1.95	2.33	—	6.61	0.89	4.78	1.55	12.98	0.36	0.79
Changxing	+215	47.4	172.0	91.0	14.8	1.82	2.65	11.1	6.73	1.04	2.01	1.74	4.75	0.35	1.40
Changxing	+130	35.7	98.5	49.0	8.69	0.92	0.83	5.07	4.34	0.58	2.58	0.84	5.55	0.35	1.17
Guangyuan	−45	58.3	136.4	60.3	10.73	1.32	1.75	9.80	6.21	0.81	3.41	1.24	6.33	0.36	1.06
Guangyuan	−120	98.2	255.0	110.0	18.2	1.32	3.65	19.2	14.5	2.30	3.39	1.11	4.57	0.20	1.16
Guangyuan	−5950	111.2	226.7	93.8	31.97	1.69	5.69	29.5	16.1	2.20	2.18	1.55	4.65	0.15	0.96
average of boundary		101.1	192.6	86.9	16.2	1.94	2.33	11.1	5.34	0.72	3.99	1.71	13.0	0.39	0.88
average of boundary		70.16	177.7	80.8	16.9	1.15	2.92	14.9	9.58	1.39	2.71	1.29	5.17	0.28	1.15
Volcano ash **		28.2	67.2	31.5	5.27	0.28	0.85	—	2.45	0.63	3.37	1.53	7.76	0.21	1.06
K/T clay ***		33.0	28.0	32.0	6.7	1.44	0.99	6.0	2.2	0.31	3.10	1.98	10.11	0.66	0.39

* Concentration of REE in ppm, with an error of less than 5%.
** Sample of volcanic ash taken from Triassic of South China.
*** Data for Cretaceous–Triassic (K/T) boundary clay from Kyte, F. T. *et al.*, 1980.

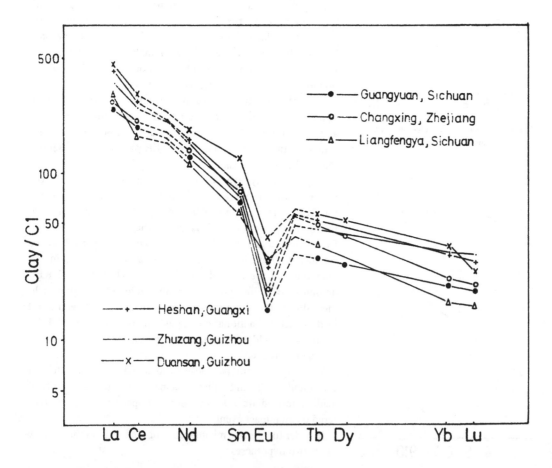

Fig. 14.13. **Rare-earth elements abundances in Permo-Triassic boundary clays of South China.**

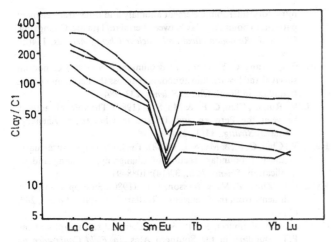

Fig. 14.14. Rare-earth elements abundances in nonboundary clays near the Permo-Triassic boundary in South China.

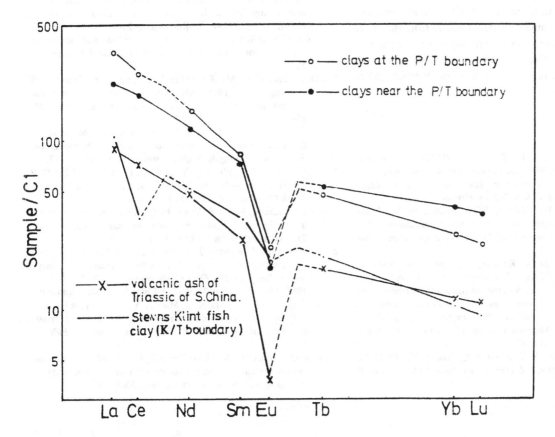

Fig. 14.15. Rare-earth elements abundances in clays at and near the Permo-Triassic (P/T) boundary and in samples of Triassic volcanic ash from South China and clay from the Cretaceous-Tertiary (K/T) boundary of Denmark. (Data for Cretaceous-Tertiary boundary clay from Kyte et al., 1980.)

2 that the Ce index of the boundary clay is affected only by the spattering of upper crustal matter and subsequent alteration on the sea floor. There is no significant fractionation of Ce in volcanic matter during the processes of eruption and alteration, which is proved, in fact, by the REE patterns of the non-boundary P/T clays and the K/T boundary clays.

With the above assumptions, we can calculate two independent mixing coefficients. The negative Eu anomaly provides the mixing ratio of the upper crustal matter (29%) to the acidic-intermediate volcanic matter (71%). Results of the negative Ce anomaly, on the other hand, are 30% to 70%. Ratios from Eu and Ce are thus so close that they support our mixing model.

Conclusion

Elemental geochemistry of rocks at the P/T boundary in South China indicates that the boundary event was a very complicated one, related not only to volcanic eruption, but also to a likely extraterrestrial impact. The mixing model proposed on the basis of our geochemical study is supported by evidence from petrology, mineralogy, and the study of microspherules. In fact, we hold that both volcanic and cosmic processes operated in formation of boundary and non-boundary clays at or near the P/T boundary in South China, and that the two processes are not necessarily antagonistic. A mixing model that holds that P/T boundary clays of South China consist of both acidic-intermediate volcanic ash and impact-spattered matter from the upper crust best explains the experimental results reported here. It is likely that the impact either triggered large-scale volcanic eruptions, or that it came just at the end of the Permian, when volcanic eruption was very active.

References

Alekseev, A. S., Barsukova, L. D., Kolesov, G. M., Nazarov, M. S. & Grigoryan, A. G. (1983). The Permian-Triassic boundary event: geochemical investigation of the Transcaucasia section. *Lunar and Planetary Science*, XIV, NASA, Houston, Texas.

Asaro, F., Alvarez, L. W., Alvarez, W. & Michel, H. V. (1982). Geochemical anomalies near the Eocene/Oligocene and Permian/Triassic boundaries. *Geol. Soc. America Spec. Papers*, **190**: 517–28.

Brandner, R., Donofrio, D. A., Krainer, L., Mostler, H., Nazarov, M. A., Resch, W., Stingl, V. & Weissert, H. (1986). Events at the Permian-Triassic boundary in the Southern and Northern Alps. In *Field Conference on Permian and Permian-Triassic Boundary in the South-Alpine Segment of the Western Tethys, and Additional Regional Reports*, Abstracts vol., pp. 15–16. Pavia: Soc. Geol. Ital. and IGCP Project 203.

Chai, C. F., Ma, S. L., Mao, X. Y., Zhou, Y. Q., Sun, Y. Y., Xu, D. Y., Zhang, Q. W. & Yang, Z. Z. (1987). Neutron activation studies of refractory siderophile element anomaly and other trace element patterns in boundary clay between Permian/Triassic, Changxing, China. *J. Radioanalytical and Nuclear Chem., Articles*, **114**(2): 293–301.

Clark, D. L., Wang, C. Y., Orth, C. J. & Gilmore, J. S. (1986). Conodont survival and low iridium abundances across the Permian-Triassic boundary in South China. *Science*, **233**: 984–6.

He, J. W., Rui, L., Chai, C. F., & Ma, S. L. (1987). The volcanic activity between the Permian and Triassic in Changxing, Zhejiang, China. *J. Stratig.*, **11**(3): 194–9.

He, J. W., Chai, C. F., & Ma, S. L. (1988). The high-temperature quartz in the P/T boundary section of Changxing, Zhejiang and its implication. *Science Bull.*, **33**(14): 1088–91.

Kyte, F. T., Zhou, Z. M. & Wasson, J. T., (1980). Sideropile-enriched sediments from the Cretaceous–Tertiary boundary. *Nature*, **288**: 651–6.

Oddone, M. & Vannucci, R. (1986). Geochemical stratigraphy at the P/T boundary in the Southern Alps. In *Field Conference on Permian and Permian-Triassic Boundary in the South-Alpine Segment of the Western Tethys, and Additional Regional Reports*, Abstracts vol., p. 44. Pavia: Soc. Geol. Ital. and IGCP Project 203.

Oddone, M. & Vannucci, R. (1988). PGE and REE geochemistry at the B–W boundary in the Carnic and Dolomite Alps (Italy). *Mem. Soc. Geol. Ital.*, **34**: 129–40.

Sheng, J. Z., Chen, C. Z., Wang, Y. G., Rui, L., Liao, Z. T., Bando, Y., Ishii, K., Nakazawa, D., & Nakamura, K. (1984). Permian-Triassic boundary in middle and eastern Tethys. *J. Fac. Sci, Hokkaido Univ.*, Ser. V, **21**(1): 133–81.

Sun, Y. Y., Chai, C. F., Ma, S. L., Mao, X. Y., Xu, D. Y., Zhang, Q. W. & Yang, Z. Z. (1984). The discovery of iridium anomaly in the Permian-Triassic boundary clay in Changxing, Zhejiang, China and its significance. In *Developments in Geoscience*, ed. Tu Guangzhi, pp. 235–46. Beijing: Science Press.

Turekian, K. K. & Wedepohl, K. H. (1961). Distribution of the elements in some major units of the Earth's crust. *Bull. Geol. Soc. America*, **72**: 175.

Xu, D. Y., Chai, C. F., Ma, S. L., Mao, X. Y., Sun, Y. Y., Zhang, Q. W. & Yang, Z. Z. (1985). Abundance variation of Ir and trace elements at the P/T boundary at Shangsi in China. *Nature*, **314**: 154–6.

Xu, G. R., Zhang, K. X., Hunag, S. J., Wu, S. B. & Bi, X. M. (1988). On the upper Permian and event stratigraphy of Permo-Triassic boundary in Huangshi, Hubei. *Earth Science*, **13**(5): 521–7.

Yang, Z. Y., Yin, H. F., Wu, S. B., Yang, F. Q., Ding, M. H. & Xu, G. R. (1987). *Permian-Triassic Boundary Stratigraphy and Fauna of South China*. Beijing: Geol. Publ. House.

Yin, H. F., Zhang, K. X. & Yang, F. Q. (1988). A new scheme of biostratigraphic delimitation between marine Permian and Triassic. *Earth Science*, **13**(5): 511–19.

Zhou, Y. Q., Chai, C. F., Mao, X. Y., Ma, S. L. & Orth, C. J. (1987a). On the catastrophic environment of the Permo-Triassic boundary in South China. In *The Final Conference of IGCP-203, 5–9 Sept. 1987, Abstracts vol.*, p. 38. Beijing: China IGCP Nat. Committee.

Zhou, Y. Q., Chai, C. F., Ma, S. L., Mao, X. Y, Sun, Y. Y. & He, J. W. (1987b). The impact events records at the Permian/Triassic boundary of Changxing, Zhejiang, China. *Science Bull.*, **32**: 1655–6.

Zhou, Y. Q., Chai, C. F., Ma, J. G., Kong, P. & Fu, G. M. (1988). Studies of the ferric microspherules at the Permian/Triassic boundary clay of Guangyuan, Sichuan. *Science Bull.*, **33**(5): 397–8.

15 Permo-Triassic orogenic, paleoclimatic, and eustatic events and their implications for biotic alteration

J. M. DICKINS

Introduction

It has long been recognized that a major change in fauna and flora was associated with a major regression in the latest Permian and earliest Triassic (see Dickins, 1983). The nature of this regression (or eustatic change) and its relationship to orogenic and tectonic developments in the Upper Paleozoic and Lower Mesozoic are discussed in this paper, which also includes comparisons with the Cretaceous–Tertiary boundary sequence. The climate of the Permo-Triassic boundary interval is also considered and its nature and influence amplified.

Volcanic rocks are widespread in Permo-Triassic boundary sequences but, as indicated elsewhere in this volume, their importance has only recently been recognized. The volcanism indicated by these rocks was related to tectonic and magmatic developments, which, together with the eustatic change and climate, must have had a profound effect on the fauna and flora of the time. These factors and apparently associated changes in the composition of seawater, the atmosphere, and the nature of the land, seem more than adequate to explain changes in the flora and fauna. The latter, in turn, are important potential guides to understanding development of the earth below the crust.

Tectonic and magmatic development

The Permo-Triassic boundary interval includes evidence of a remarkable regression, which represents a large-scale eustatic fall in sea level. Newell (1962, 1967a, 1967b, 1973) advanced strong reasons for recognizing this worldwide regression, which he thought contributed importantly to the change in life that marks the difference between the Paleozoic and Mesozoic. In the decade in which Newell's papers appeared, however, coeval worldwide transgressions and regressions were not accepted in geology, as they are today. Indeed, data suggesting such a conclusion were given scant attention, as is all too frequently the case with data and ideas that do not fit in with current orthodoxy.

The reality of coeval worldwide transgression and regression has now been established by much work, including the results of seismic stratigraphy (see Vail *et al.*, 1977), which it is beyond the scope of this paper to review. It is interesting to note, however, that the Permo-Triassic boundary interval shows up little in

seismic stratigraphic work (see Haq, Hardenbol & Vail, 1987). Reasons for this are not clear, but may involve the apparent scarcity of latest Permian deposits on the present shelves. It is also interesting that reasons for the now well-established phenomenon of coeval transgression and regression are poorly understood. This may reflect a reluctance to accept data that do not fit current orthodoxy as, for example, those that suggest worldwide, coeval tectonic activity.

Although the well-established late Permian and Cretaceous–Tertiary regressions represent the two major regressions of the Phanerozoic (Dickins, 1991a,b) their underlying cause remains obscure. The former, the latest phase in a progressive regression during the Upper Permian, was associated with extremely widespread tectonism and volcanic activity, and culminated at the end of the Changxingian (or Dorashamian), which the author regards as the latest stage of the Permian, but which is thought to be earliest Triassic by others (cf. Sweet, Chapter 11).

Marine deposits of Changxingian (or Dorashamian) age have been identified thus far only in the Tethyan region. If ultimately discovered outside this region, it is likely that they will be relatively limited in occurrence. These features are illustrated in the accompanying Upper Permian correlation chart (Fig. 15.1).

At most places within the Tethyan region the top of the Changxingian is marked by a hiatus. Even in places such as South China in which there may have been more or less continuous marine sedimentation, youngest Permian rocks indicate regression and earliest Triassic strata transgression (see elsewhere in this volume). In addition, there may be evidence at the top of the Changxingian of movements that led to a series of hiatuses or to angular unconformity. This information may indicate that shelves, as we know them, may have disappeared toward the end of the Permian.

The regression that culminated at the end of the Permian is best considered in association with a discussion of Upper Permian tectonic development. The Lower Permian was marked by widespread tension, graben formation, and associated dominantly basic and intermediate magmatic activity (Dickins, 1987, 1988a, 1988b, 1991a,b). The Hercynian or Variscan orogenic phase (or orogeny) is best considered as having been completed at the end of the Carboniferous; that is, it terminated with inception of the tensional phase. The Hunter-Bowen (or Indosinian) Orogenic Phase, which ended at the end of the Triassic, began at the beginning of the Upper Permian (using the tra-

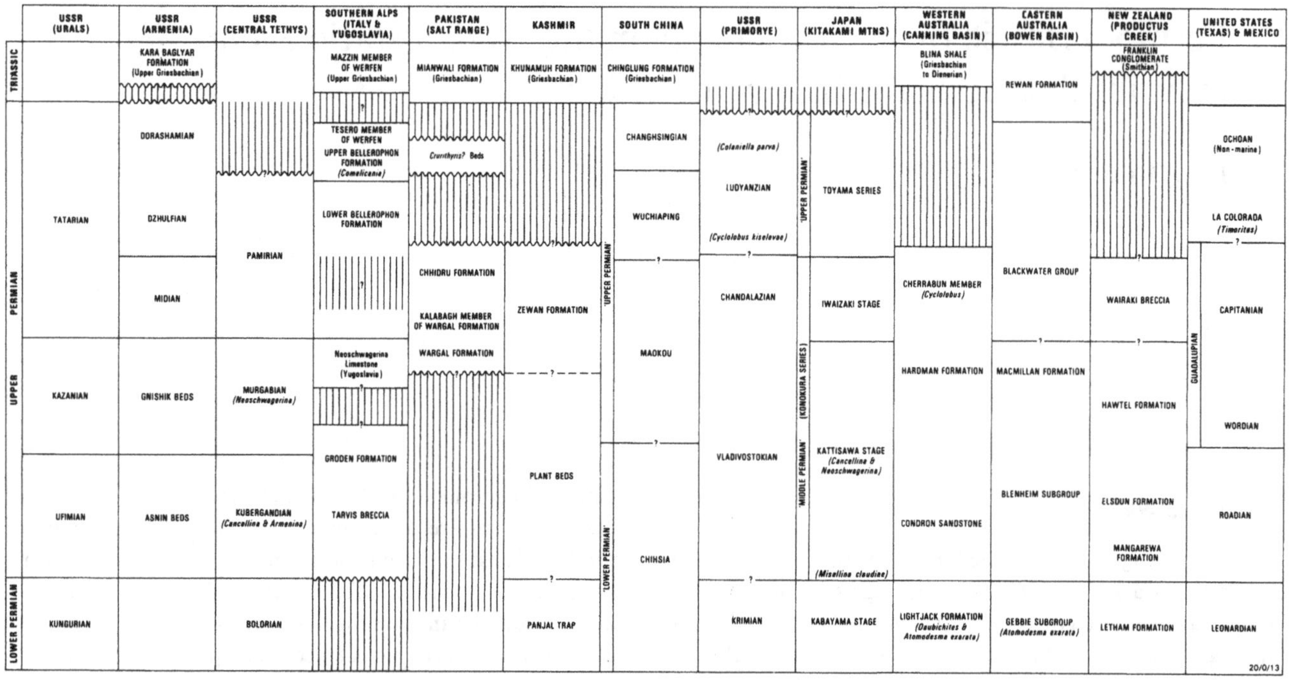

	USSR (URALS)	USSR (ARMENIA)	USSR (CENTRAL TETHYS)	SOUTHERN ALPS (ITALY & YUGOSLAVIA)	PAKISTAN (SALT RANGE)	KASHMIR	SOUTH CHINA	USSR (PRIMORYE)	JAPAN (KITAKAMI MTNS)	WESTERN AUSTRALIA (CANNING BASIN)	EASTERN AUSTRALIA (BOWEN BASIN)	NEW ZEALAND (PRODUCTUS CREEK)	UNITED STATES (TEXAS) & MEXICO
TRIASSIC	KARA BAGLYAR FORMATION (Upper Griesbachian)			MAZZIN MEMBER OF WERFEN (Upper Griesbachian)	MIANWALI FORMATION (Griesbachian)	KHUNAMUH FORMATION (Griesbachian)	CHINGLUNG FORMATION (Griesbachian)		TOYAMA SERIES	BLINA SHALE (Griesbachian to Dienerian)	REWAN FORMATION	FRANKLIN CONGLOMERATE (Smithian)	
PERMIAN (UPPER)	TATARIAN	DORASHAMIAN	PAMIRIAN	TESERO MEMBER OF WERFEN / UPPER BELLEROPHON FORMATION (Comelicana)	Crunchyrus? Beds		CHANGHSINGIAN	(Colaniella parva)	IWAIZAKI STAGE (Cyclolobus)	CHERRABUN MEMBER (Cyclolobus)	BLACKWATER GROUP	WAIRAKI BRECCIA	OCHOAN (Non-marine)
		DZHULFIAN		LOWER BELLEROPHON FORMATION	CHHIDRU FORMATION		LUOYANZIAN / WUCHIAPING	(Cyclolobus kiselevae)					LA COLORADA (Timorites)
		MIDIAN			KALABAGH MEMBER OF WARGAL FORMATION	ZEWAN FORMATION		CHANDALAZIAN			MACMILLAN FORMATION		CAPITANIAN
UPPER	KAZANIAN	GNISHIK BEDS	MURGABIAN (Neoschwagerina)	Neoschwagerina Limestone (Yugoslavia)	WARGAL FORMATION		MAOKOU		KATTISAWA STAGE (Cancellina & Neoschwagerina)	HARDMAN FORMATION		HAWTEL FORMATION	WORDIAN
	UFIMIAN	ASNIN BEDS	KUBERGANDIAN (Cancellina & Armenina)	GRODEN FORMATION	PLANT BEDS		VLADIVOSTOKIAN		CONDRON SANDSTONE	BLENHEIM SUBGROUP	ELSDUN FORMATION	ROADIAN
				TARVIS BRECCIA			CHIHSIA		(Misellina claudiae)			MANGAREWA FORMATION	
LOWER PERMIAN	KUNGURIAN		BOLORIAN		PANJAL TRAP		KRIMIAN	KABAYAMA STAGE (Monodiexodina & Atomodesma exarata)	LIGHTJACK FORMATION (Daubichites & Atomodesma exarata)	GEBBIE SUBGROUP (Atomodesma exarata)	LETHAM FORMATION	LEONARDIAN

20/0/13

Fig. 15.1. Correlation of Upper Permian rocks, illustrating the end-Permian regression and continentality. (After Dickins *et al.*, 1991.) (NB: Recent re-examination of faunas from the Kashmir sequence indicates that E1, the basal member of the Khunamuh Formation, may be Changxingian, and that there may be a break both below and above this member.)

ditional twofold division of the Permian). Dickins *et al.* (1991) show that the end of the Lower Permian was marked by widespread regression and that the Upper Permian was initiated by transgression. Rocks of early Late Permian age represent a widespread transgressive overlap and are superimposed on earlier strata (including Lower Permian rocks) that were folded, metamorphosed, and subjected to erosion. The Lower–Upper Permian boundary also marks the beginning of a period of strong acidic and intermediate intrusion and volcanism, which took place at about 270 ma in the radiometric scale (Archbold & Dickins, 1991).

Although a number of major transgressive and regressive phases can be recognized in the Upper Permian sedimentary record, Upper Permian tectonic development was for the most part characterized by progressive regression, which culminated in the uppermost Changxingian and was followed in the Triassic by progressive transgression. Progressive Upper Permian regression involved gradual isolation of marine basins and restricted the distribution of marine strata of Changxingian age.

Schopf (1974) illustrated the Upper Permian regression in a series of maps that show the extent of shelf seas. Accuracy is difficult to obtain, however, and the Upper Permian correlation chart (Fig. 15.1) of the present report may be more useful in showing the nature of the regression. This method has also been employed by Holser & Magaritz (1987, fig. 5), who used Anderson's (1981) correlation chart graphically to illustrate the regression. In their fig. 5, Holser & Magaritz show a mid-Permian regressive-transgressive phase, the same one referred to in this report. However, as a result of a new, apparently more accurate correlation, Dickins *et al.* (1991) place this phase at the end of the Kungurian, rather than at the beginning of this stage, as shown by Holser & Magaritz (1987).

The Upper Permian regression was so closely associated with development of the orogenesis that there can hardly be any doubt that the two are linked. Orogenic movements were very widespread, although their results have been overlooked to some extent in areas subsequently affected by the Alpine–Himalayan orogenic phase, which began no earlier than mid-Cretaceous. In Europe, Upper Permian movements have sometimes been ascribed to the end of the Hercynian phase (epi-Hercynian) or to the beginning of the Alpine phase (pre-Alpine). Considering the world as a whole, however, there can be no real doubt that the Upper Permian and Triassic orogenic movements were a separate and very important phase, distinguishable from both the Hercynian and Alpine–Himalayan phases.

Because glacial eustacy can be eliminated, and a large loss of water to the oceans seems unlikely, the simplest explanation for the worldwide regression is that during the Upper Permian relief increased on the surface of the earth and culminated at the end of the Permian. Schopf (1974) suggested that the heightened relief and regression indicate an interval of slower sea-floor spreading, during which oceanic ridges are thought to have been less active and thus to have occupied a smaller volume of the ocean basins (see also Larson & Pitman, 1972). However, Schopf's explanation seems unlikely beceause the strong orogenic movements documented in this interval might be expected to indicate more rapid, rather than slower, sea-floor spreading. Holser & Magaritz (1987, p. 169) also doubt that the Upper Permian regression could have been caused by mid-ocean ridge activity. Association of the regression with the Hunter-Bowen (Indosinian) orogenic phase, on the other hand, offers a fairly straightforward explanation.

Associated with development of the Hunter-Bowen (Indosinian) orogenic phase was a period of intense, predominantly intermediate and acidic, magmatic and volcanic activity. The widespread occurrence of granitic batholiths is well known. However, the extremely widespread, almost universal, occurrence and very large volumes of Upper Permian and Triassic volcanics and volcanic detritus is only now being realized. In many cases, this widespread distribution indicates that the activity was not confined to limited zones and was far distant from what might be regarded as the orogenic or subduction zones of the time. At present, with few exceptions, volcanic activity is limited to zones associated with oceanic ridges and belts of active orogenic folding and compressional faulting.

At the Seventh Gondwana Symposium, widespread acidic volcanics were recorded (Coutinho, 1988) and demonstrated on a field excursion to the east side of the Parana Basin of Brazil. The grain size of these volcanics indicates they were derived from a source close at hand, not from the far-distant Andean region. Also at the Seventh Gondwana Symposium, widespread acidic and/or intermediate volcanics were reported to occur in the Upper Permian of the Karroo Basin of South Africa (for correlation see Cole & McLachlan, 1988).

My recent examination of the Phosphoria Formation of the western United States indicates that some, if not most or all, of its bedded chert represents siliceous volcanic ash, and tuff or slightly reworked tuff is widespread. It is possible that the faunally distinct Boreal chert-evaporitic Upper Permian Province, which stretches from mid-United States through British Columbia, Arctic Canada, and Greenland to Spitzbergen, reflects this volcanic activity, much of which is referred to in Dickins (1987, 1988a, 1988b, and earlier papers).

The boundary sequences

Changes in the flora and fauna of Permo-Triassic boundary sequences are treated extensively in other parts of this volume, where the significance of these changes is also assessed. These changes have been discussed in many papers and in 1983 I reviewed the situation as it was at that time (Dickins, 1983). It is sufficient to say here that these biotic changes began long before the end of the Permian, although very special changes are recorded in the boundary sequences. Already by the end of the Changxingian the fauna had lost its Paleozoic character. Some elements regarded as Paleozoic, such as productid brachiopods, are found in post-Changxingian 'transitional beds', and these and some other elements are the basis for arguing that the Permo-Triassic boundary should be placed at a higher position (Newell, 1988). The major geological change, however, coincides with the top of the Changxingian Stage, and the coincidence at this level of unconformity, disconformity, paraconformity, hiatus, or terrestrial deposition is illustrated in Fig. 15.1.

Although explosive volcanism was widespread during the Upper Permian, a particularly strong, widespread episode marked the end of the Changxingian. From the nature of the Upper Permian regression, and the presence of a paraconformity at the top of the Changxingian, vertical movement was also

a distinctive feature. Changes are tabulated in Fig. 15.2, in the context of other late Paleozoic and early Mesozoic changes.

Fig. 15.3 provides additional information associated with the Carboniferous–Permian boundary. It is of interest to note that the most nearly complete sections across the Carboniferous–Permian boundary are in present-day mid-latitude regions, as is also the case with continuous sections in the Permo-Triassic boundary interval. The nature of tectonic developments and vertical movements, and similarity with the Cretaceous–Tertiary boundary interval has been discussed elsewhere (Dickins, 1991a, b) and is summarized from these papers in Figs. 15.3 and 15.4. The evidence indicates that the Permo-Triassic boundary interval records a time of special tectonic and magmatic-volcanic activity.

Climate

Evidence of glaciation has been found in the Upper Carboniferous and is widespread in the lowest stage of the Permian, the Asselian. Above the Asselian, however, there is no reliable evidence of glaciers, although cold-climate ice activity is known. This is summarized in Dickins (1985a). There was probably a chilling of world climate in the mid-Permian, but glacial activity has not been proved.

After the mid-Permian, world climate became steadily warmer until in the latest Permian and earliest Triassic a universally hot climate, substantially warmer than at present, prevailed. This is detailed in a number of papers (see Dickins, 1984, 1985b, 1985c). Climatic changes for the Carboniferous and Permian are shown in Fig. 15.6.

After the mid-Permian, warming of the climate is shown especially by the increasingly cosmopolitan occurrence of faunas judged to represent warm water because of their diversity and their inclusion of certain brachiopods, algae, corals, fusulinids, and conodonts. Warm waters are also indicated by development in the sedimentary sequence of reefs, desert deposits, fine-grained red beds, and evaporites. Above the mid-Permian and through the Triassic, there is not a single authenticated glacial deposit or even a deposit indicating ice action. These data have been tabulated by Dickins (1985a) and are illustrated and discussed further by Dickins (1983, 1984).

The hot climate continued through the Triassic (Dickins, 1985c). The Lower Triassic climate was quite unusual by comparison with that of the present. Warm-climate red beds were virtually universal, and desert conditions and diverse amphibian and reptilian faunas were very widespread. No humid zones can be discerned in the sedimentary record.

The hot climate of the latest Permian and earliest Triassic, together with marine regression, widespread volcanism, and tectonic instability, would have subjected the fauna and flora to extremely rigorous conditions and would no doubt have been sufficient to effect a great change in the biota.

A number of causes have been postulated for the change in climate through the Upper Permian. In scientific discussions and in various publications, generally in a rather vague way as in Holser & Magaritz (1987), it is supposed that assembly of Pangaea and/or latitudinal shifts of land sometime in the late Paleozoic had an important effect on the climate. Convincing

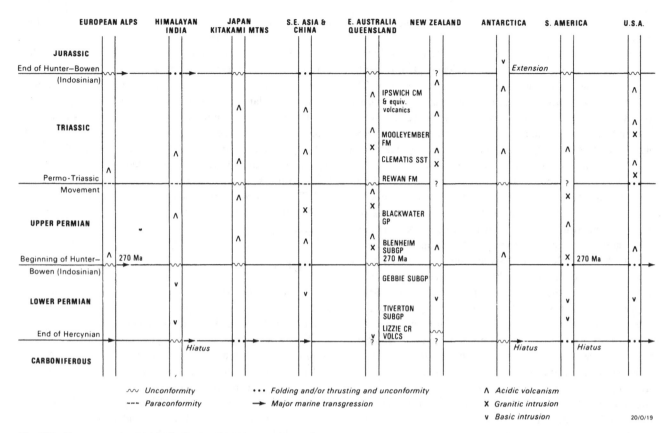

Fig. 15.2. Events associated with the Carboniferous-Permian, the mid-Permian, the Permo-Triassic, and the Triassic–Jurassic boundaries.

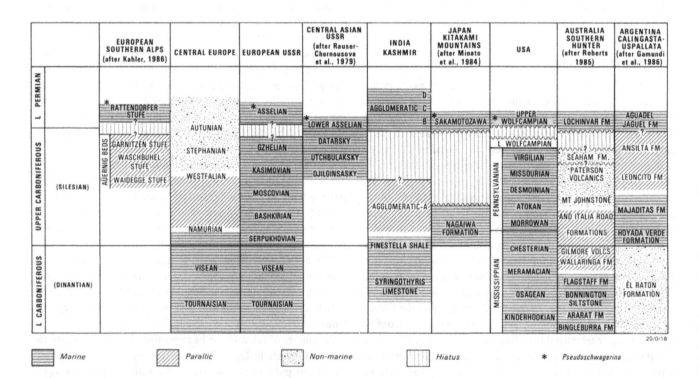

Fig. 15.3. Representative Carboniferous and Lower Permian sequences, showing worldwide regression followed by transgression in the earliest Permian.

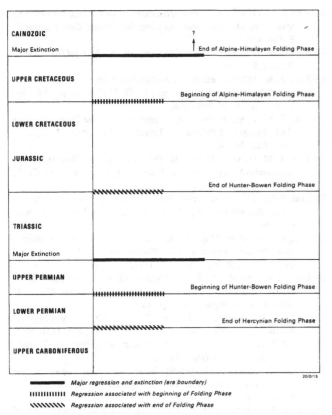

CAINOZOIC	
Major Extinction	? ↑ End of Alpine-Himalayan Folding Phase
UPPER CRETACEOUS	Beginning of Alpine-Himalayan Folding Phase
LOWER CRETACEOUS	
JURASSIC	
	End of Hunter-Bowen Folding Phase
TRIASSIC	
Major Extinction	
UPPER PERMIAN	Beginning of Hunter-Bowen Folding Phase
LOWER PERMIAN	End of Hercynian Folding Phase
UPPER CARBONIFEROUS	

▬▬▬ Major regression and extinction (era boundary)

IIIIIIIIIIII Regression associated with beginning of Folding Phase

\\\\\\\\ Regression associated with end of Folding Phase

Fig. 15.4. Major regressions and transgressions from Upper Carboniferous to lower Cenozoic and relationships with major tectonic-magmatic events and biotic extinctions.

FEATURES OF PERMIAN-TRIASSIC BOUNDARY	FEATURES OF CRETACEOUS-TERTIARY BOUNDARY
Paraconformity	Paraconformity
Strong vertical uplift	Strong vertical uplift
Very strong volcanism	Very strong volcanism
Hot climate	Apparent cooling
Major regression	Major regression
No shelves	?
Shallow water faunal change	Shallow water faunal change

Fig. 15.5. Comparison of important features of the Permian–Triassic and Cretaceous–Tertiary boundaries.

detailed studies are lacking, however, and reconstructions such as those of Smith, Hurley & Briden (1981) or Scotese (1986) seem to offer little support for such conclusions. That is, the climate changed from very cold to very hot during the Late Paleozoic and Early Mesozoic, but there were no marked changes in Pangaea that might explain this.

More likely causes for climatic change may be fluctuations in solar energy because of changes in solar emission; changes in the planetary or galactic system; change in the angle of the earth's axis; changes in the earth's atmosphere, probably as a result of magmatic and volcanic activity; or some combination of these factors. The amount of carbon dioxide and other gases in the atmosphere may have had a greenhouse effect, or the ozone layer may have been affected. It seems unlikely, however, that a simple carbon dioxide effect led to the dry climate of the Lower Triassic. An increasing area of land surface may also have had an important effect on absorption of incident solar energy. None of these factors has thus far been examined sufficiently to allow a reliable explanation.

Conclusions

A great deal of convincing evidence suggests that regression and emergence, coupled with hot climatic conditions and strong volcanic activity had a profound, adverse effect on the existing biota of the Permo-Triassic boundary interval. There is no substantial evidence of impact by extraterrestrial bodies. The factors enumerated were certainly accompanied by changes in

composition of seawater and the atmosphere, and by changes in terrestrial environments.

The lower Lower Permian, the Upper Permian, and the Triassic were times of large-scale deposition of evaporites. The effect of this on the biota has been discussed by Dickins (1983), Holser & Magaritz (1987), and Meyerhoff (1973). It has been suggested that the accumulation of evaporites might have had an effect on the salinity of the oceans and thus on life, but a direct association is not obvious. Changes in strontium- and sulphur-isotopic ratios might be related, however (see Holser & Magaritz, 1987; Meyerhoff, 1973).

Magmatic and volcanic activity almost certainly had a significant effect on the composition of the oceans and atmosphere, although the magnitude of the effect is at present difficult to tabulate. The distribution and volume of oxygen, carbon dioxide, and perhaps ozone were probably affected, and widespread Upper Permian chert reflects the amount of silica contributed directly and indirectly to the sea. Volcanic activity, both aerial and submarine, would have contributed other important materials.

Javoy & Courtillot (1989) ascribe the high strontium-isotopic ratio at the Cretaceous–Tertiary boundary to intense acidic volcanism, and the anomaly at the Permo-Triassic boundary may have a similar explanation (see Fig. 15.5). Javoy & Courtillot also suggest that anomalies in the distribution of iridium, lead, shocked minerals, and carbon isotopes might also be results of this intense volcanic activity. An iridium anomaly does not seem to be a feature of the Permo-Triassic boundary, but there is an important carbon-isotopic anomaly at this level (see Holser & Magaritz, 1987).

During the Upper Permian there was also a marked change in magnetic polarity. This change, from reversed to mixed normal and reversed (the Illawara Reversal), has been recognized in a number of parts of the world (Menning, 1986). Although in some places the position of the Illawara Reversal in relation to the stratigraphic scale is unclear, it does appear possible that it represents a single event. What this means in relation to the development of polarity during the Upper Paleozoic and Lower Mesozoic remains to be elucidated, and it is not clear if the polarity change is related to the tectonic and magmatic changes mentioned elsewhere in this report.

From a consideration of the data referred to in this report, it seems justified to conclude that the distinctive tectonic movements and the intense, peculiar magmatic-volcanic activity associated with the Permo-Triassic boundary interval (and also,

Late Permian	*Warm and hot worldwide* *No glacials*
early Late Permian	*Cool fluctuation* *No moraine identified*
Remainder of Early Permian	*No moraine identified* *"Dropstones" in Australia*
Sakmarian (Tastubian)	*Eustatic sea level rise*
Asselian	*Main glaciation* *Widespread terrestrial glaciation*
Late Carboniferous	*Glaciation limited* *Mainly montaine, some moraine*
Early Carboniferous	*Warm, no authenticated glaciation*

20-4/75

Fig. 15.6. Main changes in climate during the Carboniferous and Permian.

apparently, with the Cretaceous–Tertiary boundary interval) indicate strongly that there were significant physical and chemical changes in the subcrustal Earth at these times.

References

Anderson, J. M. (1981). World Permo-Triassic correlations: Their biostratigraphic basis. In *Gondwana Five*, eds. M. Cresswell & P. Vella, pp. 3–10. Rotterdam: A. A. Balkema.

Archbold, N. W. & Dickins, J. M. (1991). Australian Phanerozoic time scales: Permian, a standard for the Permian System in Australia. *Bur. Min. Res., Australia, Rec.*, 1989/36.

Cole, D. I. & McLachlan, I. R. (1988). Oil potential of the Permian Whitehill Shale Formation in the main Karroo Basin, South Africa. *Seventh Gondwana Symposium, Sao Paulo, Brazil, Abstracts*, 22.

Coutinho, J. M. V. (1988). Ashfall-derived vitroclastic tuffaceous sediments in the Permian of the Parana Basin and its provenance. *Seventh Gondwana Symposium, Sao Paulo, Brazil, Abstracts*, 78.

Dickins, J. M. (1983). Permian to Triassic changes in life. *Assoc. Australasian Palaeont., Mem.* **1**: 297–303.

Dickins, J. M. (1984). Evolution and climate in the upper Palaeozoic. In *Fossils and Climate*, ed. P. Brenchley, pp. 317–27. John Wiley & Sons Ltd.

Dickins, J. M. (1985a). Late Palaeozoic glaciation. *BMR J. Australian Geol. Geophys.*, **9**: 163–9.

Dickins, J. M. (1985b). Late Paleozoic climate with special reference to invertebrate faunas. *Neuvieme Cong. Int. Stratig. Geol. Carbonif.*, **5**: 394–402.

Dickins, J. M. (1985c). Climate of the Triassic. *N. Zealand Geol. Surv., Record*, **9**: 34–6.

Dickins, J. M. (1987). Tethys – a geosyncline formed on continental crust? In *Shallow Tethys 2*, ed., K. G. McKenzie, pp. 149–58. Rotterdam: A. A. Balkema.

Dickins, J. M. (1988a). The world significance of the Hunter/Bowen (Indosinian) mid-Permian to Triassic folding phase. *Mem. Soc. Geol. Ital.*, **34**: 345–52.

Dickins, J. M. (1988b). The world significance of the Hunter-Bowen (Indosinian) orogenic phase. *Proc. 22d Symposium, Dept. Geol., Univ. Newcastle*, 69–74.

Dickins, J. M. (1991a). Major sea level changes, tectonism and extinctions. *Eleventh Int. Cong. Carbonif. Stratig. Geol., Compte rendu.* (In press.)

Dickins, J. M. (1991b). Major sea level changes, tectonism and extinctions. In *Deep structure of the Pacific Ocean and its continental surroundings.* Blagoveshchensk, USSR. (In press.)

Dickins, J. M., Archbold, N. W., Thomas, G. A. & Campbell, H. J. (1991). Mid-Permian correlation. *Eleventh Int. Cong. Carbonif. Strat. Geol., Compte rendu.* (In press.)

Haq, B. U., Hardenbol, J. & Vail, P. R. (1987). Chronology of fluctuating sea levels since the Triassic. *Science*, **235**: 1156–67.

Holser, W. T. & Magaritz, M. (1987). Events near the Permian–Triassic boundary. *Modern Geol.*, **11**: 155–80.

Javoy, M. & Courtillot, V. (1989). Intense acidic volcanism at the Cretaceous-Tertiary boundary. *Earth and Planetary Sci. Letters*, **94**: 409–16.

Larson, R. L. & Pitman, W. C., III (1972). World-wide correlation of Mesozoic magnetic anomalies and its implications. *Geol. Soc. America Bull.*, **83**: 3645–62.

Menning, M. (1986). Zur Dauer des Zechsteins aus magnetostratigraphischer Sicht. *Z. Geol. Wiss., Berlin*, **14**(4): 395–404.

Meyerhoff, A. A. (1973). Mass biotal extinctions, world climatic changes and galactic motions: possible interrelations. *Can. Soc. Petrol. Geol. Mem.* **2**: 745–8.

Newell, N. D. (1962). Paleontological gaps and geochronology. *J. Paleont.* **36**: 592–610.

Newell, N. D. (1967a). Revolutions in the history of life. *Geol. Soc. America Spec. Papers*, **89**: 63–91.

Newell, N. D. (1967b). Paraconformities. In *Essays in Paleontology and Stratigraphy* eds. C. Teichert & E. L. Yochelson, 349–67. (Dept. Geology, Spec. Pub. 2.) Lawrence, Kansas: Univ. Kansas.

Newell, N. D. (1973). The very last moment of the Paleozoic Era. *Can. Soc. Petrol. Geol. Mem.* **2**: 1–10.

Newell, N. D. (1988). The Paleozoic/Mesozoic erathem boundary. *Mem. Soc. Geol. Ital.*, **34**: 303–11.

Schopf, T. J. M. (1974). Permo-Triassic extinctions: relation to sea-floor spreading. *J. Geol.*, **82**: 129–43.

Scotese, C. R. (1986). Phanerozoic reconstructions: a new look at the assembly of Asia. *Univ. Texas, Inst. Geophys. Tech. Rept.*, **66**: 54 pp.

Smith, A. G., Hurley, A. M. & Briden, J. C. (1981). *Phanerozoic paleocontinental maps.* Cambridge Earth Science Series. Cambridge: Cambridge University Press.

Vail, P. R., Mitchum, R. M., Jr., Todd, R. G., Widmier, J. M., Thompson, S., III, Sangree, J. B., Bubb, J. N. & Hatfield, W. G. (1977). Seismic stratigraphy and global changes in sea level. *Am. Assoc. Petrol. Geol. Mem.*, **26**: 49–212.

16 Permo-Triassic boundary in Australia and New Zealand

J. M. DICKINS AND HAMISH J. CAMPBELL

Introduction

Although Upper Permian and Lower Triassic rocks are widespread, no sequences representing latest Permian and earliest Triassic marine deposition are known from Australia and no well-defined boundary sequence has been recognized in New Zealand. In the western part of Australia, where marine deposits are present, the uppermost Permian and the lowermost Triassic are represented by a hiatus. In central and eastern Australia only terrestrial deposits are known both in the Upper Permian and the Lower Triassic.

The floral changes associated with the boundary in Australia have been considered by Balme & Helby (1973) who concluded that a very marked change took place between the Permian and the Triassic. Dickins (1973) who summarized Australian boundary sequences, concluded that an important regression was apparent.

Western part of Australia

Considerable hiatus is present between the Permian and the Triassic (Fig. 16.1) with generally some, but not very distinct, angular structural discontinuity (Archbold & Dickins, 1991; Dickins, 1976; Banks, 1978).

In the Perth Basin, a gap is present between the Lower Permian marine Carynginia Formation of oldest Baigendzhinian (Upper Artinskian) age and the apparently nonmarine Wagina Sandstone, the age of which, on palynological grounds, is early Upper Permian (Lower and Upper here referring to a twofold subdivision of the Permian as used in the traditional type area in the Russian Platform-Ural Mountains region). A hiatus is then present beneath the Kockatea Shale, which contains a good open-sea marine fauna regarded as not older than Upper Griesbachian (Balme, 1963; McTavish & Dickins, 1974).

In the Carnarvon Basin no marine deposits are known that are younger than the Kungurian, the top stage of the Lower Permian. Overlying these without a distinct structural break is terrestrial Upper Permian, the age of which is not reliably known. The next youngest formation, with structural and geographical discontinuity, is the Lockyer Shale with marine beds that contain fossils of Lower–Middle Triassic age (Dolby & Balme, 1976; McTavish, 1973).

The Canning and Bonaparte basins contain deposits of Dzhulfian age, the youngest marine Permian of the Australian continent. No Dorashamian or Changxingian (last stage of the Permian) is known. The overlying Triassic in both basins is apparently paralic and/or estuarine with marine fossils few and poorly preserved. On palynological grounds it is regarded as Lower Triassic and similar in age to the Kockatea Shale of the Perth Basin. Field evidence suggests the Triassic rests in open synclines formed subsequent to the deposition of the Upper Permian.

Central part of Australia

Nonmarine Upper Permian and Triassic are found in widely dispersed basins in Central Australia. In general there is a discordance in detail in the distribution, and breaks are present in the sequences. The relationships between the Permian and the Triassic remain poorly understood and no continuous or likely continuous sequence is known.

Eastern part of Australia

The best-known sequences are in the Bowen and Sydney basins (Fig. 16.2) where the uppermost Permian and the Lower Triassic developed under terrestrial conditions. Correlation is dependent largely on palynology with some support from vertebrate evidence. In both basins the boundary sequences are affected by tectonic movement so that the structural relationships are not simple. In the deepest parts of the basins no considerable hiatus may separate the Permian and the Triassic.

In the Sydney Basin (Fig. 16.3) the Newcastle Coal Measures contain a microflora regarded by Helby (1973) and Foster (1982), who proposed the name *Playfordiaspora crenulata* Zone, as equivalent to the upper part of the Chhidru Formation of the Salt Range, Pakistan, i.e., of Dzhulfian age (chart in Dickins *et al.*, 1991). According to the work of Foster, it appears likely that the first definite Triassic forms are to be found in the *Protohaploxypinus samoilovichii* Zone, which is considerably above the base of the Rewan Formation and separated from the *Playfordiaspora crenulata* Zone of apparently Dzhulfian age by two zones – the *Protohaploxypinus microcorpus* Zone and the *Lunatisporites pellucidus* Zone. Generally these two latter zones have been regarded as of Dorashamian or Dorashamian and Lower Griesbachian age because the microfloras appear to lack sharp change at any particular level and the microflora of the overlying

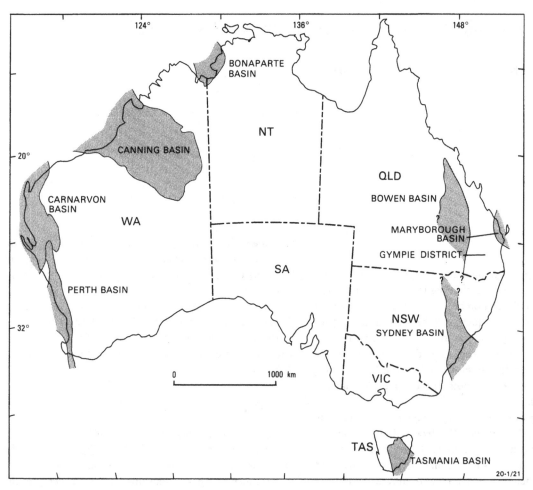

Fig. 16.2. Australian sedimentary basins with Permian and/or Triassic marine sequences.

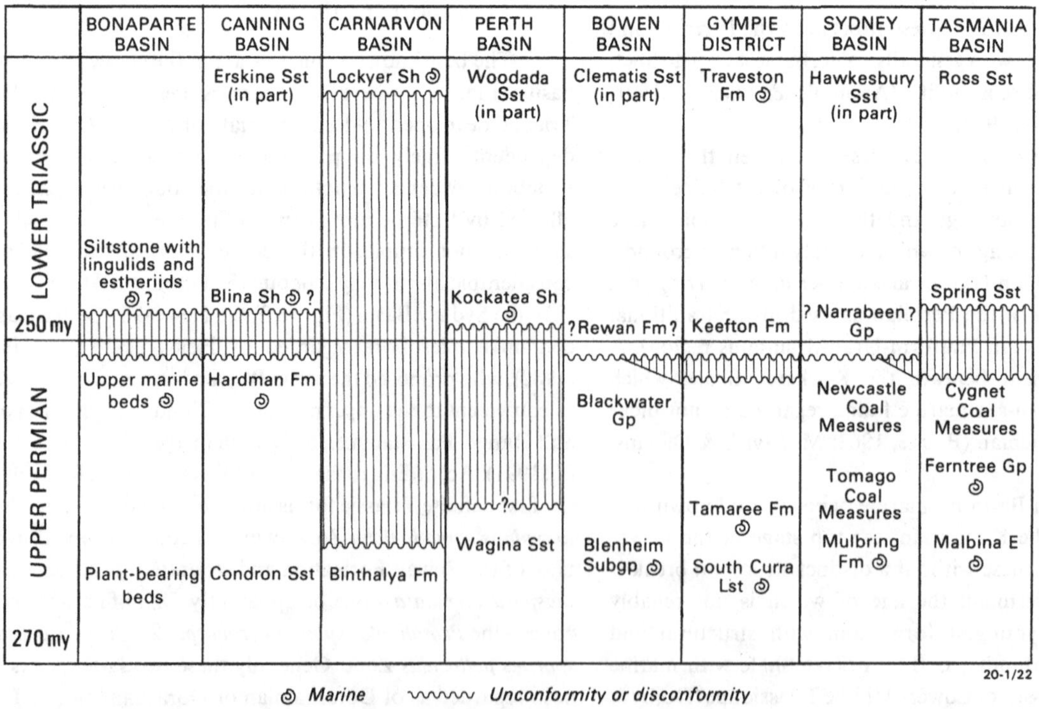

	BONAPARTE BASIN	CANNING BASIN	CARNARVON BASIN	PERTH BASIN	BOWEN BASIN	GYMPIE DISTRICT	SYDNEY BASIN	TASMANIA BASIN
LOWER TRIASSIC		Erskine Sst (in part)	Lockyer Sh ☺	Woodada Sst (in part)	Clematis Sst (in part)	Traveston Fm ☺	Hawkesbury Sst (in part)	Ross Sst
	Siltstone with lingulids and estheriids ☺ ?	Blina Sh ☺ ?		Kockatea Sh ☺	?Rewan Fm?	Keefton Fm	? Narrabeen ? Gp	Spring Sst
250 my								
UPPER PERMIAN	Upper marine beds ☺	Hardman Fm ☺			Blackwater Gp		Newcastle Coal Measures	Cygnet Coal Measures
				Wagina Sst		Tamaree Fm ☺	Tomago Coal Measures	Ferntree Gp ☺
	Plant-bearing beds	Condron Sst	Binthalya Fm		Blenheim Subgp ☺	South Curra Lst ☺	Mulbring Fm ☺	Malbina E ☺
270 my								

☺ *Marine*   〰〰〰〰 *Unconformity or disconformity*

Fig. 16.1. Sequences in Australia with marine beds in the Permian and/or Triassic. Radiometric dates from Archbold & Dickins (1991).

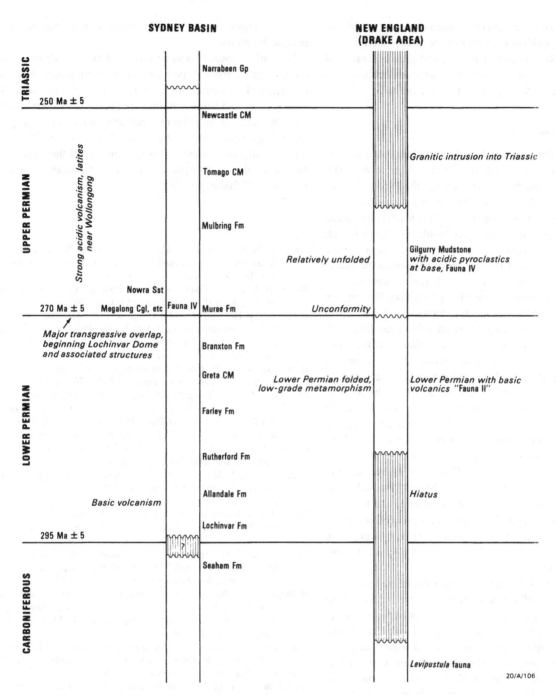

Fig. 16.3. Sequence in the Sydney Basin and New England area, northern New South Wales, showing relationships associated with Permo-Triassic boundary.

Protohaploxypinus samoilovichii Zone is related to that of the Kockatea Shale regarded as Upper Griesbachian.

The structural and sedimentational complications are, however, still poorly known. On the edges of the basins the Narrabeen and the Rewan rest with a distinct erosional discontinuity on the underlying beds (Mollan *et al.*, 1969; Herbert, 1980). Intervening beds appear in other and possibly deeper parts of the basins. At least in places there are lithological differences from the overlying beds with which they have been broadly included for want of a better solution. These intervening beds may represent the Dorashamian or some discontinuity may be pres-

ent between the Permian and Triassic. Dr N. Morris of the Geology Department of the University of Newcastle, Australia, has supplied information based on an unpublished thesis (Wilcock, 1979) that near Newcastle on the western side of Lake Macquarie, a Triassic type macroflora with *Dicroidium* and *Voltziopsis* occurs directly on top of the Newcastle Coal Measures with a *Glossopteris* flora, and the *Playfordiaspora crenulata* Zone. This may indicate that in this area, the Dorashamian is represented by a hiatus. Irrespective of this, however, it indicates a sharp change in climate from a humid, probably temperate climate to a warm, dryer climate. This is consistent with the change from the grey sediment reducing environment of the coal measures to the oxidizing, warm-climate environment of the redbeds of the Narrabeen and Rewan, and the appearance of many amphibian footprints at this level in the coalfields of both the southern and northern parts of the Sydney Basin.

New Zealand

A well-defined Permo-Triassic boundary sequence has not been recognized in New Zealand and it seems increasingly unlikely that one will be. Conventionally, the Permo-Triassic boundary has been loosely correlated with an apparent lithostratigraphic boundary between strata of the Maitai Group and the Murihiku Supergroup exposed in Southland, South Island. The basis for this interpretation lay in the fact that marine invertebrate fossils of Late Permian age are present in the Maitai Group, whereas the oldest known fossils (also marine invertebrates) within the apparently overlying Murihiku Supergroup are Early Triassic. However, only very limited fossil age control has been established. A few low-diversity fossil localities with poorly preserved fossils are sparsely distributed within thick well-bedded sequences of otherwise unfossiliferous volcaniclastic rock on either side of the Maitai-Murihiku boundary, and for the most part upper Maitai Group strata are poorly exposed. Representatives of those taxa that have been widely used in definition and recognition of the Permo-Triassic boundary elsewhere in the world have yet to be collected from New Zealand sequences (e.g., *Otoceras*, *Ophiceras*, *Claraia*, *Pseudomonotis*, conodonts, radiolarians). Hence, precise biostratigraphically established identification of the Permo-Triassic boundary is not yet possible in New Zealand. At best, 'the Permian-Triassic boundary' in New Zealand has been crudely recognized as coincident with a boundary between apparently superposed sequences with Permian fossils below (Maitai) and Triassic fossils above (Murihiku).

This conventional thinking has been challenged in recent years by a number of independent researchers (with differing specialities) and in particular with the advent of tectonostratigraphy (terrane analysis). It now seems possible that the Maitai Group and Murihiku Supergroup represent sequences within two separate tectonostratigraphic units and that their mutual boundary is tectonic. Further, on the basis of the known fossil content and age determinations it seems likely that the chronologies of the two terranes may overlap – Late Permian to Middle Triassic for the Dun Mountain-Maitai terrane (Maitai Group), and Early Triassic to Late Jurassic for the Murihiku terrane (Murihiku Supergroup). From this analysis it may be inferred that Permo-Triassic boundary strata are represented *within* the Maitai Group even though their biostratigraphic recognition is still very imprecise and only marginally more precise than previously thought.

None of the fossil taxa so far identified can provide biostratigraphic age control that permits certain determination at stage level (either Late Permian or Early Triassic) let alone zonal level. Preservation, and hence identification, of the few taxa involved is problematic. Of the better material, some appears to represent endemic species and some long-ranging benthic forms. Much of the biostratigraphic argument concerning age of these fossils is limited to the most basic concern of establishing either a Permian or Triassic age affinity.

References

Archbold, N. W. & Dickins, J. M. (1991). Australian Phanerozoic time scales: Permian, a standard for the Permian System in Australia, *Bur. Min. Res., Australia, Rec.*, 1989/36.

Balme, B. E. (1963). Plant microfossils from the Lower Triassic of Western Australia. *Palaeontology*, 6(1): 12–40.

Balme, B. E. & Helby, R. J. (1973). Floral modifications at the Permian-Triassic boundary in Australia. In *The Permian and Triassic Systems and Their Mutual Boundary*, eds. A. Logan & L. V. Hills, pp. 433–44. Can. Soc. Petrol. Geol., Mem. 2.

Banks, M. R. (1978). Correlation chart for the Triassic System of Australia. *Bur. Min. Res., Australia, Bull.*, 156C: 1–39.

Dickins, J. M. (1973). The geological sequence and the Permian-Triassic boundary in Australia and eastern New Guinea. In *The Permian and Triassic Systems and Their Mutual Boundary*, eds. A. Logan & L. V. Hills, pp. 425–32. Can. Soc. Petrol. Geol., Mem. 2.

Dickins, J. M. (1976). Correlation chart for the Permian System of Australia. *Bur. Min. Res., Australia, Bull.*, 156B: 1–26.

Dickins, J. M., Archbold, N. W., Thomas, G. A. & Campbell, H. J. (1991). Mid-Permian correlation. *11th Internat. Cong. Carb. Strat. Geol., Compte Rendu.* (In press.)

Dolby, J. H. & Balme, B. E. (1976). Triassic palynology of the Carnarvon Basin, Western Australia. *Rev. Palaeobot. Palynol.*, 22: 105–68.

Foster, C. B. (1982). Spore-pollen assemblages of the Bowen Basin, Queensland (Australia): Their relationship to the Permian/Triassic boundary. *Rev. Palaeobot. Palynol.*, 36: 165–83.

Helby, R. (1973). Review of Late Permian and Triassic palynology of New South Wales. In *Mesozoic and Cainozoic Palynology: Essays in Honour of Isabel Cookson*, pp. 141–53. Geol. Soc. Australia, Spec. Publ. 4.

Herbert, C. (1980). Depositional development of the Sydney Basin. In *A Guide to the Sydney Basin*, eds. C. Herbert & R. Helby, pp. 11–52. Geol. Surv. New South Wales, Bull. 26.

McTavish, R. A. (1973). Triassic conodont faunas from Western Australia. *N. Jb. Geol. Paläont. Abh.*, 143: 275–303.

McTavish, R. A. & Dickins, J. M. (1974). The age of the Kockatea Shale (Lower Triassic) Perth Basin – a reassessment. *J. Geol. Soc. Austral.*, 21(2): 195–202.

Mollan, R. G., Dickins, J. M., Exon, N. F. & Kirkegaard, A. G. (1969). Geology of the Springsure 1:250,000 Sheet area, Queensland. *Bur. Min. Res., Australia, Rep.*, 123: 1–119.

Index

179

Printed in the United States
By Bookmasters